FUNDAMENTAL PHYSICS OF FERROELECTRICS 2001

Previous Proceedings in the Series of Workshops on Ferroelectrics

Year	Title	Publisher	ISBN/ISSN
2000	Fundamental Physics of Ferroelectrics 2000	AIP Conf. Proceedings Vol. 535	1-56396-959-9
1999	Williamsburg Workshop (1999)	J. Phys. Chem. Solids 61 (2), 2000	0022-3697
1998	Williamsburg Workshop (5th)	AIP Conf. Proceedings Vol. 436	1-56396-730-8
1997	Williamsburg Workshop (1997)	Ferroelectrics Vol. 206 (1-4) and Vol. 207 (1-2)	0015-0193
1996	Williamsburg Workshop (4th)	Ferroelectrics Vol. 194 (1-4)	0015-0193

Other Related Titles from AIP Conference Proceedings

554 Physics in Local Lattice Distortions: Fundamentals and Novel Concepts; LLD2K
Edited by Hiroyuki Oyanagi and Antonio Bianconi, February 2001, 1-56396-984-X

550 Characterization and Metrology for ULSI Technology: 2000 International Conference
Edited by David G. Seiler, Alain C. Diebold, Thomas J. Shaffner, Robert McDonald, W. Murray Bullis, Patrick J. Smith, and Erik M. Secula, February 2001, CD-ROM included, 1-56396-967-X

551 Atomic Physics 17: XVII International Conference on Atomic Physics; ICAP 2000
Edited by Ennio Arimondo, Paolo De Natale, and Massimo Inguscio, February 2001, 1-56396-982-3

544 Electronic Properties of Novel Materials—Molecular Nanostructures: XIV International Winterschool, Euroconference
Edited by Hans Kuzmany, Jörg Fink, Michael Mehring, and Siegmar Roth, November 2000, 1-56396-973-4

507 X-Ray Microscopy: Proceedings of the Sixth International Conference
Edited by Werner Meyer-Ilse, Tony Warwick, and David Attwood, March 2000, 1-56396-926-2

500 The Physics of Electronic and Atomic Collisions: XXI International Conference
Edited by Yukikazu Itikawa, Kazuhiko Okuno, Hiroshi Tanaka, Akira Yagishita, and Michio Matsuzawa, February 2000, 1-56396-777-4

492 Simulation and Theory of Electrostatic Interactions in Solution: Computational Chemistry, Biophysics, and Aqueous Solutions
Edited by Lawrence R. Pratt and Gerhard Hummer, November 1999, 1-56396-906-8

To learn more about these titles, or the AIP Conference Proceedings Series, please visit the webpage http://www.aip.org/catalog/aboutconf.html

FUNDAMENTAL PHYSICS OF FERROELECTRICS 2001

11th Williamsburg Ferroelectrics Workshop

Williamsburg, Virginia   4-7 February 2001

EDITOR
Henry Krakauer
College of William and Mary
Williamsburg, Virginia

Melville, New York, 2001
AIP CONFERENCE PROCEEDINGS ■ VOLUME 582

Editor:

Henry Krakauer
Department of Physics
College of William and Mary
P.O. Box 8795
Williamsburg, VA 23187-8795
USA

E-mail: krakauer@physics.wm.edu

The article on pp. 82-90 was authored by U.S. Government employees and is not covered by the below mentioned copyright.

Authorization to photocopy items for internal or personal use, beyond the free copying permitted under the 1978 U.S. Copyright Law (see statement below), is granted by the American Institute of Physics for users registered with the Copyright Clearance Center (CCC) Transactional Reporting Service, provided that the base fee of $18.00 per copy is paid directly to CCC, 222 Rosewood Drive, Danvers, MA 01923. For those organizations that have been granted a photocopy license by CCC, a separate system of payment has been arranged. The fee code for users of the Transactional Reporting Service is: 0-7354-0021-0/01/$18.00.

© 2001 American Institute of Physics

Individual readers of this volume and nonprofit libraries, acting for them, are permitted to make fair use of the material in it, such as copying an article for use in teaching or research. Permission is granted to quote from this volume in scientific work with the customary acknowledgment of the source. To reprint a figure, table, or other excerpt requires the consent of one of the original authors and notification to AIP. Republication or systematic or multiple reproduction of any material in this volume is permitted only under license from AIP. Address inquiries to Office of Rights and Permissions, Suite 1NO1, 2 Huntington Quadrangle, Melville, N.Y. 11747-4502; phone: 516-576-2268; fax: 516-576-2450; e-mail: rights@aip.org.

L.C. Catalog Card No. 2001092314
ISBN 0-7354-0021-0
ISSN 0094-243X
Printed in the United States of America

Contents

Preface vii
Organizing Committee and Sponsors ix

Cryogenic Electrical Studies of Manganese-doped Lead Scandium Tantalate Thin Films: Phase Transitions or Domain Wall Dynamics? 1
  M. Dawber, S. Ríos, J. F. Scott, Q. Zhang, and R. W. Whatmore
First-Principles Computation of Elasticity of Pb(Zr,Ti)O_3: The Importance of Elasticity in Piezoelectrics 11
  R. E. Cohen, E. Heifets, and H. Fu
Lattice Distortions in Pb(Zr,Ti)O_3 Alloys Near the Morphotropic Phase Boundary 23
  M. Fornari and D. J. Singh
Structure of Pb(Zr,Ti)O_3 Near the Morphotropic Phase Boundary 33
  W. Dmowski, T. Egami, L. Farber, and P. K. Davies
Synchrotron X-ray Studies of Superlattice Ordering in Pb(Mg$_{1/3}$Nb$_{2/3}$)O_3 Single Crystals Doped with PbTiO_3 45
  A. Tkachuk, P. Zschack, E. Colla, and H. Chen
Broadband Brillouin Scattering of Relaxor Ferroelectric Crystals 55
  S. Kojima and F. Jiang
Structural Properties of Sinusoidally Modulated Pb(Sc,Nb)O_3 Alloys at Finite Temperature 62
  A. M. George and L. Bellaiche
Domain Patterns, Texture and Macroscopic Electro-Mechanical Behavior of Ferroelectrics 72
  K. Bhattacharya and J. Y. Li
Prediction of the [Na$_{1/2}$Bi$_{1/2}$]TiO_3 Ground State 82
  B. P. Burton and E. Cockayne
Unique Quantum Stress Fields 91
  C. L. Rogers and A. M. Rappe
Dynamics of Relaxors 97
  R. Blinc, R. Pirc, V. Bobnar, and A. Gregorovič
Inside Dielectrics: Microscopic and Macroscopic Polarization 107
  P. Umari, A. Dal Corso, and R. Resta
Kinetic Monte Carlo Simulations of Crystal Growth in Ferroelectric Materials 118
  C. Tahan, M. Suewattana, P. Larsen, S. Zhang, and H. Krakauer
First-Principles Study of Pb$_2$MgTeO_6 Perovskite 128
  R. Caracas and X. Gonze
Why Is There an Isotope Effect on T_c in Hydrogen-Bonded Ferroelectrics Upon Deuteration But Absent Upon Replacing ^{16}O By ^{18}O? 137
  A. Bussman-Holder and N. Dalal
Model of Polar Clusters in Relaxors: Charge Transfer and Local Configuration Instability Effects 144
  V. S. Vikhnin, R. Blinc, R. Pirc, and S. Kapphan

Pressure as a Probe of Ferroelectric Properties: Quantum Regime 155
    G. A. Samara and L. A. Boatner

Bound Charge Diffusion and Polar Nanocluster Dynamics in Proton Glass Crystals 165
    V. H. Schmidt

Nonexponential Relaxation in Piezoelectric PVDF 175
    G. W. Bohannan

Computer Simulations of Domain Pattern Formation in Ferroelectrics 185
    R. Ahluwalia and W. Cao

Temperature-Dependent Behavior of $PbSc_{1/2}Nb_{1/2}O_3$ from First Principles 191
    E. Cockayne, B. P. Burton, and L. Bellaiche

First-Principles and Semi-Empirical Calculations of Atomic and Electronic Structure for the (100) and (110) Perovskite Surfaces 201
    E. Heifets, R. I. Eglitis, E. A. Kotomin, and G. Borstel

Accurate Construction of Transition Metal Pseudopotentials for Oxides 211
    I. Grinberg, N. J. Ramer, and A. M. Rappe

Ferroelectric and Piezoelectric Properties in the Presence of Compositionally Broken Inversion Symmetry 218
    N. Sai, B. Meyer, and D. Vanderbilt

New Polaronic-Type Excitons in Ferroelectric Oxides: Nature and Experimental Manifestation 228
    V. S. Vikhnin, R. I. Eglitis, E. A. Kotomin, S. E. Kapphan, and G. Borstel

Author Index 239

Preface

This volume contains papers presented at the 11th Williamsburg Workshop on Fundamental Physics of Ferroelectrics, which took place in Williamsburg, Virginia, February 4-7, 2001. This series of meetings began in 1990, and the emphasis of the workshops in alternate years had been on theoretical and experimental results. To mark the millennium and the tenth anniversary, a joint experimental and theoretical meeting was held in Aspen, Colorado, February 2000. The workshop returned to Williamsburg in 2001, continuing as in Aspen to jointly address fundamental experimental and theoretical topics.

The role of compositional ordering and symmetry breaking on piezoelectric properties was the focus of several experimental and theoretical studies. New first-principles results were presented on diverse topics including the importance of elasticity in piezoelectrics, quantum stress fields, and the connection between microscopic and macroscopic polarization. Effective Hamiltonians, based on first-principles results, continue to show great promise in predicting electromechanical properties as a function of temperature and composition in ferroelectric alloys. There was a broad range of other topics, ranging from the physics of relaxor ferroelectrics to domain pattern formation.

The workshop demonstrates that fundamental problems of complex and technologically important ferroelectrics and piezoelectrics are now becoming tractable, and we expect dramatic progress in following years.

I want to thank Sylvia Stout of the College of William & Mary for her able assistance in organizing and helping out at the workshop. The support of the Office of Naval Research and the College of William & Mary is gratefully acknowledged.

Henry Krakauer
Department of Physics
College of William & Mary
Williamsburg, Virginia 23187

ORGANIZING COMMITTEE

Henry Krakauer, College of William & Mary
Haydn Chen, University of Illinois, Urbana-Champaign

Sponsored by
The Office of Naval Research and
The College of William & Mary

Cryogenic Electrical Studies of Manganese-doped Lead Scandium Tantalate Thin Films: Phase Transitions or Domain Wall Dynamics?

M. Dawber, S. Ríos, and J. F. Scott

*Symetrix Centre for Ferroics, Department of Earth Sciences
Cambridge University, Cambridge, U.K.*

Qi Zhang and R. W. Whatmore

*Department of Industrial and Manufacturing Engineering
Cranfield University, Cranfield, U.K.*

Abstract. We have measured coercive field, remanent polarization, built-in bias field, and the dielectric constant and loss in Mn-doped lead scandium tantalate $PbSc_{1/2}Ta_{1/2}O_3$ thin films from 15 – 400 K. Evidence for four low-temperature anomalies near 50 K, 100 K, 160 K, and 233 K is summarised, and a discussion is presented of whether these are phase transitions. For comparison, ceramic samples were studied via X-ray techniques down to 20 K, and via DSC (differential scanning calorimetry) techniques near 233 K.

INTRODUCTION

Lead scandium tantalate (PST – $PbSc_{1/2}Ta_{1/2}O_3$) is a ferroelectric material of commercial importance as a room-temperature pyroelectric detector [1]. In this use its figure of merit is about five times [2] that of lead zirconate titanate (PZT), but until recently it required such a high processing temperature (>900 °C) that it could not be integrated fully. This limitation was eliminated, however, via growth on a seed layer of PZT [3]. Relatively little is known about its properties at very low temperatures. Early studies by Setter and Cross [4] assumed that above its Curie temperature of approximately $T_c = 299\pm1$ K it had a cubic perovskite paraelectric (PE) m3m structure, and that it was rhombohedral and ferroelectric (FE) below that, with a single phase transition separating the PE and FE phases. However, the situation is complicated, even at high temperatures: For a carefully annealed specimen, the Sc and Ta ions order in planes, giving Fm3m space group and a "doubled" unit cell; whereas for a disordered Sc/Ta system, the structure Pm3m with a single formula-group primitive unit cell is more likely for the paraelectric phase [5]; however, others have observed [6] that the paraelectric phase has first-order Raman spectra similar to those in $Pb(Mg,W)O_3$, which is pseudo-cubic orthorhombic $C222_1$. Extensive Raman studies [7-12] indicate that the true Pm3m phase may be reached only above 500 °C. A review is given this year by Siny et al.[13]. The paraelectric phase between 299 K and 323 K is incommensurate, as shown clearly by electron diffraction of its anti-phase boundaries (APBs), and often referred to as antiferroelectric (AFE). However, our hysteresis data reveal no double loops in this temperature range, and so we cannot directly confirm any AFE character.

Further confusion arose concerning the ferroelectric phase when Chang and Chen concluded from TEM observations that the ferroelectric phase is tetragonal [14]. The definitive structural studies to date [15-18], however, show that it most likely has at least two phase transitions (probably Fm3m-IC-R3), which is qualitatively more compatible with the two-step model (Fm3m-R$\bar{3}$m-R3m) proposed by Salje and Bismayer [19]. Above T_I = 323 K it is paraelectric, probably Fm3m, but possibly only pseudo-cubic and isomorphic with $Pb_2(Mg,W)O_6$; that is, $C222_1$. This is evidenced by nominally forbidden X-ray diffraction Bragg reflections corresponding to (1/2, 1/2, 0) and (1/2, 0, 0). For 299 K < T < 323 K it is incommensurate and possibly antiferroelectric, and so in accord with the usual notation, we denote T_I = 323 K (somewhat sample-dependent) and the incommensurate lock-in transition temperature, T_c = 299 K, the Curie temperature for a first-order transition. Below T_c = 299 K it is ferroelectric with probable space group R3 [the splitting of the (444) reflections shows it cannot be tetragonal]; and small distortions may occur at cryogenic temperatures, beginning with a rhombohedral-monoclinic transition at 233 K [20]. The space groups and the splittings and forbidden lines, as well as the exact transition temperatures, have remained controversial because of the presence of domain walls and the differences in chemical ordering (of Sc and Ta ions) for carefully annealed versus unannealed specimens. We will tentatively refer to the PST phase below 233 K as monoclinic.

The present work was undertaken with two aims: Firstly, to examine the hysteretic switching behaviour in all phases (including the proposed antiferroelectric phase) and to obtain information on its phase transitions; and secondly, to model the temperature dependence of the coercive field at low temperatures. Only half a dozen ferroelectrics have had their coercive field $E_c(T)$ measured down below liquid nitrogen temperatures [21-24], and these all exhibit qualitatively different behaviour: TGS and guanidinium sulphates diverge as absolute zero is approached [21]; bismuth titanate is non-monotonic and approaches T=0 with zero slope (in accord with the Third Law of Thermodynamics) [21]; $LiTaO_3$ is absolutely T-independent *below* a threshold temperature of 500 K but linear in T above (probably due to ionic conduction), whereas KH_2PO_4 (KDP) is T-independent *above* a threshold temperature of 50 K; KDP then rises exactly linearly [22] with decreasing T [an effect that has been described qualitatively as domain wall freezing (or "pinning")], as does potassium ferrocyanide trihydrate [22]; $BaTiO_3$ is quite linear in T [23,24]. No quantitative theories for $E_c(T)$ have been published, but a general theory of switching is given by Janovec [25]. Switching at higher temperatures was studied carefully in PST by Randall et al.[26].

EXPERIMENTAL

Pb(OAc)$_2$·3H$_2$O and Mn(OAc)$_2$ (3 mol%) were mixed with IPA with the ratio of Pb(OAc)$_2$·3H$_2$O : IPA 1 g:1 ml. This solution was stabilised with ethanolamine (MEA) to form a clear solution. Ta(OEt)$_5$ and Sc(OAc)$_3$ were mixed in IPA and also stabilised with MEA and subsequently refluxed. The Pb solution and Ta/Sc solution were combined and further refluxed. The solution was concentrated to 0.2M by distillation.

A Mn-doped PZT layer (PMZT) was first coated on a Pt/Ti/SiO$_2$/Si substrate as a seeding layer in the attempt to reduce the PST crystallisation temperature. The thickness of a single PMZT layer is ~ 70 nm. PST layers were spun onto the top of the PMZT seed layer. Each coating was subjected to a drying (200 °C, 30 s) and annealing (550 °C, 10 min.) procedure. Total film thickness was 800 nm.

Mr. Paul Osbond (Marconi Research Caswell) is gratefully acknowledged for the supply of the ceramic specimens. For precision X-ray studies (Fig.1a), the ceramic samples were powdered. This means that the presence of phase transitions in the electrical hysteresis and/or dielectric measurements, absent in the ceramics, may be indicative either of special thin-film properties or of the Mn-doping.

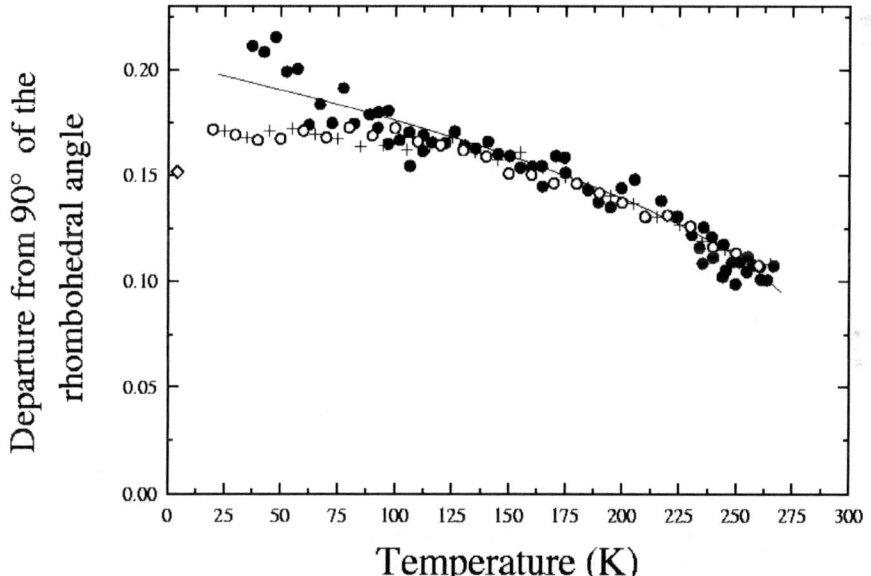

FIGURE 1a. X-ray data for pure PST powder from 4 K to 270 K: 90° minus the rhombohedral angle; solid circles are from Groves;[20] +'s (heating) and o's (cooling) are present work; diamond from Baba-Kishi and Woodward.[18] The solid curve is a least-squares mean-field fit.

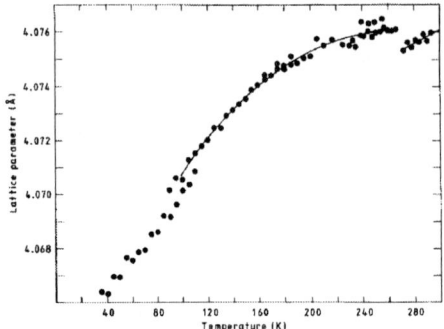

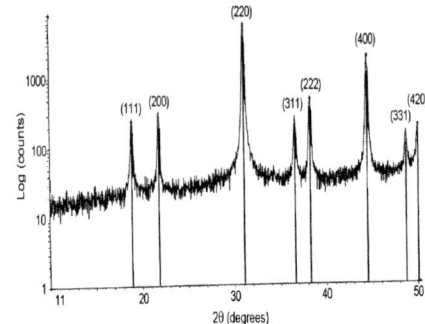

FIGURE 1b. PST lattice constant a(T) from Groves.[27] Reprinted with permission from IOP Publishing Limited.

FIGURE 1c. XRD of the oriented PST film

X-ray and Specific Heat results on Ceramic PST

X-ray data were taken on thin films (XRD in Fig.1c) and on powder from a highly ordered ceramic wafer (from GEC-Marconi). The latter gave X-ray patterns compatible with a very high degree of Sc/Ta ordering (>85%). Data collection emphasized the 50 K region, with results shown in Fig.1a. This figure shows the value of $90° - \phi$ where ϕ is the rhombohedral angle, as determined from a resolved splitting of (440) line below T_c, which lies at $2\Theta = 64.6°$. We note that there are strong differences below 50 K in the three sets of data: We do not find Groves' anomalous increase at very low T; and the ultra-precise data at 4 K of Ref.18 are outside the uncertainties of the present work. This suggests that the behavior near 50 K may be sample-dependent.

Fig.1b reproduces Groves' data on the lattice constant a(T), showing anomalous behavior near 50 K and 100 K. As discussed below, 100 K is where P_r is maximum; at 50 K P_r exhibits a thermal hysteresis.

Fig.1c shows the X-ray diffraction data on the mn-doped thin film, confirming both its single-phase character and its orientation.

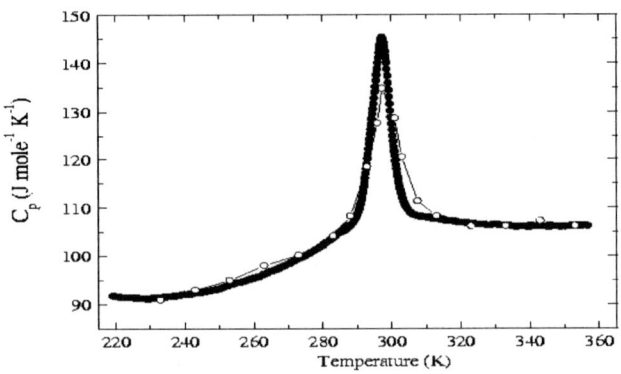

Figure 2. Specific heat (DSC) data for pure PST versus temperature (solid curve = present work). The transition entropy is nearly equal that in Setter and Cross[4] (o) and implies Sc/Ta ordering of 85%.

A Perkin-Elmer DSC apparatus was employed to look for phase transitions, including both the high temperature region up to 600 °C, where the Fm3m-Pm3m transition had been proposed, and the low temperature region near 233 K where the rhombohedral-monoclinic transition had been suggested. The sample holder, containing a slab of 144 mg of ceramic PST, was cooled down to 200 K with liquid nitrogen and then heated at 10 K/min to 360 K. Results for a ceramic sample are shown above in Fig.2. The value of entropy change ΔS at T_C was found to be 1.1 J/mol·K; by comparison with the value of 1.2 J/mol·K in Ref.4, this shows the Sc/Ta ordering to be ca. 85%. Note that the R3 ferroelectric-paraelectric phase transition at 299 K is first-order with a large latent heat; but no thermal evidence is present for any monoclinic-rhombohedral transition at 233 K; there is a very small increase near 340 K that could be a vestige of the supposed incommensurate-Fm3m transition (nominally at 323 K).

Hysteresis Measurements on Thin PST Films

The principal results from the hysteresis measurements, carried out on an Aixact TF-2000 tester with a closed-cycle CTI-Crogenics "Cryodyne" He-refrigerator, are the temperature dependences of remanent polarisation $P_r(T)$ (Fig.3), coercive voltage $V_c(T)$ (Fig.4), and built-in voltage (the asymmetric shift of the hysteresis curve from zero for positive and negative applied voltages), $V_{bi}(T)$ shown in Fig.5. In each case the data reveal the presence of the two previously known phase transitions: IC-FE at 300 K, and R3 rhombohedral to monoclinic at 233 K.

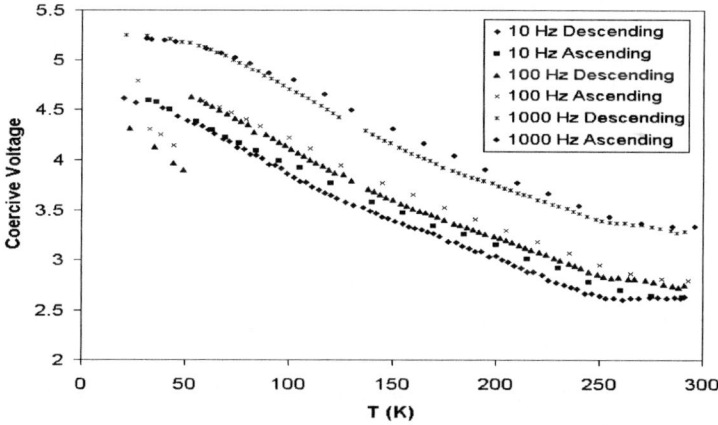

Figure 3. Coercive field $E_c(T)$ versus temperature for 800 nm thick PST:Mn film. At 100 Hz the change in slope at 50 K becomes a step discontinuity (20% drop on lowering T). This also occurs in P_r and V_{bi} at 50 K; it may be due to a resonant behaviour (it is reproducible but does not occur at 10 Hz or at 1 kHz).

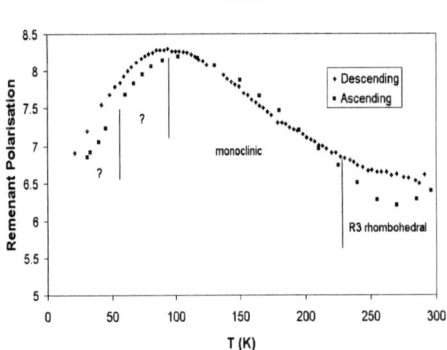
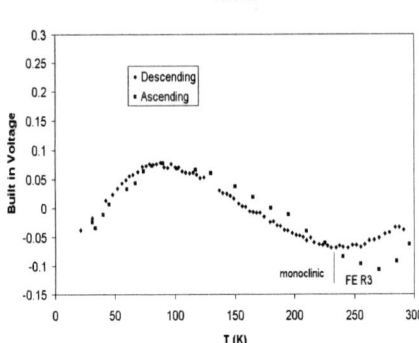

Figure 4. Remanent polarization $P_r(T)$ for PST

Figure 5. Built-in potential $V_{bi}(T)$ for a PST:Mn film.

Dielectric Measurements on Thin PST Films

The dielectric data reveal in both the real $\varepsilon(T)$ [Fig.6] and imaginary tan δ parts [Fig.7] additional phase transitions at 160 K and 50 K. The anomaly at 50 K is very distinct in both the real and imaginary parts of the susceptibility. This confirms the initial suggestion by Groves[27] that there is a phase transition at exactly 50 K. Groves' data are reproduced in Fig. 1a and show a sharp change in slope for the monoclinic distortion angle at 50 K and in Fig.1b similar anomalies in the lattice constant a(T); however, his own dielectric data (unfortunately, on a scale of thousands) showed no anomalies at this temperature, and so he concluded (erroneously, we believe) that no transition occurs near 50 K. In the Mn-doped sample there is also a strong anomaly with onset at 100 K and peak near 130 K; this anomaly may be dynamic (it is greatly enhanced in Mn-doped PST) and perhaps not signal a static structural phase transition.

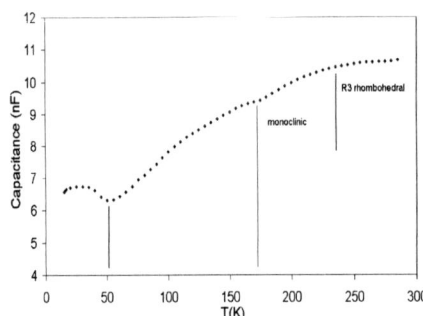
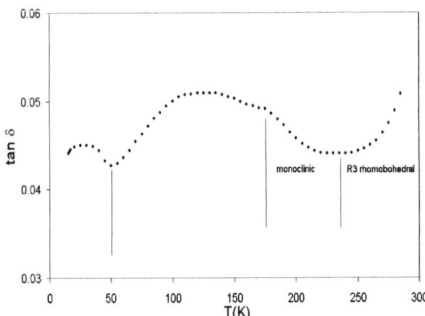

Figure 6. Real part of the dielectric constant $\varepsilon(T)$

Figure 7. Loss tangent (tan δ) for PST films at 100 kHz.

COERCIVE FIELD MODELS

All of the values reported here are from uncompensated hysteresis loops. In a lossy dielectric this will result in apparent values of P_r and E_c which are finite above the ferroelectric-paraelectric transition temperature. Nevertheless, in this initial report we have elected to use such parameters rather than compensated values, because it permits contact with raw data without any "massaging" or processing.

The behaviour of coercive field $E_c(T)$, remanent polarization $P_r(T)$, and built-in potential V_{bi}, in Figs.3-5 show that in each case anomalies occur at the same temperatures at which $\varepsilon'(T)$ and $\varepsilon''(T)$ show maxima or minima. We note in Fig.8 that the C(V) data in PST at low temperatures exhibit well-defined "butterfly loops".

Note in Fig.3 that the coercive fields are frequency dependent, and in Fig.6 that ε is also. The former phenomenon was recently rediscovered, in part by one of the present authors [28]. However, frequency dependent coercive fields were first reported by Campbell [29] and by Wieder [30].

The linear dependence of coercive field on temperature for PST in the present work is very similar to that in the KDP family. In both cases it is thought to arise from domain wall freezing. That is, domain walls experience a finite viscosity in their motion through the lattice. However, the analogy of domain wall motion in a viscous lattice to viscous fluid flow is not quantitative: The Sutherland Equation of viscosity in liquids such as water is

$$\eta = AT^{1/2}(1 + B/T)^{-1}, \qquad (1.)$$

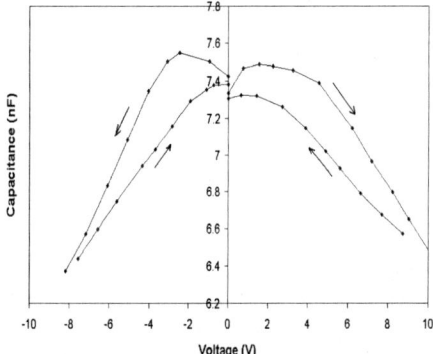

Figure 8. Capacitance versus voltage in PST:Mn.

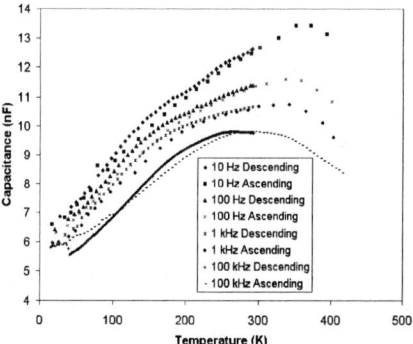

Figure 9. Frequency dependence of dielectric constant in PST thin films from 10-10^5 Hz.

an algebraic form which does not describe the temperature dependence of coercive field or domain wall reciprocal mobilities in any known ferroelectric. A $T^{1/2}$-dependence also arises for the viscosity of a dilute gas. However, for domain wall motion it is the threshold field that is of interest here, not the steady-state velocity or mobility. Consequently it is not the viscosity analogy that should be pursued (viscosity is the analogue not of the coercive field E_c but of the reciprocal of the domain wall mobility, μ_D), but rather the temperature dependence of the de-pinning energy. We emphasize that this value is not obtainable from the Landau-Devonshire free energy (which ignores domain walls and consequently yields a theoretical value of E_c ca. 10^3 too large in ferroelectrics). Our empirical result for for rhombohedral PST is E_c = constant = 3.4 MV/m; and in monoclinic PST it is (50 K < T < 233 K)

$$E_c(T) = E_c(0) - A(T-T_o), \qquad (2.)$$

where $E_c(0)$ is the value 5.7±0.2 MV/m at T = 50 K (not at zero); T_o = 50±2 K; and A = 120±10 kV/m·K. As mentioned above, both KDP and PST have a region in which E_c is temperature independent, below which it rises linearly with decreasing temperature. In KDP this "domain freezing temperature" is ca. 50 K and is thought to be unrelated to a phase transition. However, in PST it is 233 K and is the rhombohedral-monoclinic transition point. That is, in our interpretation, the coercive field is temperature independent in the R3 phase but rises linearly with decreasing temperature in the monoclinic phase(s) from 233 K to 160 K and 160 K to 50 K. An possible alternative explanation of the data is that, as in KDP, the change from T-independence to E_c linear in temperature below 233 K is not due to a structural transition, but that 233 K is a temperature at which domain wall pinning sets in.

CONCLUSIONS

The space groups of PST remain in doubt despite careful electron microscopy by Lemmens et al. [31] and X-ray studies by Bursill et al. [32,33]. The present studies do not resolve these questions but provide independent evidence for the existence of phase transitions at 50 K, 100 K, 160 K and 233 K, as summarized in Table I.. There was significant evidence for those at 50 K and 100 K in the work (Fig.1b) by Groves [27], but that was discounted because it was unconfirmed by his dielectric studies. Our work reveals anomalies in both the dielectric loss and the capacitance at those temperatures, plus anomalies in coercive voltage, remanent polarization, and built-in potential. Thus, there seems now some evidence for a monoclinic-to-unknown (triclinic?) transition at 160 K, and lower temperature transitions at 50 K and ca. 100 K that may involve unit cell multiplication; such cell multiplication seems compatible with the fact that the superlattice lines at 4.2 K in the work by Baba-Kishi and Woodward [18] are unexplainedly broadened, whereas the primary Bragg peaks are not. We note, however, that very recent ultra-precise time-of-flight neutron studies gave at 4.2 K the lattice parameter a = 0.815441(7) nm and rhombohedral angle 89.8485(3) degrees, assuming an R3 structure at such temperatures.[17] A detailed review of PST structures at temperatures near ambient has been given very recently by Bursill, Peng, Qian, and Setter [33]. In addition, infrared reflectivity down to 20 K failed to reveal any evidence of symmetry lower than R3. [34]. The present dielectric studies were carried out at 10 Hz, 100 Hz, 1 kHz, and 100 kHz. As emphasized in Ref.35, such studies are rare below 1 kHz; ref.35 measures PST down to 10^{-3} Hz.

We emphasize in this conclusion that the primary evidence for low-temperature phase transitions in our PST samples is from dynamic measurements: dielectric response and electrical hysteresis. No clear confirmation has been obtained from either of the "standard" techniques for determining phase transitions – X-rays or specific heat. Therefore it remains possible that the temperatures at which we see anomalies are not true static distortions (structural phase transitions) but instead involve subtle changes in domain-wall dynamics.

Table I. Electrical Anomalies observed in Lead Scandium Tantalate:

Temperature	$\varepsilon(T)$	$\tan \delta(T)$	$P_r(T)$	$E_c(T)$	$V_{bi}(T)$
<50 K			hysteretic	flat	
50K	minimum	minimum		maximum	sign change
50K<T<100K		hysteretic			hysteretic
100K	inflection	maximum	maximum		maximum
100K<T<160K					hysteretic
160K	minimum	maximum			sign change
160K<T<233K		hysteretic			hysteretic
233K	inflection	minimum	inflection	minimum	minimum
233K<T<299K			hysteretic	flat	
299K					
299K<T<323K					
323K	maximum	maximum			

ACKNOWLEDGEMENTS

Work at Cambridge supported by EPSRC. The thin film samples were produced through sponsorship from the Defence Research and Evaluation Agency (Malvern). We thank Prof. A. Bhalla, Prof. E. Salje, Dr. K. Z. Baba-Kishi and Dr. M. Welch for helpful discussions.

REFERENCES

1. Whatmore, R. W., Rep. Prog. Phys. **49**, 1335 (1986).
2. Shorrocks, N. M., Patel, A., R. W. Whatmore, R. W., Ferroelec. **134**, 35 (1992).
3. Huang, Z. Zhang Q., and Whatmore, R. W., J. Mat. Sci. Lett. **17**, 1157 (1998); J. Appl. Phys. **85**, 7355 (1998).
4. Setter, N. and Cross, L. E., J. Mater. Sci. **15**, 2478 (1980).
5. Setter, N. and Cross, L. E., J. Appl. Phys. **51**, 4346 (1980).
6. Galasso, F., Structure, Properties and Prep. of Perovskite-Type Compounds (Pergammon. London, 1969).
7. Smolensky, G. A., Siny, I. G., Pisarev, R. V., and Kuzminov, E. G., Ferroelectrics **12**, 135 (1976)
8. Bismayer, U., Devarajan, V., and Groves, P., J. Phys. Condens. Mater. **1**, 6977 (1989).
9. Siny, I. G., and Smirnova, T. A., Fiz. Tverd. Tela **30**, 823 (1989); Ferroelectrics **90**, 191 (1989).
10. Siny, I. G. and Boulesteix, C., Ferroelectrics **96**, 119 (1989).
11. Setter, N. and Laulicht, I., Appl. Spectrosc. **41**, 526 (1987).

12. Boulesteix, C., Caranoni, C., Kang, C. Z., Sapozhnikova, L. S., Siny, I. G., and Smirnova, T. A., Ferroelectrics **107**, 241 (1990).
13. Wang, H.-C. and Schulze, W. A., J. Am. Ceram. Soc. **73**, 1228 (1990).
14. Siny, I. G., Katiyar, R. S., and Bhalla, A. S., Ferroelec. Rev. **2**, 51 (2000).
15. Chang, Y., and Chen, Z., Ferroelec. Lett. **4**, 13 (1985).
16. Baba-Kishi,K. Z., Cressey, G., and Cernik, R. J., J. Appl. Crystal. **25**, 477 (1992).
17. Baba-Kishi, K. Z., and Barber,D. J., J. Appl. Crystal. **23**, 43 (1990)
18. Baba-Kishi, K. Z. and Woodward, P. M., paper A4b.2, AMF-3 (Hong Kong, Dec 2000), Ferroelectrics (in press, 2001); K. S. Knight and K. Z. Baba-Kishi, Ferroelec. **173**, 341 (1995).
19. Salje, E. and Bismayer, U., J. Phys. Condens. Mater. **1**, 6967 (1989).
20. Bhalla, A. S., and Randall, C. A., private communication cited in Ref.16; Randall, C. A., Barber, D. J., Whatmore, R. W., and Groves, P., J. Mat. Sci. **21**, 4456 (1986).
21. Domanski, S., Proc. Phys. Soc. (London) **B72**, 306 (1958); Holden, A. N., Merz,, W. J., . Remeika, J. P., and Matthias, B. T., Phys. Rev. **101**, 962 (1956)..
22. Sawaguchi, E. and Cross, L. E., Mat. Res. Soc. Bull. **5**, 147 (1970).
23. Barkla, H. M. and Finlayson, D. M., Phil. Mag. **44**, 109 (1953); Waku, S., Hirabayashi, H., Toyoda, H., Iwasaki, H., and Kiriyama, R., J. Phys. Soc. Jpn. **14**, 973 (1959).
24. Wieder, H. H., J. Appl. Phys. **26**, 1479 (1955).
25. Janovec,V., Czech. J. Phys. **8**, 3 (1958).
26. Randall, C. A., Barber, D. J., Groves, P., and Whatmore, R. W., J. Microsc. **145**, 275 (1987).
27. Groves, P., J, Phys. Condens. Mater. **18**, L1073 (1985).
28. Scott, J. F., Hartmann, A. J., Lamb, R. N., Ross, F. M., De Vilbis,, A., Paz de Araujo, C. A., Scott, M. C., G. Derbenwick, G., Mat. Res. Soc. Symp. Proc. **433**, 77 (1996); Scott, J. F., Ferroelectric Memories (Springer, Heidelberg, 2000), Chap.8.
29. Campbell, D. S., J. Electron. & Control **3**, 330 (1957).
30. Wieder, H. H., J. Appl. Phys. **28**, 367 (1957).
31. Lemmens, H., Richard,O., Van Tendeloo, G., and Bismayer, U., J. Electron Microsc. **48**, 843 (1999).
32. Bursill, L. A., Peng, J. L., Qian, H., and Setter, N., Physica **B205**, 305 (1995).
33. Bursill, L. A., Peng Ju-Lin, Hua, Q. and N. Setter, N., Physica B (in press, 2001).
34. Petzelt, J., Buixaderas, E., and Pronin, A., (preprint, 2001).
35. Jonscher, A. K. and Isnin, A., (preprint, 2001).

First-principles computation of elasticity of Pb(Zr,Ti)O$_3$: The importance of elasticity in piezoelectrics

R.E. Cohen[*], E. Heifets[*] and H. Fu[†]

[*]*California Institute of Technology, Pasadena, CA and Carnegie Institution of Washington, 5251 Broad Branch Rd., N.W., Washington, D.C. 20015*
[†]*Department of Physics, Rutgers University, Camden, NJ 08102*

Abstract. We have computed the elastic constant tensor for PbZrTiO$_6$, PZT 50/50, in the rhombohedral structure. We find it to be very soft elastically, and the Young's modulus to be very anisotropic. Our results for PZT are compared with literature values on other oxide perovskites, and the new experimental results on PMN-PT and PZN-PT. The elastic response is an important part of the electromechanical response, and should be studied experimentally and theoretically to help build up a systemic understanding of electromechanical materials. We find that anisotropy in Young's modulus and elastic softness may be key discriminators in evaluating or improving electromechanical properties.

INTRODUCTION

We now understand the origin of ferroelectricity in perovskite oxide ferroelectrics. Ferroelectricity results from a delicate balance of long-range and short-range forces [1–3]. The long-range forces destabilize the cubic structure, and are enhanced generally by large transverse effective charges. The large effective charges are largely the result of hybridization between the O 2p states and the B-cation d-states, which are the lowest conduction band states. Short-range overlap forces stabilize the cubic structure, and hybridization helps soften these forces so that the ions can move off-center. The electromechanical response is a result of (a) the change in polarization with strain, which depends on the effective charges and the soft-mode potential surface [4–6], (b) the elastic constants, and (c) the dielectric tensors [7, 8]. First-principles studies of effective charges and hybridization in a number of systems now show no systematics that might indicate simply which compositions will have the strongest electromechanical coupling constants [6, 9–11]. In fact, effective charges (and local structure) in perovskite oxides vary little for a given ion for different bulk compositions; e.g. Ti in PbTiO$_3$ and in PZT are very similar [6]. Here we investigate the importance of elasticity in varying the effective coupling strengths of different piezoelectrics.

A knowledge of all aspects of the electromechanical coupling in piezoelectrics would be highly desirable to optimize materials and devices for applications. Surprisingly, even for the classic ferroelectrics such as BaTiO$_3$, apparently the single crystal tensors have not been measured for each of the piezoelectric phases, that is the rhombohedral, orthorhombic, and tetragonal phases. The situation for materials such as

Pb(Zr,Ti)O$_3$ (PZT) are even more problematic, since no single crystal data exist. Fortunately, single crystal data are being measured [12] for the new high-strain piezoelectrics Pb(Mg,Nb)O$_3$-PbTiO$_3$ (PMN-PT) and Pb(Zn,Nb)O$_3$-PbTiO$_3$ (PZN-PT) [13–15], and the new tentative results will be compared with our computations for PZT and other phases here.

For PZT, phenomenological models fit to ceramic and powdered data have been successful to systematize data, but it is not clear that they have predictive power. For example, a major recent discovery of a monoclinic phase at the morphotropic phase boundary near compositions of 52/48 PZT [16, 17] was not anticipated by such models, probably because the underlying sixth order Devonshire theory could not encompass a monoclinic phase [18]. In spite of the fact that PZT can now be grown in small crystals of 10-20 microns, which are large enough for Brillouin scattering studies to extract single crystal elastic and piezoelectric constants, no such data exist. First-principles methods have been used to study the zero temperature piezoelectric strain coefficients e_{ij} in PbTiO$_3$ and e_{33} in 50/50 PZT. The temperature dependence of the piezoelectric stress coefficients d_{ij} have been studied in [8, 19–21] using model Hamiltonians fit to first-principles calculations. However, there apparently have been no studies of elasticity in ferroelectric perovskites. Here we compute the full set of single crystal elastic constants for ordered PZT 50/50 in the rhombohedral structure.

A most important parameter for piezoelectric applications, including acoustic transducers and actuators is the electromechanical coupling coefficient, k_{ij} [22]. The ratio of stored energy to input energy is given by k^2 and

$$k^2 = \frac{d^2}{\varepsilon s} \qquad (1)$$

where ε is the dielectric constant tensor and s is the compliance tensor. The piezoelectric strain tensor is

$$d_{ijk} = \frac{\partial \varepsilon_{ij}}{\partial E_k} \qquad (2)$$

$$P_i = d_{ijk}\sigma_{jk} \qquad (3)$$

where ε is the strain, σ is the stress, E is the applied electric field and P is the polarization. The elastic compliance tensor is

$$s_{ijkl} = \frac{\partial \sigma_{ij}}{\partial \varepsilon_{kl}} \qquad (4)$$

Although d_{ijk} can be determined straightforwardly from an effective Hamiltonian, it is more straightforward to compute the piezoelectric stress coefficients

$$e_{ijk} = \frac{\partial P_i}{\partial \varepsilon_{jk}} \qquad (5)$$

$$P_i = e_{ijk}\varepsilon_{jk} \qquad (6)$$

in self-consistent calculations, since electronic structure computations in periodic boundary conditions are always formally performed at constant $E = 0$ (although one can introduce a small potential perturbation that mimics an electric field). The polarization P_i can be computed from first-principles using the Berry's phase theory for polarization in bulk solids [23, 24]. By computing the polarization versus applied strains, one can compute the piezoelectric stress coefficients. These two different piezoelectric tensors are related through the elastic compliances $d_{ijk} = e_{ilm} s_{lmjk}$.

The piezoelectric, stress, strain, elastic, and compliance tensors indices are usually contracted as $11 \rightarrow 1$, $22 \rightarrow 2$, $33 \rightarrow 3$, $32 \rightarrow 4$, $13 \rightarrow 5$, $12 \rightarrow 6$, so that stresses and strains are represented by one index, and piezoelectric, elastic and compliance tensors by two indices. (See Nye [25] for details of numerical factors in some cases.) In Voigt notation, the elastic constants and compliance are represented as matrices, and are the inverse of each other.

In previous studies, we have computed the piezoelectric stress tensor for $PbTiO_3$ and the e_{33} constant for PZT 50/50 in the tetragonal structure [5, 6, 11]. Although the results for $PbTiO_3$ were in good agreement with experiment, e_{33} for PZT was of similar magnitude, and seemed too small using ceramic data for elastic compliances to estimate e_{33} for ceramics.

Whereas piezoelectricity was considered largely as a collinear effect, with the strain, electric field, and polarization in parallel, computations for polarization rotation, with the polarization direction rotated from (111) to (001) with a field along (001) showed very large piezoelectric response, consistent with the new large strain piezoelectrics.[26, 27] This effect is also consistent with the large piezoelectric coupling seen near the morphotropic phase boundary between rhombohedral and tetragonal phases in PZT, with the discovery of a monoclinic phase in this region, apparently at zero field.[28, 29] The monoclinic phase can be understood as an intermediate state between the rhombohedral (polarization 111) and tetragonal (polarization 001) phases. The large coupling in such PZT's can be thought of as a self-induced polarization rotation, as opposed to the field induced rotation in the large strain single crystals PMN-PT and PZN-PT. This effect also occurs in PZT's away from the monoclinic phase.

The question remains how much of the strong coupling is due directly to the anisotropy and softness of the elasticity, and how much is due to the underlying polarization rotation energy surface? Here we compute the elastic constants for ordered PZT 50/50 in a rhombohedral structure and examine the effects of elastic softening and anisotropy. Ordering has not been experimentally detected in PZT, and the 50/50 composition falls in the tetragonal phase field. We study the rhombohedral structure since it has the larger piezoelectric response, and is the "parent structure" of the monoclinic phase and the large coupling single crystal piezoelectrics. The use of an ordered structure is an approximation, and our results can be used as a benchmark to compare with faster, "second-principles" methods that can be used to study disordered systems.

METHOD

We have computed the energy versus strain for ordered PZT to obtain the elastic constants. We used the local density approximation and a mixed basis pseudopotential method [30]. The basis consists of the pseudoorbitals and a small number of plane

TABLE 1. Lattice and positions of atoms in zero strain (minimum energy) structure with $a = 5.78626$ Å and $\alpha = 59.2445°$ ($V = 134.63$ Å^3/10 atoms). The positions are given in terms of the lattice vectors.

lattice vectors (Å):			
	-2.86003	-1.65123	4.75126
	2.86003	-1.65123	4.75126
	0.00000	3.30247	4.75126
atomic positions:			
Pb(1)	-0.00476	-0.00476	-0.00476
Pb(2)	0.49313	0.49313	0.49313
Ti	0.26212	0.26212	0.26212
Zr	0.76231	0.76231	0.76231
O(1)	0.03100	0.03100	0.53012
O(2)	0.53012	0.03100	0.03100
O(3)	0.03100	0.53012	0.03100
O(4)	0.04493	0.51550	0.51550
O(5)	0.51550	0.04493	0.51550
O(6)	0.51550	0.51550	0.04493

waves. The pseudoorbitals, as well as the charge density and potential are expanded as plane waves, to a large cut-off wavevector G_{max}. The basis set size is much smaller, consisting of 9 orbitals per atom, plus plane waves up to a small g_{max}. The pseudopotential is a Troullier-Martins type [31] with r_{match}= 1.70, 2.00, and 1.75 bohr for $6s^1$, $6p^1$, and $5d^{10}$ for Pb, 1.50, 1.60, 1.75 bohr for $4s^2$, $4p^6$, and $4d^{0.5}$ for Zr, 1.45, 1.40, and 1.75 for $3s^2$, $3p^6$, and $3d^{0.5}$ for Ti, and 1.5, 1.5, and 2.5 for $2s^2$, $2p^4$, and $3d^{0.2}$ for O. The small g_{max} was 120 eV and the large cut-off G_{max} was 884 eV. Since the basis can be expanded in plane waves, all integrals can be computed using FFT methods. One can consider the pseudoorbitals to be contractions of large numbers of plane waves, and convergence is obtained with a relatively small basis.

The Zr and Ti ions were ordered like the rock-salt structure on the B sites. The reference structure (zero strain) was obtained by relaxing the atomic positions and rhombohedral strain in the ten atom supercell, at the experimental volume of 134.63 Å^3 [32]. The rhombohedral strain is small, the minimum energy rhombohedral angle (with relaxed atomic positions) is 59.2445°, as opposed to 60° for the fcc cubic lattice, and a=5.78626 Å. Table 1 shows the atomic positions for the unstrained structure.

There are six independent elastic constants for the rhombohedral group 3m: C_{11}, C_{33}, C_{12}, C_{13}, C_{14}, and C_{44}. Six different strains were applied to the reference structure, and the atomic positions were relaxed using analytic forces and a conjugate gradient algorithm. Table 2 shows the strains used and the linear combinations of elastic constants for each strain, and table 3 shows the elastic constants in terms of the second derivatives of the energy with respect to each strain. Note that zero strain refers to the experimental volume and the minimum energy rhombohedral strain. Internal degrees of freedom were relaxed for each strain. Figure 1 shows the energy versus strain for each case.

TABLE 2. Strains used to determine elastic constants and the second order change in energy.

	strain	ΔE
$e_1 = \varepsilon_3$	$\varepsilon_{11} = \varepsilon_{22} = -\frac{1}{2}\varepsilon_{33} = \varepsilon_3$	$(\frac{1}{4}(C_{11}+C_{12}) - C_{13} + \frac{1}{2}C_{33})\varepsilon_3^2$
$e_2 = \varepsilon_3$	$\varepsilon_{11} = \varepsilon_{22} = \varepsilon_{33} = \varepsilon_3$	$(C_{11}+C_{12}+2C_{13}+\frac{1}{2}C_{33})\varepsilon_3^2$
$e_3 = \varepsilon_4$	$2\varepsilon_{23} = \varepsilon_4$	$\frac{1}{2}C_{44}\varepsilon_4^2$
$e_4 = \varepsilon_6$	$2\varepsilon_{12} = \varepsilon_6$	$\frac{1}{2}(C_{11}-C_{12})\varepsilon_6^2$
$e_5 = \varepsilon_6$	$2\varepsilon_{12} = 2\varepsilon_{13} = \varepsilon_6$	$(C_{14}+\frac{1}{2}C_{44}+\frac{1}{4}(C_{11}-C_{12}))\varepsilon_6^2$
$e_6 = \varepsilon_3$	$\varepsilon_{33} = \varepsilon_3$	$\frac{1}{2}C_{33}\varepsilon_3^2$

RESULTS AND DISCUSSION

The energy versus strain was fit to a fourth order polynomial for each case, and the coefficient of the quadratic term derived (the B's in Table 3, and resulting elastic constants and compliances are shown in Table 4.

Note the extreme softness of the C_{44} modulus compared with the others. This is also

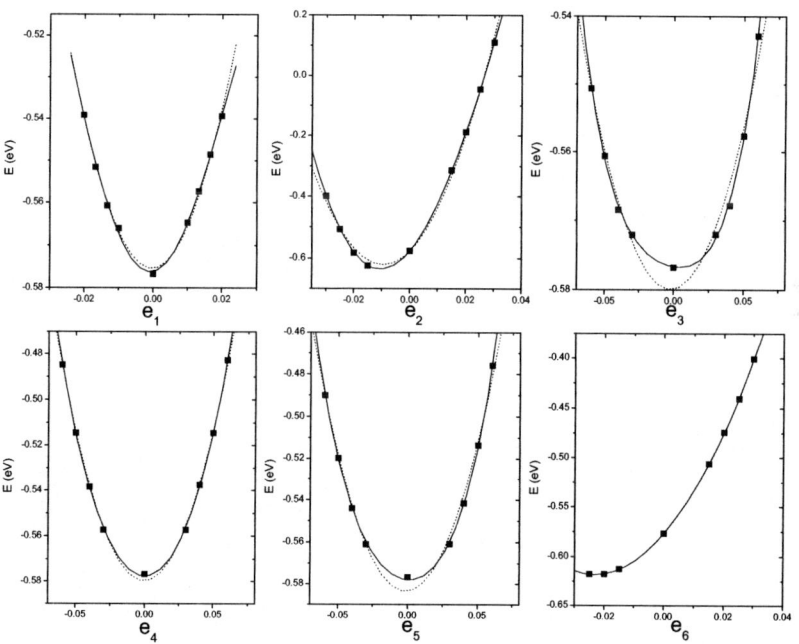

FIGURE 1. Energy versus strain for each set of strains. Strains correspond to those in Table 2. Dashed lines are from second order fits, and solid lines are fourth order fits.

TABLE 3. Elastic constants in terms of B_i, the second derivatives of the energy versus the strains e_i from table 2.

C_{11}	$(2B_1 + B_2 + 6B_4 - 3B_6)/6$
C_{33}	B_6
C_{12}	$(2B_1 + B_2 - 6B_4 - 3B_6)/6$
C_{13}	$(-4B_1 + B_2 + 3B_6)/12$
C_{14}	$(-B_3 - B_4 + B_5)/2$
C_{44}	B_3

TABLE 4. Elastic constants (GPa) and compliances ($1000\,\text{GPa}^{-1}$). Also shown is the bulk modulus and compressibility. The C's are the coefficients for the strain-energy density. The c's are for the stress-strain relations, such that the stress in the strained state T is given by $T_{\alpha\beta} = \sigma_{\alpha\beta} + c_{\alpha\beta\sigma\tau}\varepsilon_{\sigma\tau} + \sigma_{\beta\tau}\omega_{\alpha\tau} + \sigma_{\alpha\tau}\omega_{\beta\tau}$, where σ is the stress in the reference state ("unstrained"), ε is the symmetric strain relative to the reference state, and ω is the rotation (antisymmetric strain) from the reference state. The compliances are given for the stress-strain relation (c's).

C_{11}	235	c_{11}	235	s_{11}	6.75
C_{33}	180	c_{33}	180	s_{33}	6.13
C_{12}	126	c_{12}	122	s_{12}	-3.67
C_{13}	58.5	c_{13}	54.1	s_{13}	-0.93
C_{14}	-9.5	c_{14}	-9.5	s_{14}	9.4
C_{44}	8.4	c_{44}	10.5	s_{44}	111.9
K	123				

the strain that couples most strongly with the soft-modes, and is very anharmonic as can be seen in Figure 1.

Since the stress is not zero at the reference configuration, it is necessary to distinguish between the elastic constants obtained from the energy expansion (3) and those for stress versus strain [33]. The relationship between the two for hydrostatic pressures is:

$$\begin{aligned}
c_{11} &= C_{11} \\
c_{33} &= C_{33} \\
c_{12} &= C_{12} + P \\
c_{13} &= C_{13} + P \\
c_{14} &= C_{14} \\
c_{44} &= C_{44} - \frac{1}{2}P
\end{aligned}$$
(7)

where P is -4.3 GPa in our case.

We now consider various averages for application to ceramics and thin films. The elastic response of a polycrystal or ceramic depends on the texture and orientations of the crystallites [34]. The simplest, most general bounds are the Voigt bound, which assumes uniform strain, and the Reuss bound, which assumes uniform stresses. In the elastic

TABLE 5. Average bulk and shear moduli (GPa) for randomly oriented aggregates. Results are compared with experimental values for ceramic PZT 50/50 [35], and recent tentative results for PMN-33% PT and PZN-8% PT [12]. Voigt is the upper bound, Reuss the lower bound, and Hill the average estimate.

	Theory PZT	Exp. PZT	PMN-33% PT	PZN-8% PT
Voigt:				
bulk	123		112	103
shear	43		43	39
Reuss:				
bulk	117		105	101
shear	18		11	8
Hill:				
bulk	120	76	108	102
shear	31	30	27	24

TABLE 6. Comparison of computed PZT bulk modulus with experiments on other oxide perovskites [35–40].

	K (GPa)
PZT 50/50	123 (this study), 76 (exp.)
$CaTiO_3$	212, 179
$BaTiO_3$	196, 156, 139
$SrTiO_3$	179, 172
$PbTiO_3$	204
$PbZrO_3$	102
PMN-33% PT	112
PZN-8% PT	103

limit in mechanical equilibrium under load (for example in a well annealed sample under stress), the strain should approach that of the Reuss limit. The acoustic response is usually close to an average of the Reuss and Voigt limits, called the Hill estimate. We do not consider the variational Hashin-Strickman bounds here, which are narrower. The bounds are compared with experimental data on ceramic PZT 50/50 (which is probably tetragonal) in Table 5. The Hill estimate is very close for the shear modulus, but the experimental bulk modulus is significantly lower, perhaps due to porosity in the sample and/or LDA error. Actually values even for bulk moduli are hard to come by, so it is not clear if there is a problem with the theory or experiments for the bulk modulus. What is very clear though is that the large coupling materials, PZT 50/50, PMN-33% PT and PZN-8% PT are elastically very soft compared with other perovskite oxides, and the endmember compounds $PbTiO_3$ and $PbZrO_3$ (Table 6). In fact, the new experimental data on PMN-33% PT and PZN-8% PT [12] are very close to an elastic instability, which is also reflected in the large differences in the bounds for the average shear moduli, and their low values.

An important directionally dependent parameter for electromechanical applications is the Young's modulus, which gives the effective elastic constant for extension in one direction and the spontaneous strains in other directions under constant stress. We have computed these surfaces for PZT and for comparison, for $BaTiO_3$ and $PbTiO_3$ (Fig. 2). Young's modulus is soft in more directions than for the other materials. Young's modulus varies from about 20 to 163 GPa in PZT, and from 64 to 216, and 30 to 220 GPa in tetragonal $BaTiO_3$ and $PbTiO_3$, respectively. PZT appears to be generally softer than other oxide perovskites, and this is clearly one contribution to the larger strains obtainable under applied fields. We cannot separate the contributions from soft mode coupling and homogeneous and other modes coupled with strain. It is clear, however, that the soft mode coupling plays a major role, since the strain $i=3$ (which gives C_{44}) corresponds to flexing the rhombohedral angle, and couples strongly with the soft mode. This can be seen from the large change in the second derivative for a homogeneous strain, compared with the internal contributions from relaxation of the atomic positions (not shown).

We have also considered some different averages of Young's modulus, (a) corresponding to cubic c-axes parallel to (001), but randomly oriented around this axis, (b) multidomain single crystals with cubic c-axes parallel to (001), and local polarization along one of the ($\pm 1 \pm 11$) directions, (c) randomly oriented grains in a poled sample, with polarization pointing along the (111) direction closest to the average polarization direction and (Fig. 3). The shape of the anisotropy varies strongly with the texture, represented by the different averaging schemes. Note also that the total anisotropy decreases significantly with averaging, which may be one reason that oriented single crystal piezoelectrics tend

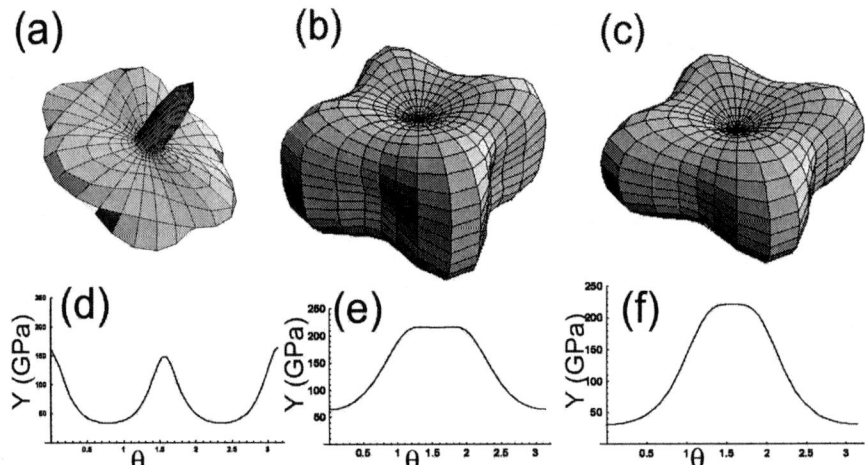

FIGURE 2. (a,d) Computed Young's modulus for rhombohedral PZT 50/50. The top figure (a) is a 3d spherical representation, with the length of the radius vector from the origin equal to the Young's modulus in that direction. The lower figure (d) shows Young's modulus as a function of direction for angles varying from the rhombohedral c-axis (cubic 111), $\theta = 0$, through the cubic (001), $\theta = 0.9553$, and around. (b,e) Experimental [38] Young's modulus for tetragonal $BaTiO_3$. (c,f) Experimental [41] Young's modulus for tetragonal $PbTiO_3$. For (e) and (f), $\theta = 0$ is the cubic (001) axis. The section along [110] is shown. PZT is much more anisotropic (and softer) than oxide perovskites.

to have much stronger coupling than poled ceramics.

We have also computed the Young's moduli for the new tentative data of Cao and Shrout [12] (Fig. 4), which are for a multidomain rhombohedral crystal. Note that it is extremely anisotropic and similar in many ways to the computed rhombohedral PZT 50/50 results, except the averaging apparently is different than our constant stress assumption. The large anisotropy in Young's modulus in these materials can be understood from the importance of polarization rotation. The (111) direction parallel to the polarization is the only direction for a Young's modulus strain that does not allow polarization rotation. The extreme anisotropy in PZN-8% PT and PMN-33% PT is a direct result of the softness for polarization rotation and the large coupling between polarization rotation and strain. Note that the bulk moduli are also soft, so that these material are soft elastically in all ways, not just with respect to shear. The PZT, PZT-PT, and PMN-PT compositions studied are all close to morphotropic phase boundaries. This is probably a key factor in the softening of the potential surface with respect to rotations. $BaTiO_3$ is also close to a phase transition, and $PbTiO3$ moderately so, but in their cases the response is not so anisotropic, and the rotational potential surface is stiffer, as was also seen in self-consistent total energy calculations for $BaTiO_3$[27].

Since elasticity and other electromechanical tensors of perovskites are often considered for the cubic cell, we have rotated our elastic constants to obtain the equivalent

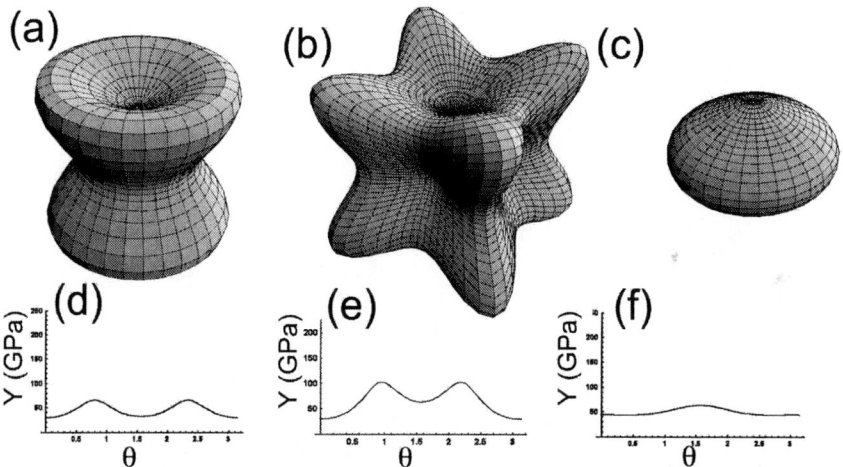

FIGURE 3. Computed Young's moduli for rhombohedral PZT 50/50 for several types of aggregates. (a,d) Rotational average around cubic (001) axis. This average is appropriate for a film or ceramic with c-axes perpendicular to the slab, and randomly oriented around this axis. See caption for fig. 2 for more details. (b,e) Multi-domain, single crystal average with cubic (001) axis vertical. This average assumes equal proportions of the four rhombohedral orientations ($\pm 1 \pm 11$) with net polarization along (001). (c,f) This average assumes random grain orientations, but for a poled sample with net polarization vertical. The assumption is that the polarization of each grain is along the rhombohedral c direction closest to the cubic (001) axis. For (e),(f), and (g) $\theta = 0$ is the cubic (001) axis. Note that the magnitude of Young's modulus is greatly decreased by the averaging process, compared with a single crystal. Also, the nature of the anisotropy is strongly dependent on the assumed texture.

TABLE 7. Computed PZT 50/50 elastic constant matrix (GPa) rotated to the cubic coordinate system.

145	112	112	3	-22	-22
112	145	112	-22	3	-22
112	112	145	-22	-22	3
3	-22	-22	61	11	11
-22	3	-22	11	61	11
-22	-22	3	11	11	61

elastic constants for the cubic axis system (Table 7). Note that the usual cubic symmetry is not present, since in fact the crystal is not cubic, but rhombohedral. Also, note that the strong anisotropy, so evident in the rhombohedral coordinate system with c along the cubic diagonal (111) direction, is hidden with c along the cubic (001) direction. The rotated constants are significantly different from the values extracted from ceramic data [35]. For example, the latter gives 111 GPa for C_{33} and 56 GPa for C_{13}, compared with 145 and 112 from Table 7. This illustrates that the elastic response of a poled ceramic cannot be considered in any way equivalent to the single crystal response. However, by modeling the texture and elastic response, it should be possible to predict ceramic behavior from the single crystal constants.

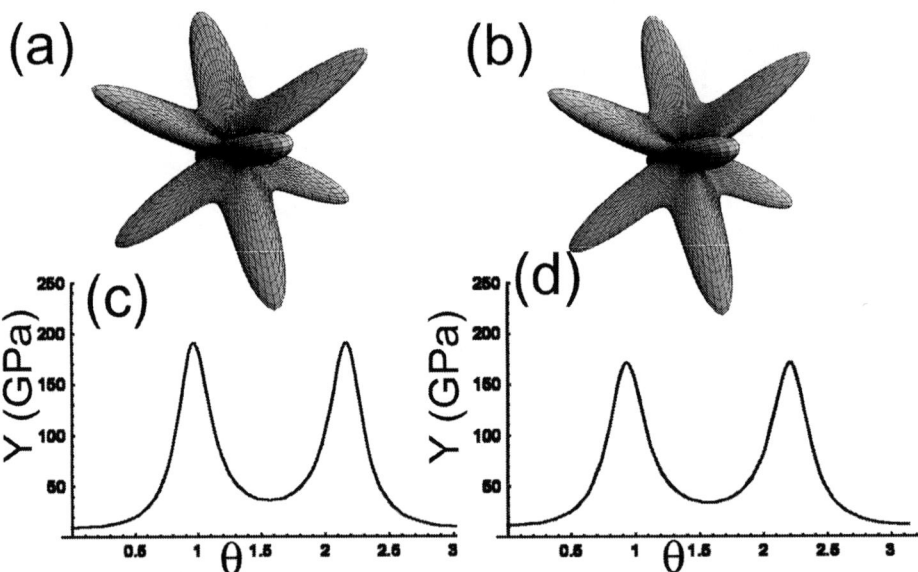

FIGURE 4. Computed Young's moduli for the tentative data [12] on (a,c) PZN-8% PT and (b,d) PMN-33% PT. The data are presented with tetragonal symmetry due to sample domains. For (c) and (d) $\theta = 0$ is the cubic (001) axis.

TABLE 8. Transformation matrix U for rotation of elastic constants in contracted Voigt notation to the cubic coordinate system.

$\frac{1}{2}$	$\frac{1}{2}$	0	0	0	$-\frac{1}{2}$
$\frac{1}{6}$	$\frac{1}{6}$	$\frac{2}{3}$	$-\frac{1}{3}$	$-\frac{1}{3}$	$\frac{1}{6}$
$\frac{1}{3}$	$\frac{1}{3}$	$\frac{1}{3}$	$\frac{1}{3}$	$\frac{1}{3}$	$\frac{1}{3}$
$\frac{\sqrt{2}}{3}$	$\frac{\sqrt{2}}{3}$	$-\frac{2\sqrt{2}}{3}$	$-\frac{1}{3\sqrt{2}}$	$-\frac{1}{3\sqrt{2}}$	$\frac{\sqrt{2}}{3}$
$\sqrt{\frac{2}{3}}$	$-\sqrt{\frac{2}{3}}$	0	$-\frac{1}{\sqrt{6}}$	$\frac{1}{\sqrt{6}}$	0
$\frac{1}{\sqrt{3}}$	$-\frac{1}{\sqrt{3}}$	0	$\frac{1}{\sqrt{3}}$	$-\frac{1}{\sqrt{3}}$	0

Since the transformation from rhombohedral to cubic coordinates will often be required as more data and theoretical elasticity computations for perovskites become available, we include for convenience the transformation U from c along the three-fold axis (cubic 111) to c along the cubic four-fold axis (cubic 001) for a fourth order tensor in Voigt notation (8). The transformation is applied as

$$C' = U^T C U. \tag{8}$$

The transformation is not orthogonal, and is equivalent to rotating the fourth rank tensor.

CONCLUSIONS

We have computed the elastic constant tensor for rhombohedral 50/50 PZT. We find that PZT is much softer elastically than other comparable oxide piezoelectrics. We find a similar picture in the new tentative data for PZN-8% PT and PMN-33% PT [12]. The high strain materials all have extreme anisotropy in the Young's modulus, and are soft in may directions. Understanding the elastic response is one of the key elements to designing better materials and using them to their best advantage in devices. These results also suggest that elasticity may be a screening diagnostic for looking for new materials and material improvements.

ACKNOWLEDGMENTS

This work was supported by the Office of Naval Research contract number N000149710052. We thank W. Cao and T. Shrout for allowing us to include their new data on PMN-PT and PZN-PT. Computations were performed on the Cray SV1 at the Geophysical Laboratory, supported by NSF EAR-9975753 and the Keck Foundation.

REFERENCES

1. Cohen, R. E., *Nature*, **358**, 136–138 (1992).
2. Cohen, R. E., and Krakauer, H., *Phys. Rev. B*, **42**, 6416–6423 (1990).
3. Cohen, R. E., and Krakauer, H., *Ferroelec.*, **136**, 65–84 (1992).
4. Bellaiche, L., and Vanderbilt, D., *Phys. Rev. Lett.*, **83**, 1347–1350 (1999).
5. Saghi-Szabo, G., Cohen, R. E., and Krakauer, H., *Phys. Rev. Lett.*, **80**, 4321–4324 (1998).
6. Saghi-Szabo, G., Cohen, R. E., and Krakauer, H., *Phys. Rev. B*, **59**, 12771–12776 (1999).
7. Bellaiche, L., and Vanderbilt, D., *Phys. Rev. B*, **61**, 7877–7882 (2000).
8. Rabe, K. M., and Cockayne, E., in *First-principles Calculations for Ferroelectrics: Fifth Williamsburg Workshop*, edited by R. E. Cohen, AIP, 1998, vol. 436, pp. 61–70.
9. Ghosez, P., Cockayne, E., Waghmare, U. V., and Rabe, K. M., *Phys. Rev. B*, **60**, 836–843 (1999).
10. Ghosez, P., Michenaud, J. P., and Gonze, X., *Phys. Rev. B*, **58**, 6224–6240 (1998).
11. Saghi-Szabo, G., Cohen, R. E., and Krakauer, H., in *First-Principles Calculations for Ferroelectrics: Fifth Williamsburg Workshop*, edited by R. E. Cohen, AIP, New York, 1998, vol. Conference Proceedings 436, pp. 43–52.
12. Cao, W., and Shrout, T. R., Personal communication (2001), unpublished.
13. Park, S. E., and Shrout, T. R., *Mat. Res. Innov.*, **1**, 20–25 (1997).
14. Park, S.-E., and Shrout, T. R., *J. Appl. Phys.*, **82**, 1804–1811 (1997).
15. Liu, S.-F., Park, S.-E., Shrout, T. R., and Cross, L. E., *J. Appl. Phys.*, **85**, 2810–2814 (1999).
16. Noheda, B., Cox, D., Shirane, G., Gonzalo, J., Cross, L., and Park, S., *Appl. Phys. Lett.*, **74**, 2059–2061 (1999).
17. Noheda, B., Gonzalo, J., Cross, L., Guo, R., Park, S., Cox, D., and Shirane, G., *Phys. Rev. B*, **61**, 8687–8695 (2000).
18. Vanderbilt, D., and Cohen, M. H., *Phys. Rev. B*, **63**, 94108–94117 (2001).
19. Bellaiche, L., Garcia, A., and Vanderbilt, D., *Phys. Rev. Lett.*, **84**, 5427–5430 (2000).
20. Cockayne, E., and Rabe, K. M., *Phys. Rev. B*, **57**, R13973–R13976 (1998).
21. Garcia, A., and Vanderbilt, D., *Appl. Phys. Lett.*, **72**, 2981–2983 (1998).
22. Uchino, K., *Acta Material.*, **46**, 3745–3753 (1998).
23. King-Smith, R. D., and Vanderbilt, D., *Phys. Rev. B*, **47**, 1651–1654 (1993).
24. Resta, R., *Rev. Mod. Phys.*, **66**, 899–915 (1994).
25. Nye, J. F., *Physical Properties of Crystals: Their Representation by Tensors and Matrices*, Oxford University Press, New York, 1985.
26. Fu, H., and Cohen, R. E., in *Fundamental Physics of Ferroelectrics 2000*, edited by R. E. Cohen, AIP, New York, 2000, vol. Conference Proceedings.
27. Fu, H., and Cohen, R. E., *Nature*, **403**, 281–283 (2000).
28. Du, X., Zheng, J., Belegundu, U., and Uchino, K., *Appl. Phys. Lett.*, **72**, 2421–2423 (1998).
29. Guo, R., Cross, L. E., Park, S., Noheda, B., Cox, D., and Shirane, G., *Phys. Rev. Lett.*, **84**, 5423–5426 (2000).
30. Gulseren, O., Bird, D. M., and Humphreys, S. E., *Surf. Sci.*, **402-404**, 827–830 (1998).
31. Troullier, N., and Martins, J.-L., *Phys. Rev. B*, **43**, 1993–2006 (1991).
32. Jaffe, H., Roth, R. S., and Marzullo, S., *J. Res. Nat. Bur. Stand.*, **55**, 239 (1955).
33. Barron, T. H. K., and Klein, M. L., *Proc. Phys. Soc.*, **85**, 523–532 (1965).
34. Anderson, O. L., *Equations of State of Solids for Geophysicists and Ceramic Science*, Oxford University Press, New York, 1995.
35. Berlincourt, D. A., Cmolik, C., and Jaffe, H., *Proceedings of the IRE*, **48**, 220–229 (1960).
36. Xiong, D. H., Ming, L., and Manghnani, M. H., *Phys. Earth Planet. Inter.*, **43**, 244 (1986).
37. Fischer, G. J., Wang, W. C., and Karato, S., *Phys. Chem. Minerals*, **20**, 97 (1993).
38. Berlincourt, D. A., and Jaffe, H., *Phys. Rev.*, **111**, 143 (1958).
39. Edwards, L. R., and Lynch, R. W., *J. Phys. Chem. Sol.*, **31**, 573 (1970).
40. Kobayashi, Y., Endo, S., Ming, L. C., Deguchi, K., Ashida, T., and Fujishita, H., *J. Phys. Chem. Sol.*, **60**, 57–64 (1999).
41. Gavrilyachenko, V. G., and Fesenko, E. G., *Soviet Phys. Cryst.*, **16**, 549 (1971).

Lattice Distortions in Pb(Zr,Ti)O$_3$ Alloys Near the Morphotropic Phase Boundary

Marco Fornari* and David J. Singh

*Center for Computational Materials Science,
Naval Research Laboratory, Washington D.C. 20375
* also at Institute for Computational Sciences and Informatics,
George Mason University, Fairfax, VA 22030*

Abstract. The lattice total energy surface and the structural instabilities that characterize rhombohedral PbZr$_x$Ti$_{1-x}$O$_3$ in the morphotropic phase boundary region near $x = 0.5$ are investigated using first principles density functional supercell calculations. As expected, we find a strong ferroelectric instability. However, we also find a substantial R-point rotational instability, close to but not as deep as the ferroelectric one. This is similar to the situation in pure PbZrO$_3$. These two instabilities are both strongly pressure dependent, but in opposite directions so that lattice compression of less than 1% is sufficient to change their ordering. Because of this, local stress fields due to B-site cation disorder may lead to coexistence of both types of instability are likely present in the alloy near the morphotropic phase boundary.

Rhombohedral PbZr$_x$Ti$_{1-x}$O$_3$ (PZT) ceramic alloys with compositions near the morphotropic phase boundary (MPB) around $x = 0.52$ form the basis of most piezoelectric transducer devices. [1] This is due to a combination of high response, realizable strains in the tenths of % range and favorable weak temperature dependencies. [2,3] Recently, there has been renewed scientific interest in these materials, driven partly by the discovery of nearly order of magnitude higher performance in related relaxor single crystals, [4,5] and partly by the fundamental understanding of them, being obtained by experimental investigation and first principles calculations. The development of effective Hamiltonian approaches, in which the number of lattice degrees of freedom is substantially reduced, has greatly helped in modelling the relevant properties of these systems; direct simulation of temperature dependent properties of these complex, strongly anharmonic materials using the full lattice Hamiltonian, *e.g.* at the density functional level, is impractical at present. Within this approach, the true Hamiltonian (or the best available approximation to it – generally the results of density functional calculations) is mapped onto a so-called effective Hamiltonian, which has a very limited number of carefully chosen degrees of freedom, and which is intended to reproduce the low energy behavior of the true Hamiltonian. Clearly, this down-folding, benefits from physical insight into the important characteristics of the true Hamiltonian.

The microscopic physics underlying the low temperature phases of the end-points is well known. In PbTiO$_3$ ferroelectricity is due to condensation of a Γ_{15} unstable phonon where the oxygen octahedra shift against the cations. The ground state structure has shifts along [001] with a tetragonal lattice strain that stabilizes this direction. A rhombohedral ferroelectric (FE) phase with [111] shifts is not favored and does not occur because of the large electronic hybridization between Pb and O, as may be seen by comparison with BaTiO$_3$, which has a rhombohedral FE ground state. [6–8]

PbZrO$_3$ has a complex anti-ferroelectric ground state [9–12] that may be viewed as arising from the ideal cubic perovskite structure by a combination of strong zone boundary instabilities. [13,14] Significantly, the ferrodistortive Γ_{15} mode that gives the FE ground state of PbTiO$_3$ is also found strongly unstable in density functional (DF) studies of cubic perovskite PbZrO$_3$ even though the ground state is not FE. Also the instabilities of cubic perovskite PbZrO$_3$ are much stronger than for PbTiO$_3$, with energies of 0.20 – 0.25 eV per formula unit. [12,15]

The actual structure arises from a delicate balance between modes characterized as octahedral rotations, octahedral distortions and off-centering. With the addition of small amounts of Ti, the PZT phase diagram shows a transition to FE behavior, in other words freezing in of the Γ_{15} instability. However, the actual situation is no doubt more complex than this. First of all, in the Zr rich part of the phase diagram, there is a low temperature rhombohedral phase in addition to the high temperature phase characterized by the pure Γ_{15} displacement. This low temperature phase is associated with a co-existence of the frozen in Γ_{15} FE instability with rotations of the oxygen octahedra. [2,16,17] Secondly, local probes indicate a complex local structure that differs substantially from the average diffraction structure in Zr rich PZT and also in the related Pb based relaxor single crystals. [5,18,17,19–21]

Complete phonon dispersions of cubic perovskite PbTiO$_3$ and PbZrO$_3$ determined by DF calculations were reported by Ghosez et al.. [22] Both materials show both FE (Γ_{15}) and rotational (R$_{25}$ type) instabilities, though in the titanate the rotational (R$_{25}$) instability is weak and occurs only in small regions of the zone around the R and M points. As mentioned, the R$_{25}$ unstable mode consists of rotation of the oxygen octahedra around the transition metal. The rapid upward dispersion away from R and M is notable; it reflects the rigidity of the octahedra. In cubic perovskite structure PbZrO$_3$ the R$_{25}$ mode is considerably more unstable and disperses upwards from R weakly, implying more deformable octahedra, a point that was associated with pressure dependencies. [23] It should, however, be noted that in materials like PbZrO$_3$ where the ideal structure is highly unstable, the size of the various instabilities is not simply related to the magnitudes of the corresponding imaginary phonon frequencies in the cubic structure. For example, the phonon frequencies indicate that the R and M point instabilities associated with octahedral rotation in PbZrO$_3$ are much more unstable than the FE Γ_{15} mode, [14,22] but, energetically, the FE mode is more unstable. [12,15]

Here, we use simple ordered supercells with composition $x = 0.5$ to investigate the relative strengths of the FE and rotational instabilities in this region. We find that the FE instability is stronger, as expected, but only marginally so, and that the balance between these may be easily switched by modest strains at the sub 1% level. Noting the difference in ionic radii of Ti and Zr (and the cell volumes of PbZrO$_3$ and PbTiO$_3$) and that there is no evidence for cation ordering in PZT near the MPB, [24] one may conclude, first of all, that these two types of instability may coexist in alloys near the MPB and, secondly, that it may be helpful to build rotational degrees of freedom into effective Hamiltonians.

We focus on the rhombohedral side of the MPB neglecting the rhombohedral strain that is known to be very small in contrast to the tetragonal phase. [25] The supercell used to model the alloy in most of the work reported here is a 10 atom FCC cell with alternating Ti and Zr layers along the [111] direction. This is the same as one of the cells used by Sághi-Szabó et al. [8] to investigate piezoelectricity on the tetragonal side of the MPB and Ramer et al. to investigate microscopic stress fields. [26,27] Some additional results are reported for a [100] ordered 10 atom supercell, which then has tetragonal P4/mmm symmetry and features alternating sheets of Zr and Ti atoms (so each B-site cation has four like B-site neighbors and two unlike B-site neighbors), and a [110] ordered 10 atom supercell. This latter cell, which also has tetragonal symmetry P4/mmm, features lines of like B-site neighbors running along the [001] direction, so each B-site cation has two like and four unlike B-site neighbors). The present DF calculations were done within the Hedin-Lundqvist local density approximation (LDA) using the general potential linearized augmented plane-wave method [28] with local orbital extensions to relax linearization errors and treat semi-core states. [29] The Brillouin zone samplings were done using $4 \times 4 \times 4$ special **k** point meshes (note this is for the doubled perovskite cell). Well converged basis sets of over 1650 functions were used with sphere radii of 2.0, 1.83, 2.25 and 1.47 a_0 for Zr, Ti, Pb and O, respectively. Forces were calculated by the method of Yu et al.. [30]

Calculations were done for three lattice parameters in 1% increments, i.e. 7.555, 7.631 and 7.708 a_0. While the ideal low temperature cubic perovskite structure cannot be accessed experimentally, Jaffe et al. [31] report an effective (by cell volume in the tetragonal phase) value of 7.73 a_0 at room temperature, while Noheda et al. [33] obtain 7.69 a_0 for both the tetragonal (slightly above room temperature) and monoclinic (20 K) phases at a slightly more Zr rich composition, $x = 0.52$. Within the LDA we obtain a slightly smaller effective lattice parameter for the supercell: 7.55 a_0, with cubic symmetry (but including the O breathing), and 7.59 a_0 with a full relaxation into the ferroelectric structure. Such $1-2\%$ smaller lattice parameters relative to experiment are typical of LDA errors in this class of materials. [12,34] In the ideal cubic structure, the [111] supercell has one internal parameter corresponding to breathing of the octahedron around Ti.

At $a = 7.631$ a_0 relaxation of this gives an energy gain of 33 mRy (all energies are for the 10 atom cell - see Table 1) with a 0.116 a_0 reduction of the Ti-O bond lengths – very close to what would be expected (0.11 a_0) based on the difference in Ti and Zr ionic radii. This symmetric breathing corresponds to the highest phonon branch

TABLE 1. Calculated energetics of the full symmetry O breathing in the [111] [001] and [110] supercells. The lattice parameters are constrained to the pseudocubic value $a = 7.631$ a_0. The relative energy is the energy of the breathed cell relative to the breathed [111] ordered cell and includes both the effects of B-site cation ordering and breathing but not ferroelectricity. Energies are given on a per 10 atom supercell basis.

Cell	Number of Breathed O	O Displacement (a_0)	Breathing Energy	Relative Energy
111	6 / 6	0.11	33 mRy	0
110	4 / 6	0.13	24 mRy	+11 mRy
001	2 / 6	0.15	14 mRy	+24 mRy

as shown in Table 2. The frequency of 805 cm^{-1} is higher than that calculated in either pure PbZrO$_3$ or PbTiO$_3$ (725 cm^{-1} and 612 cm^{-1}, respectively), [22] as can be expected from bond length considerations (note that the O is between a Zr and Ti yielding a short relaxed Ti-O bond length). The [110] and [001] supercells have larger O breathing displacements, but involving fewer atoms, for a smaller net energy gain. The result is that these cells which are already slightly disfavored energetically with respect to the [111] cell become more disfavored. It should be noted that while the [111] and [110] cells have a single full symmetry internal structural coordinate, corresponding to O breathing, the [001] cell has two such coordinates. These are the O breathing, for which calculations are shown, and the Pb height between the planes of Zr and Ti. Relaxation of both of these coordinates will lower the energy of this cell relative to the calculation with only the O breathed, shown in the table, and presumably reduce the energy difference between it and the [111] ordering. This would be consistent with results of Burton, which showed for other systems that Pb motions can effectively screen B-site cation orderings in energetics. [32] These results are summarized in Table 1. We note however, that the scale of the relative energies is not very high, consistent with the non-observation of B-site cation ordering in experiments.

We computed the remaining phonons compatible with a rhombohedral symmetry (these are the Γ_{15} and zone folded R$_{15}$ modes of the simple cubic perovskite) from atomic forces for a variety of small distortions about the breathed FCC structure. The dynamical matrix was determined by least squares fit to these, and then diagonalized to obtain the phonon eigenvectors and frequencies (given in Table 2) for the three volumes. Additionally, we calculated the energetics of the simple perovskite R$_{25}$ rotational mode (which, however, is not strictly speaking a true pure mode for the FCC supercell). We note that the ferroelectric and rotational modes and the Γ_{15} and folded R$_{15}$ modes belong to different symmetries and therefore do not mix at the harmonic level (*e.g.* the ferroelectric mode breaks inversion).

The 500 cm^{-1} mode in Table 2 involves mainly O displacements, like the breathing mode: small octahedra tilting mixed with stretching along [001]. The other modes above 300 cm^{-1} involve distortions of the O octahedra, while the lowest stable mode involves mainly transition metal off-centering in the octahedra. In rhombohedral symmetry there is also one marginally unstable and one unstable

TABLE 2. Calculated [111] supercell vibrational frequencies (cm^{-1}) of phonons compatible with rhombohedral symmetry.

a_L (a_0)	ω_1	ω_2	ω_3	ω_4	ω_5	ω_6	ω_7
7.555	125i	16i	158	326	357	538	838
7.631	122i	33i	150	334	341	500	805
7.708	143i	64i	137	317	324	465	764

mode – the R_{15} mode and the FE Γ_{15} instability, respectively. In the ideal perovskite cell and the true disordered alloy, they belong to the same phonon branch and a rough interpolation suggests that the FE instability extends over the entire Brillouin zone as expected by comparison with the full dispersion curve in PbZrO$_3$ (Ref. [22]).

Energy minimization along the FE eigenmode at $a = 7.631\ a_0$ yields an energy gain of 11 mRy (again per doubled cell). The local minimum compatible with rhombohedral symmetry, however, occurs away from this line and the relaxation provides futher energetic gain of 7 mRy. Its coordinates [35] compare well with the displacement pattern obtained by Sághi-Szabó et al. for the tetragonal FE state (note that at the harmonic level the tetragonal and rhombohedral eigenvectors are the same). [8] As shown in Fig. 1, the instability is energetically disfavored by compression and favored by expansion. Over the range of 2% in lattice parameter we find a variation of the energy gain along the FE eigenmode of about 5 mRy. Further, the variation in depth of the local minima are larger: 11, 18 and 22 mRy for $a = 7.555\ a_0$, $a = 7.631\ a_0$ and $a = 7.708\ a_0$ respectively.

Interestingly, the character of the ferroelectric instability differs from that in the classical perovskite ferroelectrics BaTiO$_3$ and KNbO$_3$ in that we obtain a significant ferroelectric instability against a rigid motion of a single cation sublattice in our PZT supercell. In particular if we hold all the other ions fixed, there is still a lattice instabilty against shifting the Pb ions off-center. This reflects the different nature of the ferroelectricity in PZT, where the primary instability arises because of the large volume available to Pb in the ideal perovskite structure.

The precise mechanism of the high electromechanical response in PZT and the related relaxor crystals is still not fully established but there are strong indications that polarization rotation from rhombohedral to tetragonal and the strong strain coupling for the tetragonal direction are the main ingredients. [36–38] Recently, it has been suggested both theoretically [38] and experimentally [39] that the monoclinic phase found at the MPB [33] can bridge the tetragonal and the rhombohedral phase favoring such polarization rotation. Bellaiche et al. [38] used an effective Hamiltonian approach adapted to the alloy with parameters determined from first principles calculations. The effective Hamiltonian used allowed strains, alloy disorder and off-centering displacements to interact, but octahedral rotations were not included. Nonetheless, excellent agreement with experimental data was obtained for the temperature and composition dependence of the piezoelectric and

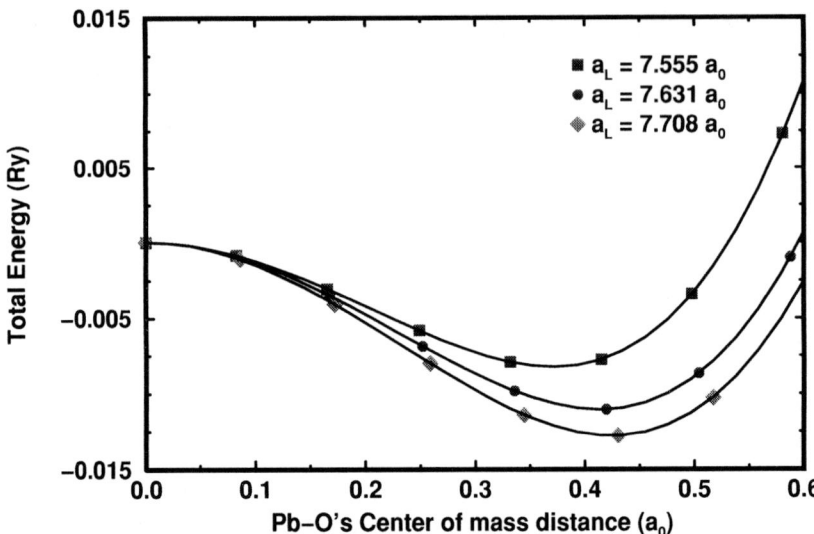

FIGURE 1. Relaxation along the FE unstable mode (energies for 10 atom supercells) for the reference volume (circles), 1% compression (squares), and 1% expansion (diamonds). The instability decreases under pressure.

structural properties in PZT, except that the temperature scale had to be uniformly adjusted downwards by approximately one third.

The microscopic mechanism that makes polarization rotation occur at modest field strengths is still somewhat uncertain though the importance of Pb chemistry has been widely recognized [8,12,22,40,41] In particular, it is not fully understood how the instabilities present in $PbTiO_3$ and $PbZrO_3$ are modified in the alloy and the extent to which they compete or perhaps coexist in real samples.

Certainly, the simple supercell discussed above is a rather severe approximation to the disordered alloy. Even though Zr and Ti have the same valence, their ionic sizes and electronic properties differ significantly. At the very least, local stresses related to Zr-Ti disorder should be expected in PZT alloys. Zr rich local regions will be under compressive stress so the volume available for the Pb ions is reduced thus lowering the local FE tendency. Ionic considerations, supported by experimental evidence, [42] predict that octahedral rotations should respond in the opposite way to compression especially if the octahedra are stiff. Consider such rotations around [001] (C_{4h} symmetry) and [111] (C_{3i} symmetry). As mentioned, these are not true modes for the supercell, but they are zone boundary rotational modes that play a key role in at least $PbZrO_3$. Both these rotations are unstable (FIG. 2).

At $a = 7.631\ a_0$, the energy gain for the [111] rotation is 9 mRy (7 mRy around [001]). It is remarkable that the sizes of the FE and rotational instabilities are so similar for this $x = 0.5$ supercell, considering that they are also quite close,

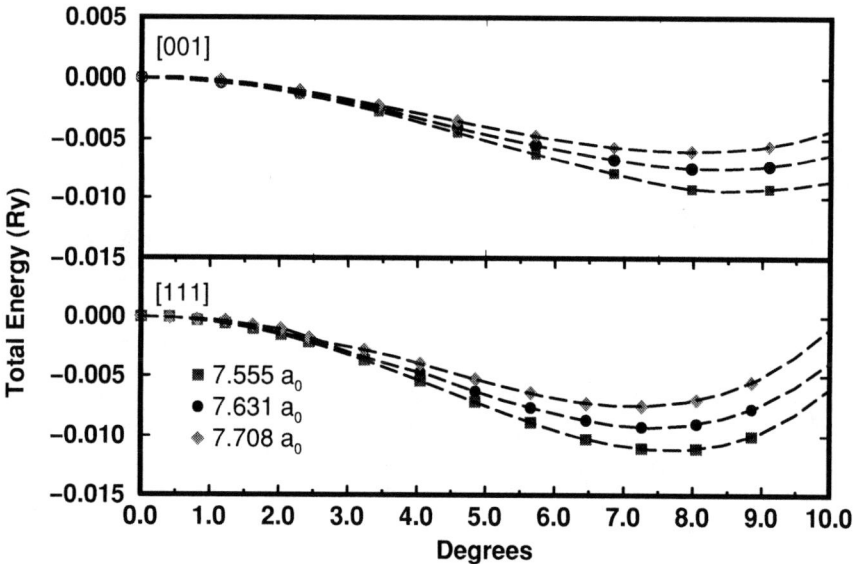

FIGURE 2. Variation of the energy (of 10 atom supercells) with "rotation" (a) around [111] and (b) around [001], for the reference lattice parameter (circles), under compression (squares) and under tensile stress (diamonds), as in Fig. 1.

again with the FE instability lower, for $PbZrO_3$ though with a much higher energy scale. Presumably, they track each other across the rhombohedral side of the phase diagram. This can be qualitatively understood in terms of ionic size effects. Unlike $BaTiO_3$ or $KNbO_3$, [43,44] the ferroelectric mode in these Pb based materials is best described as a Pb off-centering with respect to the surrounding O with a smaller transition metal displacement in the same direction. This is clearly seen, *e.g.* in the Γ_{15} phonon eigenvector. Thus the FE instability is controlled by the volume available to the Pb ion; the rotational mode also involves changes in Pb-O bond length, again controlled by the same distances.

However, a 1% compression (to $a = 7.555\ a_0$) increases the rotational instabilities to 11 mRy about [111] and 9 mRy around [001]. Qualitatively, this is related to the stiffness of the O octahedra, especially around Ti; under pressure the octahedra size and geometry can only be retained by rotation, which is made at the expense of the relatively soft Pb-O interaction. This driving force is clearly absent for the FE instability. Experimental evidence of opposite stress dependencies of rotational (R_{25}) and FE (Γ_{15}) instabilities was discussed for $PbZrO_3$ early on. [42] Stress-field response calculations [26] for $PbZr_{0.5}Ti_{0.5}$ supercells suggest that competition between different distortions may be important when uniaxial stress is not parallel to the FE distortion. Considering the different experimental cell volumes of the end-points (effective $a(PbZrO_3) = 7.7883\ a_0$ and $a(PbTiO_3) = 7.5028\ a_0$) [45,46], the existance of local stresses sufficient to tip the order of the FE and rotational instabilities seem quite reasonable in the disordered alloy. As noted, the FE instability occurs at the Γ and R points and probably over most or all of the zone. Thus it can occur on a very local scale in real space. The rotational instability is no doubt more localized in reciprocal space, so several octahedra must rotate in concert within a region, but this region may be realizably small, as the ZrO_6 octahedra may be soft as in $PbZrO_3$ providing weak connections in the interlinked network. In this scenario, coexistence of rotational and FE distortions in the alloy is expected. The resulting disorder might increase response away from the MPB, though at the expense of maximum attainable response at the best composition – this would be favorable for obtaining a desirable weak temperature dependence of the response.

In any case, it is clear that rotational degrees of freedom are in the low energy space along with the FE mode, and presumably should be considered in the construction of effective Hamiltonians, either via explicit additional coordinates or renormalization of the existing FE and strain related coordinates. This could possibly improve the temperature scale. Because the balance between the rotational and ferroelectric instabilties is very strongly volume dependent it may well be that the thermal expansion of the lattice will play a more significant role when both of these types of distortion are included.

We thank R.E. Cohen, L. Bellaiche, T. Egami and B. Burton for helpful discussions. The code FINDSYM by H. T. Stokes and L. Boyer was used for some symmetry analysis. This work is supported by ONR and the DoD ASC computer center.

REFERENCES

1. K. Uchino, *Piezoelectric Actuators and Ultrasonic Motors* (Kluwer Academic, Boston, 1996).
2. B. Jaffe, W. R. J. Cook, and H. Jaffe, *Piezoelectric Ceramics* (Academic Press, New York, 1971).
3. M. E. Lines and A. M. Glass, *Principles and applications of ferroelectrics and related materials* (Clarendon, Oxford, 1977).
4. S.-E. Park and T. R. Shrout, J. Appl. Phys. **82**, 1804 (1997).
5. T. Egami, W. Dmowski, M. Akbas, and P. K. Davies, in *First Principles Calculations for Ferroelectrics*, edited by R. Cohen (AIP, New York, 1998), Vol. 436, p. 1.
6. R. E. Cohen, Nature **358**, 137 (1992).
7. U. V. Waghmare and K. M. Rabe, Phys. Rev. B **55**, 6161 (1997).
8. G. Sághi-Szabó, R. E. Cohen, and H. Krakauer, Phys. Rev. B **59**, 12771 (1999).
9. H. Fujishita and S. Katano, J. Phys. Soc. Jpn. **66**, 3484 (1997).
10. H. Fujishita and S. Katano, Ferroelectrics **217**, 17 (1998).
11. S. Teslic and T. Egami, Acta Cryst. B **54**, 750 (1998).
12. D. J. Singh, Phys. Rev. B **52**, 12559 (1995).
13. H. Fujishita and S. Hoshino, J. Phys. Soc. Jpn. **53**, 226 (1984).

14. W. Zhong and D. Vanderbilt, Phys. Rev. Lett. **74**, 2587 (1995).
15. D. J. Singh, Ferrroelectrics **194**, 299 (1997).
16. D. Viehland, J. F. Li, X. H. Dai, and Z. Xu, J. Phys. Chem. Sol. **57**, 1545 (1996).
17. D. L. Corker, A. M. Glazer, R. W. Whatmore, A. Stallard, and F. Fauth, J. Phys.: Condens. Matter **10**, 6251 (1998).
18. W. Dmowski, M. K. Akbas, P. K. Davies, and T. Egami, J. Phys. Chem. Solids **61**, 229 (2000).
19. T. Egami, W. Dmowski, M. Akbas, and P. K. Davies, Ferroelectris **436**, 1 (1998).
20. S. Teslic, T. Egami, and D. Viehland, J. Phys. Chem. Sol. **57**, 1537 (1996).
21. X. Dai, Z. Xu, and D. Viehland, J. Appl. Phys. **80**, 1919 (1996).
22. P. Ghosez, E. Cockayne, U. V. Waghmare, and K. M. Rabe, Phys. Rev. B **60**, 836 (1999).
23. E. Cockayne and K. M. Rabe, J. Phys. Chem. Solids **61**, 305 (2000).
24. C. A. Randall, A. S. Bhalla, T. R. Shrout, and L. E. Cross, Ferroelectr. Lett. **11**, 103 (1990).
25. R. W. Whatmore and A. Glazer, J. Phys. **C12**, 1505 (1979).
26. N. J. Ramer, E. J. Mele, and A. M. Rappe, Ferroelectrics **296**, 31 (1998).
27. N. J. Ramer and A. M. Rappe, J. Phys. Chem. Sol. **61**, 315 (2000).
28. D. J. Singh, *Planewaves, pseudopotentials and the LAPW method* (Kluwer Academic, Boston, 1994).
29. D. Singh, Phys. Rev. B **43**, 6388 (1991).
30. R. Yu, D. J. Singh, and H. Krakauer, Phys. Rev. B **43**, 6411 (1991).
31. B. Jaffe, R. S. Roth, and S. Marzullo, J. Res. Nat. Bur. Stand. **55**, 239 (1955).
32. B.P. Burton, J. Phys. Chem. Solids **61**, 327 (2000).
33. B. Noheda, D. E. Cox, G. Shirane, L. E. Cross, and S.-E. Park, Appl. Phys. Lett. **74**, 2059 (1999).
34. N. Marzari and D. J. Singh, Phys. Rev. B **62**, 12724 (2000), and refs. therein.
35. The atomic positions of the atoms in the relaxed FE cell are: (0.0, 0.0, 0.0) and $(1.0+\delta_{Pb}, 1.0+\delta_{Pb}, 1.0+\delta_{Pb})$ for Pb's, $(1.5+\delta_{Zr}, 1.5+\delta_{Zr}, 1.5+\delta_{Zr})$ for Zr, $(0.5+\delta_{Ti}, 0.5+\delta_{Ti}, 0.5+\delta_{Ti})$ for Ti, $(0.5+\delta_{O1}^x, 0.5+\delta_{O1}^x, 0.0152+\delta_{O1}^z)$, $(0.5+\delta_{O1}^x, 0.0152+\delta_{O1}^z, 0.5+\delta_{O1}^x)$, $(0.0152+\delta_{O1}^z, 0.5+\delta_{O1}^x, 0.5+\delta_{O1}^x)$, $(0.5+\delta_{O2}^x, 0.5+\delta_{O2}^x, 0.9848+\delta_{O2}^z)$, $(0.5+\delta_{O2}^x, 0.9848+\delta_{O2}^z, 0.5+\delta_{O2}^x)$, $(0.9848+\delta_{O2}^z, 0.5+\delta_{O2}^x, 0.5+\delta_{O2}^x)$ for the 6 O's. The energy minimun occurs for δ_{Pb}=-0.006, δ_{Zr}=0.032, δ_{Ti}=0.032, δ_{O1}^x=0.069, δ_{O1}^z=-0.084, δ_{O2}^x=0.069, δ_{O2}^z=0.089. The coordinates are in terms of the lattice parameter a_L= 7.631 a_0.
36. S. E. Park and T. R. Shrout, J. Appl. Phys. **82**, 1804 (1997).
37. H. Fu and R. E. Cohen, Nature **281**, 403 (2000).
38. L. Bellaiche, A. Garcìa, and D. Vanderbilt, Phys. Rev. Lett. **84**, 5427 (2000).
39. R. Guo, L. E. Cross, S.-E. Park, B. Noheda, D. E. Cox, and G. Shirane, Phys. Rev. Lett. **84**, 5423 (2000).
40. L. Bellaiche, J. Padilla, and D. Vanderbilt, Phys. Rev. B **59**, 1834 (1999).
41. B. P. Burton and E. Cockayne, Phys. Rev. B **60**, 12542 (1999).
42. G. A. Samara, Phys. Rev. B **1**, 3777 (1970).
43. R. E. Cohen and H. Krakauer, Phys. Rev. B **42**, 6416 (1990).
44. D. J. Singh and L. L. Boyer, Ferroelectrics **136**, 95 (1992).

45. T. Mitsui and al., *Numerical data and functional relationships in science and technology* (Springer-Verlag, New York, 1981).
46. E. Sawaguchi, J. Phys. Soc. Japan **8**, 615 (1953).

Structure of Pb(Zr,Ti)O₃ Near the Morphotropic Phase Boundary

W. Dmowski, T. Egami, L. Farber and P.K Davies

*Department of Materials Science and Engineering, University of Pennsylvania,
Philadelphia, PA 19104-6272, USA*

Abstract. The atomic structure of Pb(Zr,Ti)O₃ solid solutions near the morphotropic phase boundary (MPB) was examined using time-of-flight neutron diffraction. In addition to the conventional crystallographic refinement the atomic pair distribution function (PDF) analysis was used. Crystallographic analysis suggests that the average structure changes significantly through the MPB. However, the PDF analysis, which is more accurate in describing the local structure, shows that changes are more gradual. In particular the PDF suggests that the local environment of each element remains relatively invariant of composition. Ti is always ferroelectrically active, while Zr is not. Pb is always displaced against oxygen atoms and forms short bonds at about 2.45 Å. What changes most with the Ti/Zr ratio is the distribution in the direction of Pb displacement. It is suggested that the population of local Pb displacements changes between <100> and <110> in the pseudocubic notation with the Ti/Zr ratio, and the MPB is a crossover point with maximum disorder.

INTRODUCTION

The solid solution of PbZrO₃ (PZ) and PbTiO₃ (PT), Pb(Zr$_{1-x}$Ti$_x$)O₃ (PZT), is widely applied because of its high dielectric permittivity and excellent electro-mechanical coupling. On the Zr-rich side PZT is antiferroelectric below 250°C and up to 7% of Ti. With higher amounts of Ti it becomes ferroelectric. The PT-PZ phase diagram is characterized by the almost vertical morphotropic phase boundary (MPB) at $x = 0.52$, which separates the Zr-rich rhombohedral phase (R3m or R3c depending on temperature and Ti content) from the tetragonal Ti-rich phase (P4mm) [1]. At this phase boundary PZT exhibits exceptionally high piezoelectric response, which is stable over a wide range of temperature. The extremely high electromechanical coupling observed in this region is related to the presence of the MPB. In the rhombohedral phase the ferroelectric polarization is along the pseudo-cubic <111>$_c$ direction, while in the tetragonal phase it is along the <001>$_c$ direction. Near the MPB the free energies of the two phases are close to each other, making it easy for the polarization to rotate from one direction to another when the external electric field is applied. Intuitively this is the basis for the high electromechanical coupling. However, it is not obvious why the height of the energy barrier between the two phases is so low.

Recently a previously unknown monoclinic phase Cm(4) was discovered at low temperatures along the MPB by the high-resolution x-ray diffraction study of PZT [2]. Apparently this phase provides the bridge between the tetragonal and rhombohedral phases and facilitates polarization rotation. It was suggested [2] that this phase results from the "freezing out" of the local displacements of Pb from the average crystallographic sites, either in the $<110>_c$ directions in the tetragonal or $<100>_c$ in the rhombohedral phase. However, the real structure of PZT is much more complex than the crystal structure suggests. For instance electron microscopy (TEM) studies by Viehland et al. [3], Dai et al. [4], and Ricote et al. [5] reported extra reflections, not consistent with average rhombohedral structure, suggesting the presence of small domains with ordered cation displacements. In addition conventional crystallographic studies encounter many problems when there are local atomic displacements from the high symmetry positions [6]. Since the results of TEM studies and anisotropic temperature factors of Pb atoms cannot be explained in terms of the average structure, a model with domain-type local structure had been proposed [3-6]. The earlier study of PZT in the rhombohedral phase using the pulsed neutron pair distribution function (PDF) analysis [7] suggested that the direction of local Pb polarization is closer to the $<110>_c$ even though the average polarization is along $<111>_c$. In the present work we studied the local structure of PZT across the MPB using the pulsed neutron PDF analysis as well as the crystallographic analysis. In the PDF analysis the interatomic distances are obtained by the direct Fourier transformation of the normalized structure factor, without the assumption of unit cell symmetry. It, therefore, can describe the local structure of complex systems with much higher accuracy than the conventional crystallographic methods [8,9].

EXPERIMENTAL DETAILS

We have examined three compositions of PZT with 60%-Ti (tetragonal phase, P4mm), 48%-Ti (monoclinic phase, Cm) and 40%-Ti (rhombohedral phase, R3c, the so called "low temperature rhombohedral" phase), in addition to pure PT, at the temperature of 10 K. Powder samples were prepared by the standard solid state ceramic processing method. Powder diffraction measurements were carried out at the GPPD spectrometer at the IPNS of the Argonne National Laboratory using the pulsed neutron time-of-flight method at 10 K. The Rietveld analysis was carried out using the GSAS program [10]. The high-resolution ($\Delta d/d =0.26\%$) 145 deg detector bank was examined with the d-spacing range of 0.45 - 2.5 Å. We refined the unit cell parameters, atomic positions and temperature factors in addition to the background, scale factor and particle broadening. The same data sets after corrections and normalization were used to obtain the atomic pair distribution functions.

RESULTS OF THE RIETVELD ANALYSIS

The neutron diffraction Rietveld refinement results on the pulsed neutron diffraction data are presented in the Table 1. Similar to Ref. 6 we have allowed the positions of Zr and Ti atoms to be refined independently, even though they occupy the same crystallographic sites, resulting in the improvement of the agreement factors. Most critical for the refinement were the Pb and O positions and temperature factors. In general the Rietveld factors, R_{wp} and R_P, are small and comparable to those of pure PT (shown also in the Table 1) and PZ [11]. Notably for PZT samples, even at 10K, thermal factors of Pb and oxygen atoms are large, which indicates significant atomic displacements from the average positions, in agreement with other results [2,6].

Table 1. Rietveld Results for PZT Solid Solution

$PbTiO_3$, P4mm, a=3.90044 c=4.15115, wR_p=0.07, R_p=0.049, T=300 K.					
Atom	X	Y	Z	Fraction	Uiso [Å^2]
Pb	0.5	0.5	0.5	1.0	0.005
Ti	0.0	0.0	0.0337	1.0	0.006
O	0.5	0.0	0.1123	1.0	0.006
O	0.0	0.0	0.6088	1.0	0.007
$PbZr_{0.4}Ti_{0.6}O_3$, P4mm, a=3.992 c=4.1758, wR_p=0.079, R_p=0.053, T=10 K.					
Pb	0.5	0.5	0.5	1.0	0.010
Zr	0.0	0.0	0.0446	0.4	0.003
Ti	0.0	0.0	0.0304	0.6	0.003
O	0.5	0.0	0.1310	1.0	0.009
O	0.0	0.0	0.6095	1.0	0.010
$PbZr_{0.52}Ti_{0.6}O_3$, Cm, a=5.7154 b=704 c=4.1378, β=90.493 wR_p=0.088, R_p=0.061, T=10 K.					
Pb	0.0	0.0	0.0	1.0	0.015
Zr	0.51	0.0	0.4404	0.48	0.011
Ti	0.4840	0.0	0.4433	0.6	0.005
O	0.5423	0.0	-0.0899	1.0	0.007
O	0.2891	0.2595	0.3777	1.0	0.014
$PbZr_{0.6}Ti_{0.46}O_3$, R3c, a=5.756 c=14.242, wR_p=0.093, R_p=0.063, T=10 K.					
Pb	0.0	0.0	0.2746	1.0	0.011
Zr	0.0	0.0	0.0080	0.6	0.003
Ti	0.0	0.0	0.0180	0.4	0.003
O	0.1937	0.3445	0.0739	1.0	0.013

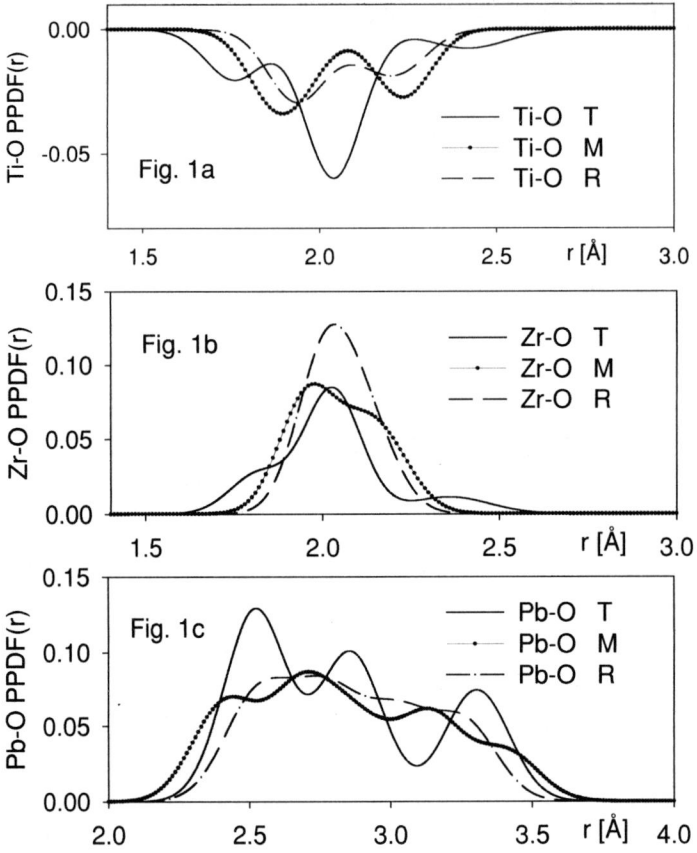

FIGURE 1. Partial pair distribution functions calculated using the crystallographic Rietveld refinement: a) for Ti-O, b) for Zr-O and c) for Pb-O. T, M and R denote tetragonal (60% Ti), monoclinic (42% Ti) and rhombohedral (40% Ti) phases, respectively.

Using the results from the Table 1 we have calculated the partial pair distribution functions (PPDF) weighted by the neutron scattering lengths, which can be compared with the experimental results. Figure 1 shows the model calculations based on the Rietveld refinement. Since Ti has a negative neutron scattering length any correlation with an atom that has a positive neutron scattering length (O, Pb, Zr) produces negative peaks in the PPDF. The plot for Ti-O distribution (Fig. 1a) indicates that in the tetragonal (T) phase Ti atom is displaced towards the apex oxygen atom, along $[001]_c$, in a similar way as in pure PT. This move creates three peaks in the plot corresponding to the split of the Ti-O bonds into three groups, and indicates local Ti polarization. For the monoclinic (M) and rhombohedral (R) phases the Ti-O distances separate into two groups resulting from the displacement of Ti toward the face of the oxygen octahedron, along <111>$_c$. Fig 1b reveals that according to the Rietveld analysis Zr in the T phase is polarized by as much as 0.2 Å in the [001]$_c$ direction in a similar way to Ti. In the M phase displacements of Zr against the O atoms are of the order of 0.1 Å and along the triad axis, <111>$_c$. In the R phase there is a single Zr-O peak indicating small distribution of Zr-O distances and negligible polarization.

The picture is more complex in the case of Pb-O correlation as shown in Fig 1c. The Rietveld result suggests simple Pb displacements along the [001]$_c$ direction in the T phase. The distribution of the Pb-O distances is broader in the M phase and somewhat reduced in the R phase. The direction of Pb displacements is dictated by the symmetry in the M and R phases. In the Rietveld analysis the effect of static displacements of atoms from the high symmetry positions shows up in the artificially large thermal factors. It is interesting to compare the thermal factors of Pb for different Ti compositions including pure PT and PZ. Figure 2 shows the Pb thermal factor as obtained from the neutron Rietveld refinement at 10K with the exception of PT which was measured at 300K. We can note that Pb thermal factors are much smaller in PZ and PT (at 300K !) than in the PZT samples. In addition the maximum is observed at the MPB. This indicates that disorder manifested by displacements from the average high symmetry positions is largest at the morphotropic phase boundary, in agreement with the recent theoretical study [12].

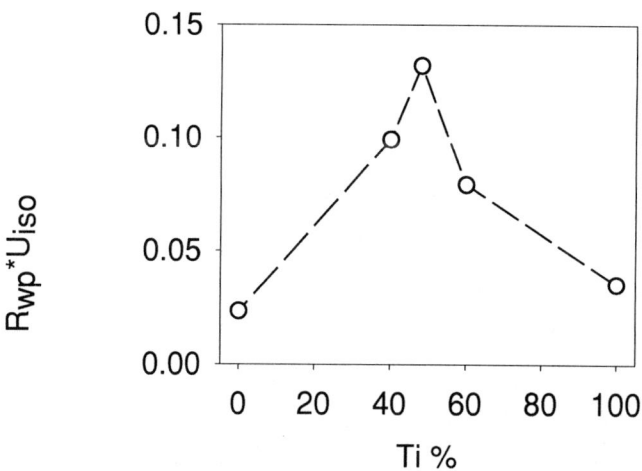

FIGURE 2. Thermal factors of Pb obtained from the crystallographic refinement at 10K. The values are multiplied by the agreement factor (R_{wp}) for each phase.

Large temperature factors suggest that the Rietveld results may provide an inaccurate, or wrong, picture of the local atomic structure and the microscopic changes through the MPB. A better description of the local structure is provided by the atomic pair distribution function, which is obtained without assumption of the crystal symmetry. Before examining PDF's of the PZT samples it is interesting to review the data for PZ and PT.

RESULTS OF THE PDF ANALYSIS

Pure PT

$PbTiO_3$ (PT) is one of the simplest ferroelectric materials [13]. At the temperature of 766 K it exhibits a displacive transition from the cubic to tetragonal phase. From the structural studies it was found that in PT both Ti and Pb are displaced against slightly distorted oxygen octahedron [14]. This contribution from both Ti and Pb provides the basis for strong polarization of PT. PT is a good reference point in the studies of other Pb based perovskite ceramics. Structural parameters are listed in Table 1. They are in good agreement with the original work [14]. Here we emphasize some important features evident from this table. The refined thermal factors are small and isotropic, and their magnitudes are reasonable for room temperature. These values provide a reference for thermal factors in other samples. In the ideal perovskite structure Pb is surrounded by 12 equally distanced oxygen atoms, but in the tetragonal PT Pb forms four

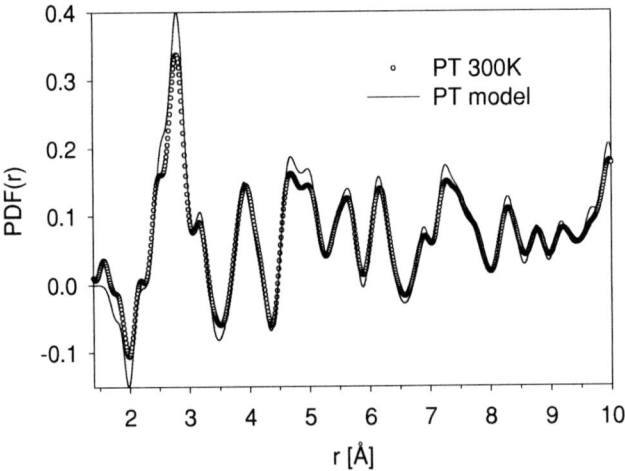

FIGURE 3. Model PDF of pure PT calculated using results of the Rietveld refinement (solid line) compared with the experimental data (circles).

short Pb-O bonds in a similar way as in PbO [14]. As a result 12 Pb-O distances are split into 3 groups of short (2.54 Å) and longer (2.8, 3.2 Å) bonds. The oxygen octahedron around Ti is slightly distorted with the O-O distances of 2.78 and 2.85 Å, 2.8 Å being ideal.

The neutron PDF of PT is shown in the Figure 3. The experimental data are shown by circles while the solid line represents the calculated PDF based on the atomic positions and temperature factors from the Rietveld refinement. Since the neutron scattering length of Ti is negative, correlations involving Ti and other atoms show up as negative peaks. The first negative peak corresponds to Ti-O correlations and shows a clear split, reflecting off-centering of Ti in the oxygen octahedron. The three-way split of Pb-O distances is seen at 2.5, 2.8 and 3.2 Å. It is seen that the calculated PDF is in very good agreement with the experimental PDF. This indicates that indeed atoms in PT are in the high symmetry positions as refined by the Rietveld method and the local atomic structure is the same as the average structure.

Pure PZ

At the other end of the PZ-PT diagram, $PbZrO_3$ is considered to be a classic example of an antiferroelectric material. The structure of PZ is more complex and have been reexamined several times [e.g. 15,16,11 and references there in]. The most recent [11] study of powder neutron diffraction used both the Rietveld and PDF analyses. The structure of PZ is well characterized at low temperatures by an orthorhombic Pbam (#55) symmetry. The agreement between the experimental PDF and the PDF calculated from the Rietveld refinement is very good [11]. Pb atoms are arranged in double rows, and are displaced in the $[110]_c$ (pseudo-cubic) directions in the antiparallel fashion. The oxygen octahedra are tilted about the $[1\bar{1}0]_c$ axis. As a result of Pb displacements and O_6 tilts the Pb environment is very different from that in the ideal perovskite. In fact both the PDF and Rietveld refinement indicate that Pb-O distances are split into three groups of ~2.45, 3.08 and 3.4 Å. Similarly to PT, Pb forms 4 short bonds with oxygen ions and is off-centered, but in more complex structural rearrangements. This off-centering results in strong Pb polarization.

The average Zr-O distance is 2.09 Å with a dispersion of 0.024 Å. The Zr-O bondlength is too long to be accommodated within the O_6 octahedron, since the ideal B-O distance is 2.00 Å. This results in a small distortion of the octahedron and bending of the O-Zr-O bond. Off-centering of Zr from the center of mass of the O_6 octahedron is 0.20 Å, and produces some ZrO_6 polarization. However, this polarization is *transverse* with respect to the Zr-O bond (due to bending of the O-Zr-O bond), and not *longitudinal*. So the polarization must be purely electrostatic and must not involve electronic activity. The agreement between experimental and calculated PDF is very good [11] despite the fact that the structure is much more complex than PT, confirming that the average structure is the same as the local structure.

PZT Solid Solutions

Figure 4 presents the PDF's calculated from the Rietveld analysis (4a) and those obtained directly from the experimental data (4b). The Rietveld results (4a) show that short Pb-O and B cation-oxygen distances change significantly through the MPB. In addition there is a significant overlap of the Ti-O and Zr-O peaks, resulting in cancellation of each other, since according to the average structure both Ti and Zr are equally polarized. However, the experimental PDF's show much smaller changes with composition, despite the phase changes. In particular the position of the short Pb-O peak at 2.45 Å is almost independent of composition. The Ti-O and Zr-O peaks in the

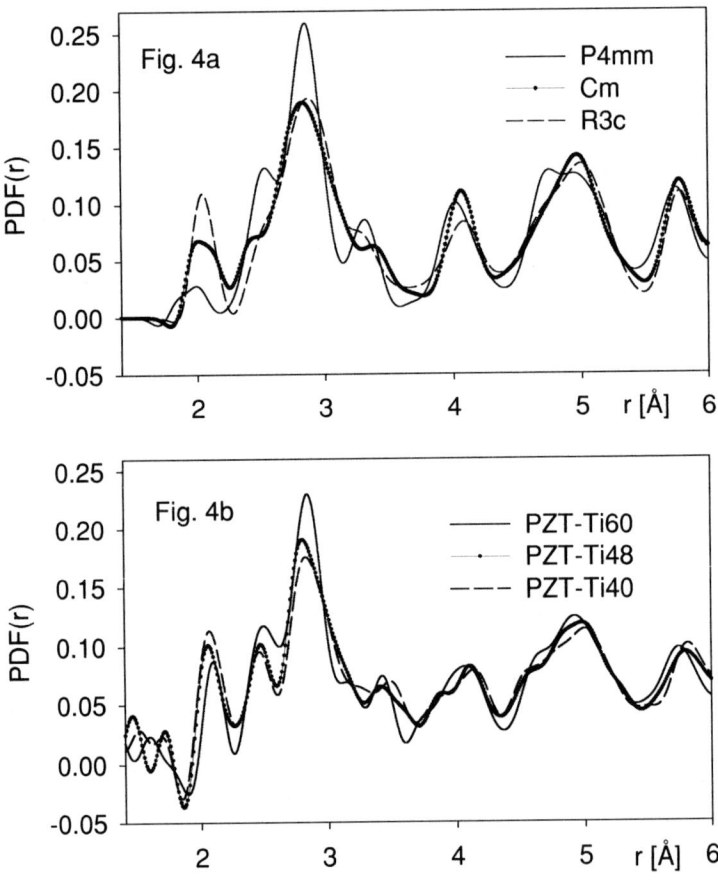

FIGURE 4. Top figure shows PDF's calculated using structural parameters obtained from the Rietveld analysis. Experimental PDF's are displayed below.

experimental PDF are clearly separated with small overlap. Overall agreement between the calculated and experimental PDF's is the best for the T phase and the worst for the R phase. The experimental and calculated PDF's in the short range region are compared in Figure 5 for three phases. They show that the atomic positions determined by the Rietveld analysis are incorrect.

For the T phase (Fig. 5a) the Ti-O distance suggested by the Rietveld analysis is 2.0 Å, while in the experimental PDF the main Ti-O peak is at 1.9 Å. The Ti-O peak is similar to that in PT, suggesting [001] polarization of Ti. In the M phase (Fig 5b) and the R phase (Fig. 5c) a single Ti-O peak is seen in the experimental PDF at 1.86 Å (T)

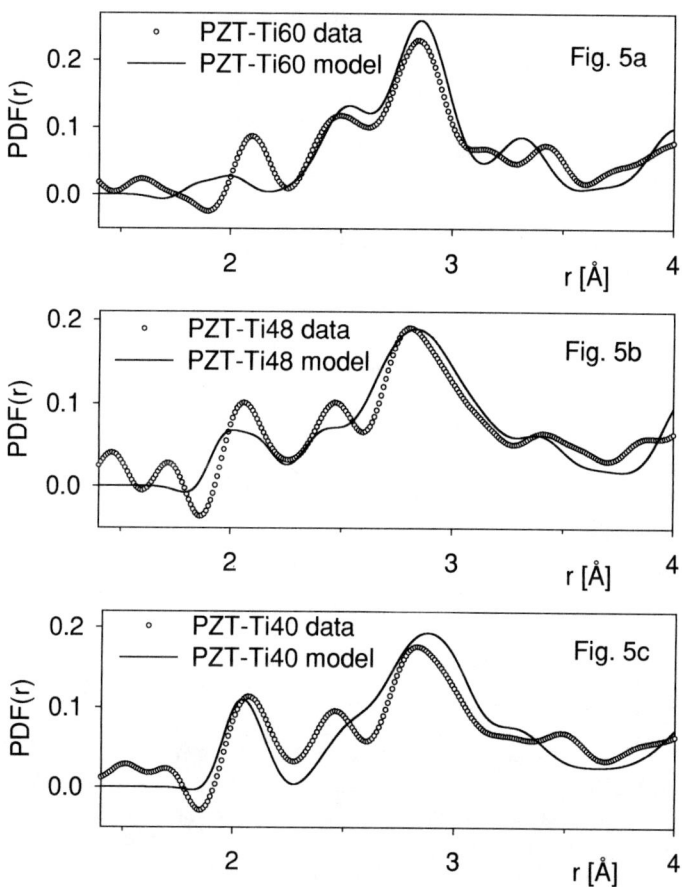

FIGURE 5. Comparison of the short range atomic order for three phases at 10K. Each figure presents calculated model basing on crystallographic parameters (average structure) and experimental PDF.

and 1.85 Å (R). This may appear to indicate the unpolarized Ti in the O_6 octahedron. However the bondlength is too short for the unpolarized Ti-O bond, which is expected to be about 2.0 Å, and the area under the peak is to small to include six Ti-O bonds. Simple Gaussian modeling suggests that it is most likely that the Ti-O peak is split into two groups by a displacement of Ti toward the face of the octahedron, by 0.2 Å in the $[111]_c$ direction. The long Ti-O peak overlaps with the positive Zr-O peak. Thus we conclude that Ti is displaced in the M and R phases along the triad axis. The Rietveld analysis suggests the correct Ti polarization, but incorrect Ti-O bondlengths.

For the T phase (Fig. 5a) the Rietveld results give the average Zr-O distance of 2.0 Å, much shorter than that in PZ, and the Zr-O peak should split into two, reflecting the $[111]_c$ cubic displacement. However, the experimental PDF gives a Zr-O distance of 2.1 Å, very close to that in PZ. The Zr-O peak is not split, again similar to that in PZ. In fact the Zr-O peak strongly resembles that in PZ in position and width, suggesting that, just as in PZ, the longitudinal Zr polarization is very small. In the R phase (Fig. 5c) both the experimental and the Rietveld PDF's show a single Zr-O peak, also similar to PZ. Thus in all three phases the Zr-O peak is not split, suggesting that the Zr-O bond is not ferroelectrically active. Therefore the suggestion of the Rietveld results that Zr is ferroelectrically active in the T and M phases is incorrect.

For the T phase (Fig. 5a) the Pb-O peaks are split in three-ways in the measured PDF, and are in reasonable agreement with the Rietveld PDF, except that the peak at 2.45 Å is better defined in the measured PDF. For the M phase (Fig. 5b) and R phase (Fig. 5c) the Pb-O peak in the measured PDF disagrees with the Rietveld result in both position and magnitude. In the measured PDF the Pb-O peak at 2.45 Å is very well defined in all three phases, as in PZ and PT. This means that the Pb atom bonds to 4 near oxygen ions in PZT as in PZ and PT. In PT these four Pb-O bonds are created by [001] displacements of Pb. In PZ they are created by Pb moving in the cubic $[1\bar{1}0]_c$ direction coupled with rotation of the octahedra around $[1\bar{1}0]_c$ axis. In the R3c or R3m symmetry ideally Pb should become displaced along the pseudo-cubic $<111>_c$ directions towards 3 oxygen atoms, while the experimental PDF shows that Pb is bonded to four O atoms. Therefore the displacement of Pb cannot be the along $<111>_c$ directions, and is not consistent with the R3m or R3c symmetry. Indeed our earlier PDF study of PZT up to 40%-Ti concluded that the Pb polarization in the R phase is closer to the $<110>_c$ rather than the $[111]_c$ direction [7]. The Rietveld analysis is inaccurate in describing the environment of Pb, and grossly underestimates the ferroelectric activity of Pb.

Thus we arrive at the following picture of the Pb displacement in PZT. We suggest that Pb is locally displaced approximately in either the $<110>_c$ or $<100>_c$ directions. If the environment of a particular Pb is rich in Ti due to statistical fluctuation Pb will be polarized along $<100>_c$ as in PT. In a Zr rich environment Pb is displaced in $<110>_c$ as in PZ. Since the Pb polarizations are ferroelectrically coupled, there are some local deviations from these directions toward the average polarization direction. What changes with changing the Ti/Zr ratio is the population of local Pb displacements in $<110>_c$ or $<100>_c$ directions. In the R phase there are more $<110>_c$ displacements, since there are more Zr rich environment, while in the T phase $<100>_c$ displacements are the majority. The MPB is a crossover point. PZT in the vicinity of MPB is therefore characterized by a maximum disorder in the Pb polarization. The crystal structure in the monoclinic phase at the MPB represents an "average" structure, but not a true microscopic structure. As a result of the strong randomness in the Pb polarization and local deviations from the easy polarization directions (non-collinear state) the anisotropy energy between the $<111>_c$ and $<100>_c$ directions [17] is averaged out. Therefore the rotation energy of the macroscopic polarization is small. This explains high dielectric permittivity and large piezoelectric response observed near the MPB [18].

SUMMARY

We have examined the structure of three PZT samples near the MPB at 10 K using the Rietveld refinement and the pair distribution function (PDF) analysis. The Rietveld results describe the "average" structure, while the actual local structure is significantly different. In particular the PDF shows that the Zr displacement against oxygen is very small and the Zr-O bond is not ferroelectrically active, as in PZ, while the Rietveld analysis incorrectly suggests ferroelectrically active Zr in the T and M phases. Ti is polarized in the [001] direction in the T phase as in PT, and in the M and R phases Ti displaces toward the octahedron face in the cubic $<111>_c$ direction. Pb displaces toward four oxygen atoms and creates short Pb-O distances of 2.45 Å in all samples as in pure PT and PZ. Disorder in the Pb displacements is largest near the morphotropic phase boundary, in agreement with a recent theoretical study [12]. We propose a model in which Pb displacements are in either the cubic $<110>_c$ or $<100>_c$ directions and the population of each changes with the Ti/Zr ratio. In this model the resulting non-collinear structure leads to lowering of the anisotropy energy [17] and easy rotation of the polarization vector, thus explaining why piezoelectric coupling is high near the MPB [18].

ACKNOWLEDGMENTS

This work was supported by the Office of Naval Research grants N00014-98-1-0584 and N00014-98-1-0583. Useful discussions with B. Noheda, G. Shirane, L. Bellaiche, D. Vanderbilt and K. Rabe are acknowledged. The pulsed neutron diffraction data were obtained at the IPNS of the Argonne National Laboratory which is supported by the Department of Energy, Division of Basic Energy Sciences, under contract W-31-109-Eng-38.

REFERENCES

1. Jaffe, B., Cook, W. R. and Jaffe, H., *Piezoelectric ceramics*, London, Academic, 1971.
2. Noheda, B., Cox., D. E., Shirane, G., Gonzalo., J. A., Cross, L. E. and Park, S-E., *Appl. Phys. Letters* **74**, 2559 (1999).
3. Viehland, D. et al., *Phys. Rev.* **B52**, 778 (1995).

4. Dai, X., Li, J-F. and Viehland, D., *J. Am. Ceram. Soc.* **76**, 2815 (1995).
5. Ricote, J., Corker, D. L., Wahtmore, R. W., Impey, S. A., Glazer, A. M., Dec, J. and Roleder, K.., *J. Phys.: Condens. Matter* **10**, 1767 (1998).
6. Corker, D. L., Glazer, A. M., Whatmore, R. W., Stallard, A. and Fauth F., *J. Phys.: Condens. Matter* **10**, 6251 (1998).
7. Teslic, S., Egami, T. and Viehland, D., *Ferroelectrics*, **194,** 271 (1997).
8. Egami, T., in *Structure and Bonding*, **vol. 98**, ed. J. B. Goodenough (Springer-Verlag, Berlin, 2001) p. 115.
9. Egami, T. and Billinge, S. J. L., *Underneath the Bragg Peaks: Structural Analysis of Complex Materials*, (Elsevier Science, Oxford, 2001), to be published.
10. Larson, A. C. and R. Von Dreele. R., GSAS Program Manual, Los Alamos National Lab., unpublished
11. Teslic, S. and Egami, T., *Acta Cryst.* **B54**, 750 (1998).
12. Bellaiche, L., Garcia, A. and Vanderbilt, D., cond-mat/0102254.
13. Lines, M., E., Glass, A. M., *Principles and Applications of Ferroelectrics and Related Materials*, Clarendon Press, Oxford, 1977
14. Shirane, G. ,Pepinsky, R. and Frazer, B. C., *Acta Cryst.* **9**, 131 (1956).
15. Jona, F., Shirane, G., F. Mazii and Pepinsky, R., *Phys. Rev.* **105**, 849 (1951).
16. Fujishita, H., Shiozaki, Y., Achiwa, N., and Sawaguchi, E., J. Phys. Soc. Jpn. 51, 3583, (1982).
17. Egami, T., *Ferroelectrics*, **222,** 163 (1999).
18. Egami, T., Ferroelectrics, in press.

Synchrotron X-ray Studies of Superlattice Ordering in Pb(Mg$_{1/3}$Nb$_{2/3}$)O$_3$ Single Crystals Doped with PbTiO$_3$

Andrei Tkachuk[*], Paul Zschack[*], Eugene Colla[†] and Haydn Chen[*]

[*]*Department of Materials Science and Engineering and Frederick Seitz Materials Laboratory, University of Illinois, Urbana, Illinois 61801, USA*

[†]*Department of Physics, University of Illinois, Urbana, Illinois 61801, USA*

Abstract. The temperature dependence of the superlattice reflections: a) F spots and b) α spots in a lead magnesium niobate (PMN) single crystals containing 0% and 6% of PbTiO$_3$ (PT) has been studied using synchrotron x-ray scattering techniques. (No superlattice reflections were found in PMN doped with 32% PT). Analysis of the temperature dependence of the α spots suggests the existence of the correlated anti-parallel atomic displacements that form nanoregions different from the chemical nanodomains. While the correlation length is temperature independent, the magnitude of these displacements increases on cooling below the freezing temperature T_f. Intensities of the α spots above this temperature become indistinguishable from the background. Our results show that value of T_f for each composition is very close to the one obtained from a Vogel-Fulcher fit to the frequency dependence of the dielectric constant maximum T_m. The relation of these correlated anti-ferrodistortive fluctuations to polar ferroelectric nanodomains and relaxor behavior needs further study.

INTRODUCTION

High dielectric constant of the relaxor ferroelectrics, which can exceed 50,000, is an attractive feature for applications in the capacitor industry. Relaxor properties can be tuned by different atomic substitutions. This was shown to increase material's electrostriction and piezoelectric coefficients, thereby expanding the applications to the area of very effective actuators and nanopositioners [1]. Among other interesting characteristics of the ferroelectric relaxors, the following defining properties necessitate better understanding: a) broad dielectric constant peak over a wide temperature range, b) strong dispersion of the complex susceptibility as a function of frequency of the applied electric field. These features which distinguish the relaxors from the "normal ferroelectrics" are presently under extensive investigation. As a result, several models have been proposed in publications to explain the origin of the relaxor behavior [2-5]

Pb(Mg$_{1/3}$Nb$_{2/3}$)O$_3$ (PMN) can be considered as a model ferroelectric relaxor. The temperature dependence of the dielectric constant exhibits a broad peak near T_m=265 K (AC f=1 kHz). Its position and magnitude strongly depend on the frequency of the applied electric field [3]. Substitution of Ti on the B sites (Mg/Nb) of the ABO$_3$

perovskite structure of the PMN, was shown to change the dielectric properties of the material. In general, this type of doping reduces the frequency dispersion of the dielectric constant and moves the position of the peak's maximum (T_m) to higher temperatures. Relaxor properties disappear for concentrations of Ti apporaching 34% which marks the morphotropic phase boundary on PMN-PT diagram.

Present understanding of relaxors assumes the existence of small regions of the ferroelectric polar clusters, which may interact with each other by means of dipolar interactions [3]. The size of these polar regions is on the nanometer scale. They were postulated to exist at temperatures much higher than T_m and grow in size during cooling [3]. Polar regions are imbedded into a chemically disordered matrix of Mg and Nb atoms. At the same time Nb and Mg are also ordered on the nano-scale, forming chemically ordered nanodomains. The random layer model, proposed by Davis et al [6], suggests the most probable type of chemical ordering. Chemical disorder is believed to prevent polar nanodomains from growing into regular micro-size ferroelectric domains [7,8]. At the sufficiently low freezing temperature (T_f), the polar clusters become randomly frozen in space without forming long-range ordered ferroelectric phase [2]. This scenario is very similar to the magnetic spin glasses where the structural disorder coupled with competing interactions accounts for observed glassy behavior [9]. However, phenomena related to disorder in the dielectrics are more complicated than in magnetic systems. Electric charge interactions for example, may cause large structural distortions, which do not take place in magnetic spin glasses. In magnetic systems the source of competing interactions is due to frustration when both ferromagnetic and anti-ferromagnetic interactions occur simultaneously. It is still an open question if there is a source of competing interactions in the Pb containing relaxors. From our measurements, there is evidence of correlated anti-parallel distortions forming small nanoclusters. These exist in pure PMN and PMN-6%PT on a short-range scale. The magnitude of these distortions grows on cooling below the T_f. Therefore, the distortions might be considered a source of intractions competing with the establishment of ferroelectric order. However, the aforementioned is still somewhat speculative.

Local disorder and local fluctuations (chemical and dipolar) in PMN and other related relaxors, was demonstrated to be intimately related to relaxor ferroelectric behavior [3,7,10-13]. Correlation in these fluctuations is responsible for the formation of very weak and broad superlattice reflections. They were studied by utilization of synchrotron x-ray scattering techniques. The focus of this study was to understand superlattice reflections as a function of temperature and limited chemical compositions currently available to us.

EXPERIMENTAL

Single crystal of PMN was grown by the Chochralsky technique. PMN-6%PT and PMN-32%PT single crystals were grown by the melted flux method. Sizes of the single crystals were 3x5x1 mm^3 for pure PMN, 5x5x0.5 mm^3 for PMN-6%PT and 7x7x0.5 mm^3 for PMN-32%PT with its surface normal perpendicular to 001 crystallographic plane. All samples were thoroughly studied and characterized with dielectric measurements of the complex dielectric constant as a function of frequency and temperature. We studied the crystals with synchrotron x-ray radiation at beamline X-18A at the National Synchrotron Light Source (NSLS), Brookhaven National Laboratory and beamline 33-ID at the Advanced Photon Source (APS), Argonne National Laboratory. Crystals were studied at the NSLS with the incident x-ray energy tuned to 10 keV by two Si (111) single crystals and subsequently focussed by the cylindrical mirror. Measurements were performed in the reflective geometry with KEVEX solid state detector. At APS the x-ray energy was tuned by a double-crystal monochromator using Si 111 and focussed vertically by two Rh coated mirrors. Diffracted radiation was measured with an Oxford scintillation detector. Additional studies near Pb L_{III} absorption edge (13.35 keV) and Nb K edge (18.99 keV) were performed at APS on pure PMN. In both NSLS and APS, harmonics in the incident x-ray beam were suppressed to the level that did not cause any observable data contamination. Low temperature measurements were performed in a closed cycle He compressed gas cryostat, which was mounted on the standard four-circle diffractometer. The sample height was regularly adjusted to compensate for the temperature contraction of the sample holder inside the cryostat.

RESULTS AND DISCUSSION

The temperature T_m does not mark any structural phase transition. Numerous electron, x-ray and neutron scattering experiments reported in the literature were unable to find any clear evidence of macroscopic phase transition in a wide 15-600 K temperature range in powder and single crystals. These results were obtained from studies of fundamental Bragg reflections. Bragg reflections are mostly sensitive to the average long-range structure. Superlattice short-range order peaks and diffuse scattering are more sensitive to the local structure. Two types of very weak superlattice reflections were found to exist in both pure PMN and PMN-PT for small concentrations of Ti: a) F peaks and b) α peaks [14]. The Cubic reciprocal unit cell is shown in Figure 1 where F and α spots are marked.

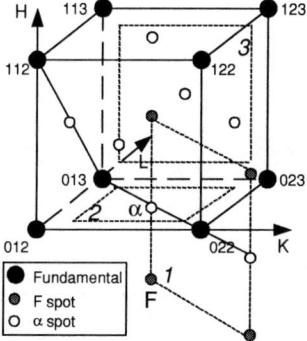

FIGURE 1. Reciprocal unit cell: corners-Bragg peaks; body centers- F spots and face centers-α spots. Reciprocal planes, discussed in the text, are labeled 1, 2 and 3. Dashed lines mark their perimeters.

Fundamental Bragg peaks, F spots and α spots are indicated by circles in Figure 1. Doubling of the unit cell in the real space leads to the half integer indexing of the superlattice reflections in the reciprocal space. Terminology for calling these peaks F and α spots is adapted from electron microscopy studies of the related ferroelectric relaxors [14]. Superlattice reflections were found initially in pure PMN and PMN-PT by transmission electron microscopy (TEM) techniques [14,15]. Later Vakhrushev et al. [16] found both F and α spots from the x-ray synchrotron measurements, but no temperature dependence of these reflections was reported. Zhang et al. [17] have studied the temperature dependence of the F spots with a rotating anode x-ray generator, but they were unable to observe any α spots. This is not surprising since α spots are weaker and more diffused than F spots so that the incident intensity from a conventional x-ray source might not be high enough to resolve them from the background. Synchrotron radiation is much brighter than conventional x-ray sources and therefore more useful to study these weak peaks. Gosula et al. [18] were the first to study the temperature dependence of F and α spots in pure PMN using synchrotron radiation. Work reported here extends the studies from pure PMN to PMN-6%PT and PMN-32%PT compositions.

In this work we have obtained the 3D intensity maps in sections of the reciprocal space. Figure 2 depicts the intensity distribution near the base of the 022 fundamental Bragg peak in the plane marked 1 in Figure 1. F and α spots are clearly seen in pure PMN and PMN-6%PT at 15 K. They are very weak in comparison to 022 Bragg peak, whose intensity is more than 8 orders of magnitude higher. We found the F peaks in PMN-6%PT to be weaker and broader than in pure PMN (see Figure 2). The width of the α spots for both compositions is identical, but the temperature dependence of the integrated intensity is different and will be discussed later in this section. No superlattice reflections of any kind were found in PMN-32%PT (T=15-300 K).

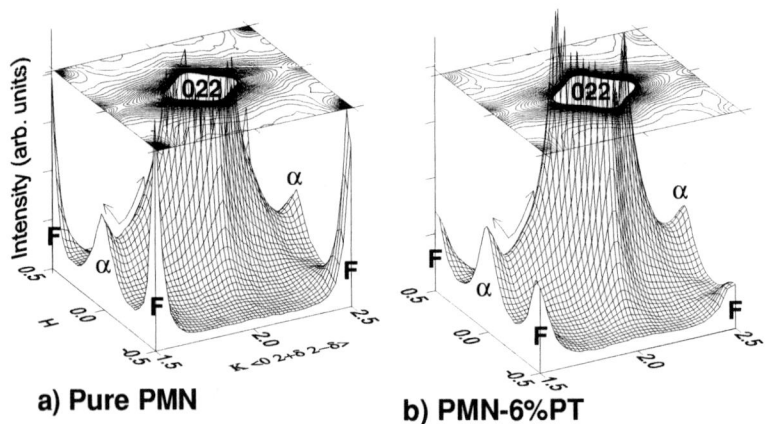

FIGURE 2. Section of reciprocal space marked 1 (Fig 1) in the vicinity of 022 Bragg peak. Figure indicates the presence of F and α spots for both PMN and PMN-6%PT at 15 K. The double-ended arrow indicates direction of the diffuse scattering ridge.

The main contribution to the F peaks is coming from the Nb/Mg chemical short-range ordering in [111] directions [14,15,17-19]. Additional contributions may come from the corresponding atomic displacements, as was proposed by Zhang et al. [17]. However, the effect is much smaller compared to that due to chemical ordering. It appears that PbTiO$_3$ doping destroys the Mg/Nb short range ordering. The estimated size of the chemically ordered regions in PMN-6%PT was obtained from the full width at half maximum (FWHM) using the well-known Scherrer equation for crystallite size determination. It was found on average to be 24 Å. This is similar to the size obtained from the electron diffraction dark field imaging in PMN doped with 7% Ti, where the estimated size was reported to be close to 30 Å [14]. The size of chemically ordered domains in pure PMN is about 50 Å. The intensity of the α peaks was found to be strongly temperature dependent in contrast to TEM studies of Hilton et al. [14]. This can be explained by the fact that additional intensity from the diffuse scattering ridge, like the one marked by arrow in Figure 2, contributes to the intensity of the α spots as a background. Therefore, any cross section of the ridge will appear as a peak. In Figure 1 these ridges are shown schematically by solid lines that stretch between the corners of the cube along their face diagonals. A different cross section of the same ridge in plane 2 (see Fig. 1) is shown in Figure 3. Figure 3a clearly shows coexistence of the α spot with a diffuse ridge at low temperature. It is clear that the α spot can be completely separated from the diffuse scattering only along the direction of the ridge. Linear scans across the ridge can be mistaken for a peak even if the actual α spot is no longer present (Figure 3b). This may be why earlier TEM measurements reported α spots to be temperature independent [14]. Figure 3b shows that the 1/2(035) α peak in PMN-6%PT cannot be resolved from the diffuse intensity ridge at 300 K. As discussed by H. You et al. [20], the origin of the diffuse scattering is attributed to the pure transverse soft phonon modes. Acoustic modes contribute to the intensity of the diffuse ridges near Bragg peaks with H+K+L=even, and optic modes contribute when H+K+L=odd [20].

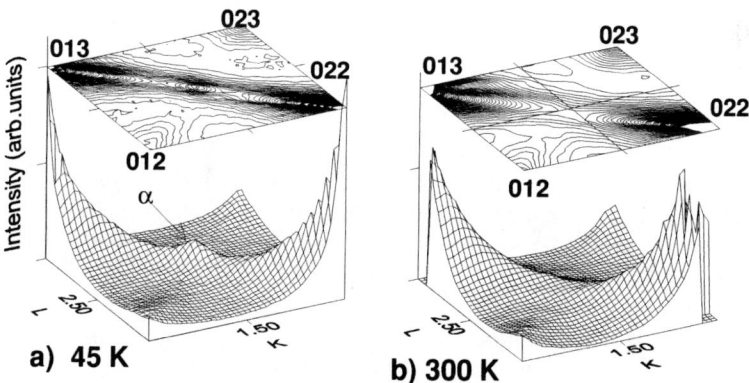

FIGURE 3. Section of reciprocal space marked 2 (Fig 1) in the vicinity of 022 Bragg peak. Figure indicates α spot at low temperature a) 45 K, which is indistinguishable from the diffuse intensity ridge at b) 300 K. Both 0% and 6% PT samples behave this way, but α spots disappear at different temperatures.

It has been shown that in the immediate vicinity of the Bragg peaks, the intensity of the diffuse scattering increases on cooling [16,20]. However, we found that closer to the center of the zone where α spots are located, ridges are not changing with temperature up to 800 K. Figure 4 depicts the temperature dependence of the integrated intensity of some α spots from the linear scan along the diffuse ridge. Different α spots of both pure PMN and PMN-6%PT are shown for comparison on the same plot normalized to their corresponding intensity at 15 K. To summarize, we found that the intensity of the α spot is decreasing on warming from 15 K, while the intensity of the ridge remains constant.

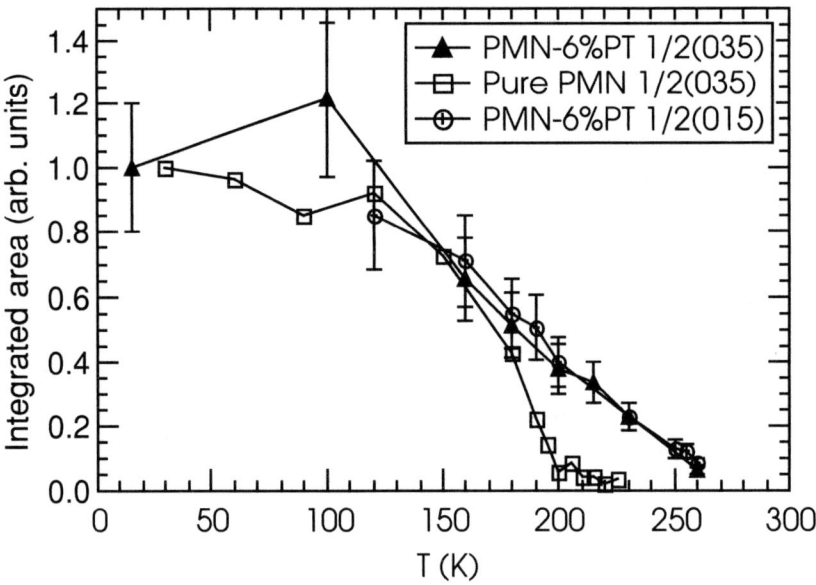

FIGURE 4. Temperature dependence of integrated intensity of α spots in PMN and PMN-6%PT.

Glazer [21] studied the origins of different superlattice reflections in perovskites. According to his proposed classification scheme, the origin of the α spots can be attributed to correlated in-phase oxygen octahedra tilts. The analysis of the structure factor has shown that α spots with two identical odd indices (e.g. 1/2(011), 1/2(033) etc.) cannot occur due to the oxygen tilting alone. In our experimental study we were able to observe such reflections. For example, 1/2(055) a spot is shown in Figure 5 This means that correlated displacements of atoms other than oxygen must be coupled to the oxygen octahedra tilts. For example, the presence of the correlated displacements of these atoms in 011 type directions can explain the appearance of the α spots. In order to double the unit cell, these displacements must be arranged in some kind of anti-parallel fashion.

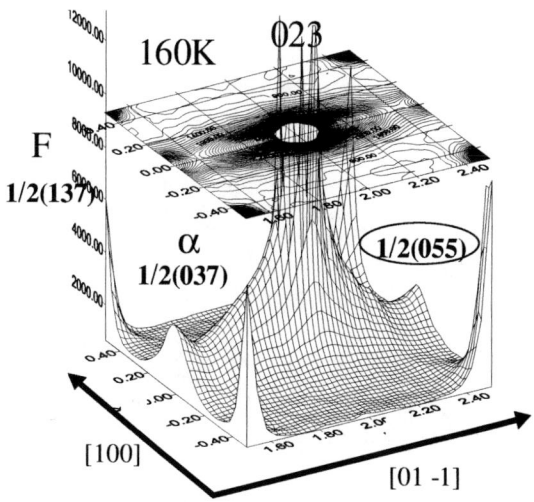

FIGURE 5. Superlattice peaks near the 023 Bragg peak. 1/2(055) α spot is circled. α spots with two odd Miller indices the same cannot occur due to the oxygen tilting alone [21].

These displacements are anti-ferrodistortive in nature. In a related $PbIn_{1/2}Nb_{1/2}O_3$ (PIN), α spots were also found and their temperature dependence was directly correlated with anti-ferroelectric behavior [22]. For example, the intensity of the α spots in ordered PIN sharply increased just below the Neel temperature of the anti-ferroelectric phase transition. Assuming that in PMN-PT α spots are also due to anti-ferrodistortions, we can make some interesting conclusions from the analysis of PMN and PMN-6%PT. These correlated anti-ferrodistortions form clusters whose size is temperature independent and differ from the chemical nanodomains. FWHM along the ridge (see Figure 3a) of pure PMN and PMN-6%PT is both identical and temperature independent at temperatures where α peaks can be distinguished from the diffuse background, as shown in Figure 6.

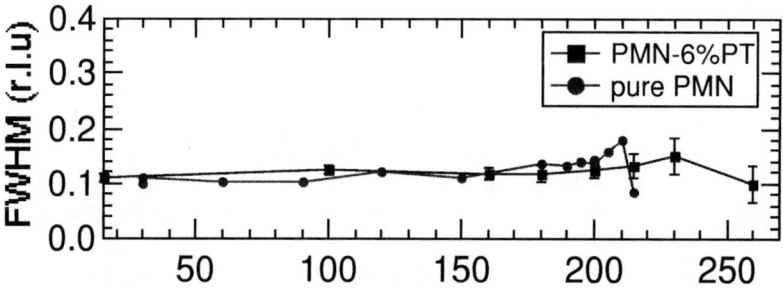

FIGURE 6. FWHM for 1/2(035) α spot along the ridge for PMN and PMN-6%PT.

Figure 6 suggests the size of the clusters formed by anti-ferrodistortive fluctuations is an intrinsic property independent of these two different compositions. The size was estimated to be near 32 Å for both PMN and PMN-6%PT from the Scherrer formula.

On the plot of the integrated area versus temperature (see Figure 4) the interpolated temperature, where intensity goes to zero, was found close to T_f=260 K in PMN-6%PT and T_f=220 K for pure PMN. Interestingly, these are the freezing temperatures that are very close to the ones obtained from the Vogel-Fulcher fit of the frequency dependent dielectric response. Vogel-Fulcher fits from the dielectric measurements are shown in Figure 7. From Figure 4 it is also evident that intensity of the α peaks in PMN-6%PT decreases more gradually than in pure PMN. Long tails are often indicative of the short-range order effects.

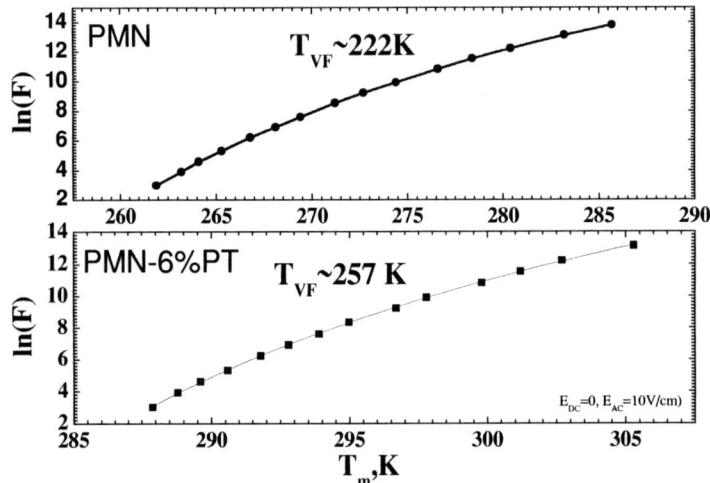

FIGURE 7. Plots of the frequency dependence of the dielectric constant maximum T_m for PMN and PMN-6%PT. Temperature (T_{VF}), indicated on the plots, was used as an adjustable parameter in Vogel-Fulcher fitting equation.

CONCLUSIONS

The temperature dependence of superlattice reflections for three compositions of PMN (with 0%, 6% and 32 % $PbTiO_3$) has been determined by utilizing synchrotron scattering techniques. No superlattice peaks of any kind were found for the 32% composition. We have found the temperature dependence of α spots by separating them from the temperature independent anisotropic diffuse scattering background. By doing so, we have found that their integrated intensity increases on cooling below the freezing temperature T_f, while FWHM of the α spots is temperature independent. This means that magnitude of the correlated anti-parallel atomic distortions increases on cooling below the freezing temperature, while the size of the clusters that are formed by these distortions is temperature independent. The size of these clusters is about 32 Å, and is also identical for both 0% and 6% PT doped compositions. However, the size of the chemically ordered nanodomains is twice as large for pure PMN compared to PMN-6%PT (50 Å and 24 Å respectively). This suggests that chemical ordering is independent of the clusters that are responsible for α spots. The temperature T_f obtained from our x-ray studies is close to the phenomenological T_f freezing temperature obtained from the Vogel-Fulcher fit of the frequency dependence of the T_m maximum of the dielectric constant. It is not clear what the connection is between the anti-polar clusters and polar nanodomains expected in relaxor ferroelectrics, because the size of anti-ferrodistortive clusters we found to be temperature independent. On the other hand, the polar nanodomains are expected to grow in size from embryo size at T_d=600 K to the cryogenic temperatures, reaching couple of hundreds of Å in size. We can only speculate that these anti-polar displacements correlated on a 30 Å scale, may compete with establishment of the parallel alignment of electric dipoles which is required to catalyze a ferroelectric phase. Further studies are underway in effort to determine which atoms, apart from the oxygen, contribute to the structure factor of the α spots from the anomalous scattering synchrotron techniques.

ACKNOWLEDGMENTS

This material is based upon work supported by the U.S. Department of Energy, Division of Materials Sciences under Award No. DEFG02-ER9645439, through the Frederick Seitz Materials Research Laboratory at the University of Illinois at Urbana-Champaign.

We would like to thank the staff of both the UNICAT ID-33 beamline and the MATRIX X-18A beamline at National Synchrotron Light Source, Brookhaven National Laboratory for assistance during the experiments.

The UNICAT facility at the Advanced Photon Source (APS) is supported by the Univ of Illinois at Urbana-Champaign, Materials Research Laboratory (U.S. DOE, the State of Illinois-IBHE-HECA, and the NSF), the Oak Ridge National Laboratory (U.S. DOE under contract with UT-Battelle LLC), the National Institute of Standards and Technology (U.S. Department of Commerce) and UOP LLC. The APS is supported by the U.S. DOE, Basic Energy Sciences, Office of Science under contract No. W-31-109-ENG-38.

Research carried out (in part) at the National Synchrotron Light Source, Brookhaven National Laboratory, which is supported by the U.S. Department of Energy, Division of Materials Sciences and Division of Chemical Sciences.

REFERENCES

1. Nomura, S. and Uchino, K., *Ferroelectrics* **41**, 117-132 (1982).
2. Viehland, D., Li, J.-F., Jang, S.J., Cross, L.E., and Wutting, M., *Physical Review B* **43**, 8316-8320 (1991).
3. Cross, L.E., *Ferroelectrics* **151**, 305-320 (1994).
4. Blinc, R., Dolinsek, J., Gregorovic, A., Zalar, B., Filipic, C., Kutnjak, Z., and Levstik, A., *Journal of Physics and Chemistry of Solids* **61**, 177-183 (2000).
5. Kleemann, W. and Lindner, R., *Ferroelectrics* **199**, 1-10 (1997).
6. Davies, P.K. and Akbas, M.A., *Journal of Physics and Chemistry of Solids* **61** (2000).

7. Westphal, V., Kleemann, W., and Glinchuk, M.D., *Physical Review Letters* **68**, 847-850 (1992).
8. Chen, I.-W., Ping, L., and Wang, Y., *Journal of Physics and Chemistry of Solids* **57**, 1525-1536 (1996).
9. Colla, E.V., Chao, L.K., Weissman, M.B., and Viehland, D.D., *Physical Review Letters* **85** (2000).
10. Setter, N. and Cross, L.E., *Journal of Materials Science* **15**, 2478-82 (1980).
11. Egami, T., Dmowski, W., Teslic, S., Davies, P.K., Chen, I.-W., and Chen, H., "Nature of Atomic Ordering and Mechanism of Relaxor Ferroelectric Phenomena in PMN," in *ONR Williamsburg Workshop on Ferroelectrics. Feb 2-5,* , 1997, .
12. Egami, T., Teslic, S., Dmowski, W., Viehland, D., and Vakhrushev, S., *Ferroelectrics* **199**, 103-113 (1997).
13. Davies, P.K. and Akbas, M.A., *Ferroelectrics.* **221**, 27-36 (1999).
14. Hilton, A.D., Barber, D.J., Randall, C.A., and Shrout, T.R., *Journal of Materials Science* **25**, 3461-3466 (1990).
15. Husson, E., Chubb, M., and Morell, A., *Materials Research Bulletin* **23**, 357 (1988).
16. Vakhrushev, S., Naberezhnov, A., Sinha, S.K., Feng, Y.P., and Egami, T., *Journal of Physics and Chemistry of Solids* **57**, 1517-1523 (1996).
17. Zhang, Q.M., You, H., Mulvihill, M.L., and Jang, S.J., *Solid State Communications* **97**, 693-698 (1996).
18. Gosula, V., Tkachuk, A., Chung, K., and Chen, H., *Journal of Physics and Chemistry of Solids* **61**, 221-227 (2000).
19. Fanning, D.M., Robinson, I.K., Jung, S.T., Colla, E.V., Viehland, D.D., and Payne, D.A., *Journal of Applied Physics* **87**, 840-8 (2000).
20. You, H. and Zhang, Q.M., *Physical Review Letters* **79**, 3950-53 (1997).
21. Glazer, A.M., *Acta Cryst.* **A 31**, 756 (1975).
22. Nomura, K., Yasuda, N., Ohwa, H., and Terauchi, H., *Journal of the Physical Society of Japan* **66**, 1856-1859 (1997).

Broadband Brillouin Scattering of Relaxor Ferroelectric Crystals

Seiji Kojima and Fuming Jiang

Institute of Materials Science, University of Tsukuba, Tsukuba, Ibaraki, 305-8573, Japan

Abstract. The dynamical properties of relaxor ferroelectrics have been studied by broadband Brillouin scattering. In 0.65PMN-PT an intense central peak (CP) was observed in a wide temperature range below about 250 °C. The relaxation time determined from CP shows slowing down above the cubic to tetragonal transition temperature T_{ct} = 158 °C. Under the assumption that the origin of CP is the dynamical behavior of polar micro regions, this behavior was well reproduced by the extended superparaelectric model above T_{ct}. The intense CP of uniaxial relaxor 0.61SBN with the dielectric maximum temperature T_m = 70 °C was also observed in a wide temperature range below about Tc + 400 °C. The anisotropy of CP intensity indicates that CP of 0.61SBN is originated mainly from fluctuating polar clusters along the c-axis. However, the slowing down of the relaxation time determined from CP stopped at about T_m + 100 °C. This fact may indicate growing interaction between PMRs in 0.61SBN.

INTRODUCTION

Relaxor ferroelectrics are characterized by a diffuse and frequency dependent dielectric peak with a relaxation spectrum being much broader than the Debye type. The maximum temperature of a low-frequency dielectric peak T_m obeys the Vogel-Fulcher law of disordered system, like a spin glass and a glass-forming material. These non-Debye and non-Arrhenius behaviors of relaxor ferroelectrics have received much attention as common natures in complex disordered system, and many theoretical models were proposed. However, their physical mechanism is not yet confirmed.

Up to now, low-frequency dielectric properties were extensively studied in many relaxor ferroelectric materials. Such a low frequency dispersion have been analyzed by some theoretical models related to a polar micro region (PMR), for example, the superparaelectric model, the dynamically correlated domain model [1,2,3]. In these models, PMRs fluctuate independently in the high-temperature paraelectric phase. Recently the importance of the interaction between PMRs was recognized near T_m. The interaction was discussed on the basis of random bonds, or a distribution of local fields [4,5]. On the other hand, high frequency behaviors, especially in a gigahertz range, were not much studied. According to the temperature variation of linear birefringence, the appearance of PMRs was detected by the change of slope. This temperature is called Burn's temperature T_B, which is usually a few hundred degrees above T_m. The other experimental evidence of PMRs is the observation of strong 1st order Raman scattering in a cubic phase of lead perovskite relaxors [6]. In the high-temperature phase, the potential energy of PMR is described by a double well or multiwell structure [5]. Consequently, hopping or flipping between wells may cause the high-frequency relaxation. Therefore, it is important to study the dynamical behaviors of PMRs in a high-frequency range, especially above 1 GHz.

In this paper, we report the broadband Brillouin scattering study of two types of relaxor ferroelectric crystals to make clear the high-frequency dynamical properties. One is a lead perovskite relaxor, and the other is a tungsten-bronze uniaxial relaxor. Recently large crystals were grown in PMN-PT, PZN-PT, etc. However, since relaxor crystals have some kinds of inhomogeneity, the micro-Brillouin scattering technique [7-9] has been employed in this study.

LEAD PEROVSKITE RELAXOR : PMN-PT

Since most of relaxors are mixed perovskite oxides of which A site is occupied by Pb^{2+} ions, at first we studied $0.65Pb(Mg_{1/3}Nb_{2/3})O_3$-$0.35PbTiO_3$ (0.65PMN-PT) with perovskite structure at a composition near the morphotoropic phase boundary. Because of a huge piezoelectric effect, large single crystals are available for 0.65PMN-PT. With the decrease of temperature it undergoes a cubic to tetragonal phase transition at T_{ct} = 158 °C and then undergoes a tetragonal to rhombohedral transition at about T_{tr} = 60 °C. Our previous study of micro-Brillouin scattering indicates the hetero structure below T_{tr} [8]. Therefore, by choosing a homogeneous area in a (100) plate, the broadband Brillouin scattering was measured in the temperature range from 300 °C to -190 °C. The typical broadband Brillouin spectrum is shown in Fig. 1. The strong central peak (CP) appears as broad Rayleigh wings and the sharp doublet is a Brillouin component of a longitudinal acoustic mode.

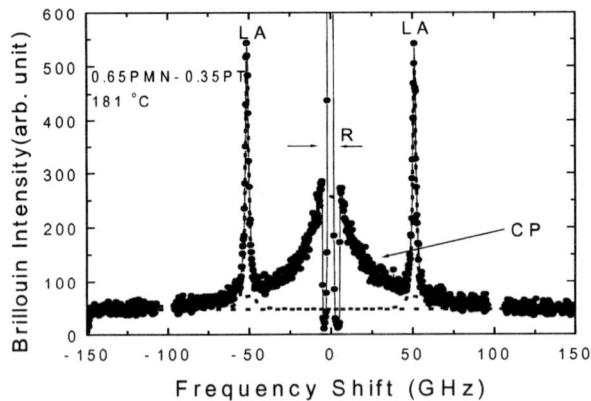

FIGURE 1. Broadband Brillouin spectrum of 0.65PMN-0.35PT at 181 °C.

Normal ferroelectrics with translational invariance show CP only in the vicinity of the Curie temperature T_c. The marked narrowing of CP is caused by the critical slowing down. While in relaxor ferroelectrics, translational symmetry is destroyed by the disorder in the B site. According to the high-frequency dielectric measurements, the relaxation time does not diverge towards T_{ct}, but shows the maximum around T_{ct} [10].

Therefore, the origin of CP in relaxors may be related to the intermediate-order or local PMRs. The notable result in 0.65PMN-PT is the appearance of CP in a large temperature range of a few hundreds degree. It has the strong correlation with the appearance of intense 1st order Raman scattering by optical modes which can be explained by assuming the local B-site order with Fm3m symmetry [6].

The temperature dependence of relaxation time for 0.65PMN-PT is shown in Fig.2. The relaxation time shows slowing down with decreasing temperature. However, the relaxation time does not diverge like a normal order-disorder ferroelectric transition. At high temperatures, PMRs fluctuate independently as predicted by the superparaelectric model [2]. Since PMRs appear at the Burn's temperature T_B and grow gradually on cooling, we extended the superparaelectric mode by the assumption that the temperature variable potential barrier height for polarization switching is given by $H = H_0(1-T/T_B)$ for $T < T_B$ and $H = 0$ for $T > T_B$, where H_0 is the barrier height at 0 K. We also assume that the temperature dependence of relaxation time τ is given by $\tau = \tau_D \exp(-H/k_B T)$, where τ_D is the inverse of Debye frequency determined from sound velocity. By fitting line-shape of CP, we obtained reasonable values of $T_B = 852$ K ($> T_{ct} = 441$ K) and $H_0 = 6136$ K. In this extended superparaelectric model, the interaction between PMRs is not included. However, for 0.65PMN-PT, the dielectric dispersion is relatively small above T_{ct} and the deviation from the Curie-Weiss law is also small. Therefore, the interaction can be negligible for CP. Another notable result is the fact that CP exists below T_{ct} and its intensity gradually decreases on further cooling. It means that parts of PMRs are still dynamical below T_{ct} and freeze at low temperature.

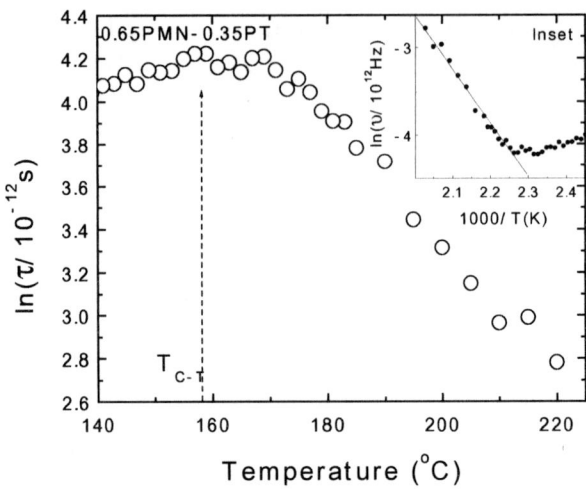

FIGURE 2. The temperature dependence of relaxation time in 0.65PMN-0.35PT. The solid line in Inset denotes the calculation by the extended superparaelectric model.

TUNGSTEN-BRONZE RELAXOR : SBN

Strontium barium niobate $Sr_xBa_{1-x}Nb_2O_6$ has a tetragonal tungsten-bronze structure. The long chains of oxygen octahedra along the c-axis resemble those in perovskite ferroelectrics, while normal to this axis the structure consists of slightly puckered sheets of oxygen atoms. At high Sr contents, it shows relaxor behavior, and spontaneous polarization appears only along the c-axis at least above room temperature. Therefore, it is called a uniaxial relaxor. We studied $Sr_{0.61}Ba_{0.39}Nb_2O_6$ (0.61SBN) with $T_m = 70°C$. The broadband Brillouin scattering was measured in a 0.61SBN single crystal. The broad and intense CP is clearly observed as shown in Fig. 3. We measured CP for different scattering geometries and found that the strong CP appears only at the geometry for $A_1(z)$ symmetry. This anisotropy of CP intensity indicates that the polarization fluctuation mainly occurs along the c-axis. Such intensity anisotropy was not observed in 0.65PMN-PT.

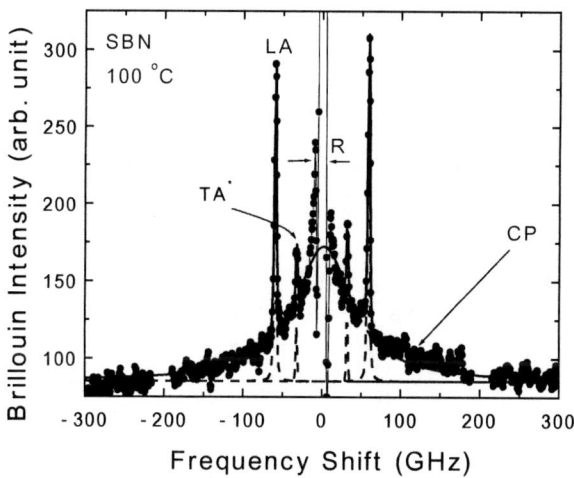

FIGURE 3. Broadband Brillouin scattering spectrum of SBN at 100 °C.

The intense CP was observed, as in the case of 0.65PMN-PT, in the large temperature range below about $T_m + 400$ °C. This may indicate that dynamical PMRs appear at very high temperatures. The temperature dependence of CP was measured, and the relaxation time was determined from the line width of CP as shown in Fig. 4. In cooling from the high temperature, the relaxation time gradually increases at first above $T_m + 100$ °C. However, the slowing down of the relaxation time stopped at about $T_m + 100$ °C, and with further cooling the relaxation time decreases again. At this temperature the deviation of the CP shape from Lorentzian increases. This fact may indicate growing interaction between PMRs or long-range order in 0.61SBN. The recent theory on CP of disordered ferroelectrics predicts that the 2nd order CP line shape deviates significantly from Lorentzian when the temperature approaches the critical temperature [13]. Therefore, the detailed line shape analysis is in progress.

CONCLUSION

Two types of relaxor ferroelectric crystals were studied by broadband Brillouin scattering. In perovskite relaxor 0.65PMN-PT, the intense CP was observed in the large temperature interval of about 200 °C. We consider that CP is originated from dynamical switching of randomly oriented PMRs. The temperature dependence of the relaxation time determined from CP is well reproduced by the extended superparaelectric model. This means that the interaction between PMRs does not affect the CP in a cubic phase. The existence of CP below T_{ct} indicates that the dynamical PMRs still remain in tetragonal and rhombohedral phases.

While in uniaxial relaxor SBN with tungsten-bronze structure, the marked anisotropy of CP intensity shows that the polarization fluctuation along the c-axis plays a dominant role in CP. The intense CP is also observed in a large temperature interval. However, in the temperature dependence of relaxation time above T_m maximum appears at about $T_m + 100$ °C. Such a deviation from monotonic increase on cooling may be originated from the strong interaction between PMRs, which can be closely related to the marked dielectric dispersion in the high temperature phase. To make clear the behavior of CP in relaxors, more accurate and high-resolution measurements are necessary to determine the non-Debye behavior.

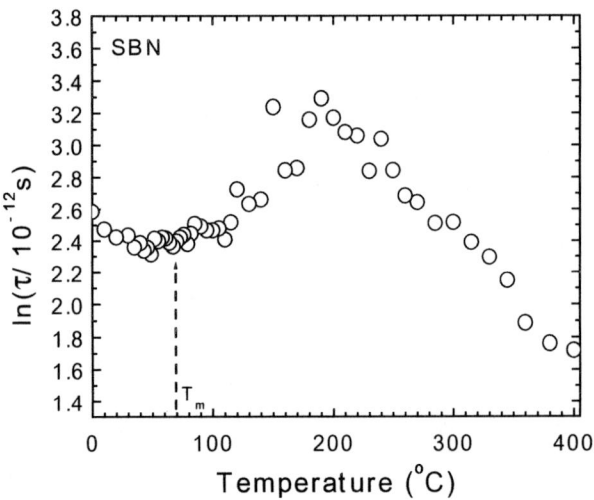

FIGURE 4. The temperature dependence of relaxation time in 0.61SBN.

ACKNOWLEDGMENTS

We acknowledge generous support from the Japanese Society for the promotion of Science (JSPS, No. 10440155), the Foundation for Advancement of International Science and the Marubun Research Foundation. One of the authors (F. M. Jiang) is grateful to the Scholarship from the Japanese Ministry of Science, Culture and Sports.

REFERENCES

1. Vieland, D., Jang, S. J., Cross, L. E., and Wuttig, M., *Phys. Rev. B* **46**, 8003-8006 (1992).
2. Cross, L. E., *Ferroelectrics*, **76**, 241-267, (1987).
3. Chamberlin, R. B., *Phys. Rev. B* **48**, 15638-15654 (1993).
4. Pirc, R., and Blinc, R., *Phys. Rev. B* **60**, 13470-13478 (1999).
5. Vugmeister, B. E., and Rabitz, H., *Phys. Rev. B* **57**, 7581-7585 (1998).
6. Siny, I., and Katiyar, R., *Ferroelectrics* **206-207**, 307-315 (1998).
7. Ahart, M., Jiang, F. M., Mikami, M., Park, I. S., and Kojima, S., *Jap. J Appl. Phys.* **38**, 3058-3061 (2000).
8. Jiang, F. M., and Kojima, S., *Appl. Phys. Letters* **77**, 1271-1273 (2000).

9. Maczka, M., Hanuza, J., Jiang, F. M., and Kojima, S., *Phys. Rev. B* (to be published).
10. Elissade, C., Ravez, J., and Gaucher, P., *Mater. Sci. Engineer., B* **22**, 303-309 (1994).
11. Lines, M.E. and Glass, A.M.., *Principles and Applications of Ferroelectrics and Related Materials*, Oxford: Clarendon Press, 1977, pp. 280-292.
12. Dec, J., Kleemann, W., Woike, Th., and Pancrath, R., *Eur. Phys. J. B* **14**, 627-632 (2000).
13. Vugmeister, B. E., Yacoby, Y., Toulouse, J., and Rabitz, H., *Phys. Rev. B* **59**, 8602-8606 (1999).

Structural properties of sinusoidally modulated Pb(Sc,Nb)O$_3$ alloys at finite temperature

A.M. George and L. Bellaiche

*Physics Department,
University of Arkansas, Fayetteville, Arkansas 72701, USA*

Abstract. A first-principles-derived approach is used to study finite-temperature properties of a Pb(Sc,Nb)O$_3$ alloy that is compositionally sinusoidally modulated along the [001] crystallographic direction. The *local* electrical polarizations — centered in the planes perpendicular to the modulation direction — are found to be spatially-dependent for any temperature. This unusual structural feature can be viewed as a direct consequence of an internal composition-modulation-induced electric-field.

INTRODUCTION

Complex insulating perovskite alloys are of great technological and fundamental interest because of their large dielectric and piezoelectric properties and because of their intriguing structural properties. Examples include (Ba$_{1-x}$Sr$_x$)TiO$_3$ which emerges as a leading potential dielectric material for the memory cell capacitors in dynamic random access memory [1], and the class of Pb(Mg$_{1/3}$Nb$_{2/3}$)O$_3$–PbTiO$_3$ (PMN–PT) and Pb(Zn$_{1/3}$Nb$_{2/3}$)O$_3$–PbTiO$_3$ (PZN–PT) alloys that exhibit remarkably large piezoelectric constants [2] — these latter thus promise dramatic improvements in the resolution and range of ultrasonic and sonar listening devices [3]. Another example of a technologically and fundamentally-interesting perovskite solid solution is Pb(Zr$_{1-x}$Ti$_x$)O$_3$ (PZT). As a matter of fact, this alloy is widely used in transducers and actuators [4] and exhibits an unexpected monoclinic phase for a narrow composition range around x=0.48 [5,6].

Recent studies have demonstrated that playing with the atomic ordering of perovskite alloys shows great promise for the development of new materials with very unusual properties [7–11]. In particular, a first-principles study [11] pointed out that atomic ordering in *heterovalent* alloy — i.e., in a alloy made of atoms belonging to different columns of the Periodic Table — can have a *drastic* effect on structural properties. The aim of this article is to predict and understand the finite-temperature structural features resulting from a particular form of atomic ordering in the heterovalent Pb(Sc$^{+3}_{1-x}$Nb$^{+5}_x$)O$_3$ (PSN) alloy. More precisely, the presently studied atomic ordering consists of a *sinusoidal* variation of the x composition along a given crystallographic direction inside the PSN material. Such peculiar ordering has recently been observed in semiconductor alloys and results in unusual electronic and optical properties (See Ref. [12] and references therein). Our main finding is that sinusoidally modulated heterovalent alloys exhibit an interesting phenomena, namely a continuous change of the electrical polarization inside the material both in the paraelectric and ferroelectric phase. This phenomena is a direct consequence of a *spatially-dependent* internal electric-field, partially resulting from the difference in valence between the alloyed atoms.

METHOD

Here, we use the numerical scheme proposed in Ref. [6,13], which consists of constructing an alloy effective Hamiltonian from first-principles calculations. The total energy E is written as a sum of three energies,

$$E(\{\mathbf{u}_i\}, \{\mathbf{v}_i\}, \eta_H \{\sigma_j\}) = E_{\text{ave}}(\{\mathbf{u}_i\}, \{\mathbf{v}_i\}, \eta_H) \\ + E_{\text{loc,s}}(\{\mathbf{v}_i\}, \{\sigma_j\}) + E_{\text{loc,m}}(\{\mathbf{u}_i\}, \{\sigma_j\}) , \quad (1)$$

where $\mathbf{u}_i$ is the local soft mode in unit cell i; $\{\mathbf{v}_i\}$ are the dimensionless local displacements which are related to the inhomogeneous strain variables inside each cell [14]; η_H is the homogeneous strain tensor; and the $\{\sigma_j\}$ characterize the atomic configuration of the alloy. That is, $\sigma_j=+1$ or -1 corresponds to the presence of a Nb or Sc atom, respectively, at lattice site j of the PSN alloy. The $\{\sigma_j\}$ parameters are incorporated into the energies $E_{\text{loc,s}}$, which takes into account the effect of atomic arrangement on the inhomogeneous strain, and $E_{\text{loc,m}}$, which characterizes the interaction between the local soft modes and the alloy configuration. For E_{ave}, we generalize the analytical expression of Ref. [14] — proposed for simple compounds — to the study of the PSN alloy by creating a Pb$\langle$B$\rangle$O$_3$ simple system in which the $\langle$B$\rangle$ atom is a virtual atom involving a kind of potential average between Sc and Nb atoms [15]. E_{ave} thus consists of five parts: a local-mode self-energy, a long-range dipole-dipole interaction, a short-range interaction between soft modes, an elastic energy, and an interaction between the local modes and local strain [14]. The expression for $E_{\text{loc,s}}$ and $E_{\text{loc,m}}$ are [13]

$$E_{\text{loc,s}}(\{\mathbf{v}_i\}, \{\sigma_j\}) = \sum_{ij} R_{|j-i|} \sigma_j \mathbf{f}_{ji} \cdot \mathbf{v}_i \quad (2)$$

and,

$$E_{\text{loc,m}}(\{\mathbf{u}_i\}, \{\sigma_j\}) = \sum_{ij} Q_{|j-i|} \sigma_j \mathbf{e}_{ji} \cdot \mathbf{u}_i \quad (3)$$

where the sum over i runs over all the unit cells, while the sum over j runs over the mixed sublattice sites. $\mathbf{f}_{ji}$ is a unit vector joining the B-site j to the origin of displacement $\mathbf{v}_i$ –located on the A site. $\mathbf{e}_{ji}$ is a unit vector joining the B-site j to the (B-site) center of the soft mode $\mathbf{u}_i$. The alloy parameters $R_{|j-i|}$ and $Q_{|j-i|}$ only depend on the distance between i and j, and are found to decrease as this distance increases. As a result, we included the contribution up to the first neighbors for $R_{|j-i|}$ and up to the third neighbors for $Q_{|j-i|}$ — that we will denote Q_1, Q_2 and Q_3 in the following. All the parameters of Eqs. (1)-(3) are derived from first principles [15–17].

Once our effective Hamiltonian is fully specified, the total energy of Eq. (1) is used in Monte-Carlo simulations to compute structural properties of PSN alloys. We use $12 \times 12 \times 12$ supercells to obtain converged results [18]. Two different atomic configurations are generated. The first studied structure is the disordered PSN, and is mimicked by randomly distributing the $\{\sigma_j\}$ variables in our supercells as in Ref. [13]. The second structure has the chemical $Pb(Sc_{1-\nu(k)}Nb_{\nu(k)})O_3$ stochiometry with a niobium composition varying *along the [001] direction* — and thus dependent on an integer k indexing the different (001) B-planes— and given by

$$\nu(k) = 0.5 + 0.2 * \sin\frac{2\pi k}{12} \quad (4)$$

In other words, this second structure is a *sinusoidally modulated* alloy (1) with an amplitude of the modulation set at 0.20 (concentrations of scandium and niobium then range between 0.30 to 0.70), (2) with an *average* concentration equal to the charge neutrality leading composition in PSN – i.e. 50 % of scandium and 50 % of niobium, and (3) with a wavelength corresponding to twelve B-planes. (Note that the different B-planes have k integers ranging between 0 and 11.) The compositional variation along the [001] direction affects the populations of σ_j equal to +1 or −1 in the (001) B-planes, while we assume atomic disorder in these (001) planes. For each studied structure, the $\{\sigma_j\}$ variables are kept fixed during the Monte-Carlo simulations and 2×10^4 Monte-Carlo sweeps are first performed to equilibrate the system, and then 2×10^4 sweeps are used to get the various statistical averages. The outputs of the Monte-Carlo procedure are the local mode vectors $\{\mathbf{u}_i\}$ (directly related to the electrical polarization), and the homogeneous strain tensor η_H. Note that a recent study has demonstrated that our effective-Hamiltonian approach gives excellent agreement with (finite-temperature) measurements and (0 Kelvin) direct first-principles calculations for the structural properties of disordered and rocksalt-ordered PSN alloys [13].

RESULTS

Figure 1 shows the cartesian coordinates ($<u_x>$, $<u_y>$ and $<u_z>$) — along the [100], [010] and [001] directions, respectively — of the supercell average of the local mode vectors in disordered and sinusoidally modulated PSN, as a function of the temperature. Each coordinate is close to zero at high temperature. As each of the two systems is cooled down, $<u_x>$, $<u_y>$ and $<u_z>$ increase and remain nearly equal to each other. This characterizes a transition from a paraelectric cubic phase to a ferroelectric rhombohedral structure, consistent with experiments done on disordered PSN [19,20]. As in Refs [6,13,21], we linearly rescale the temperature in Fig. 1 so that the theoretical T_c in disordered PSN is forced to match the experimental value. Interestingly, Fig. 1 indicates that the average local modes ($<u_x>$, $<u_y>$ and $<u_z>$) for the sinusoidally modulated PSN are very comparable to those of the disordered PSN alloy. In particular, we predict that the sinusoidal modulation described by Eq. (4) leads to a Curie temperature nearly identical to that of disordered PSN.

In fact, structural effects that are characteristic of the studied sinusoidal modulation can be seen in Fig 2. This figure shows the cartesian coordinates $u_x(k)$,

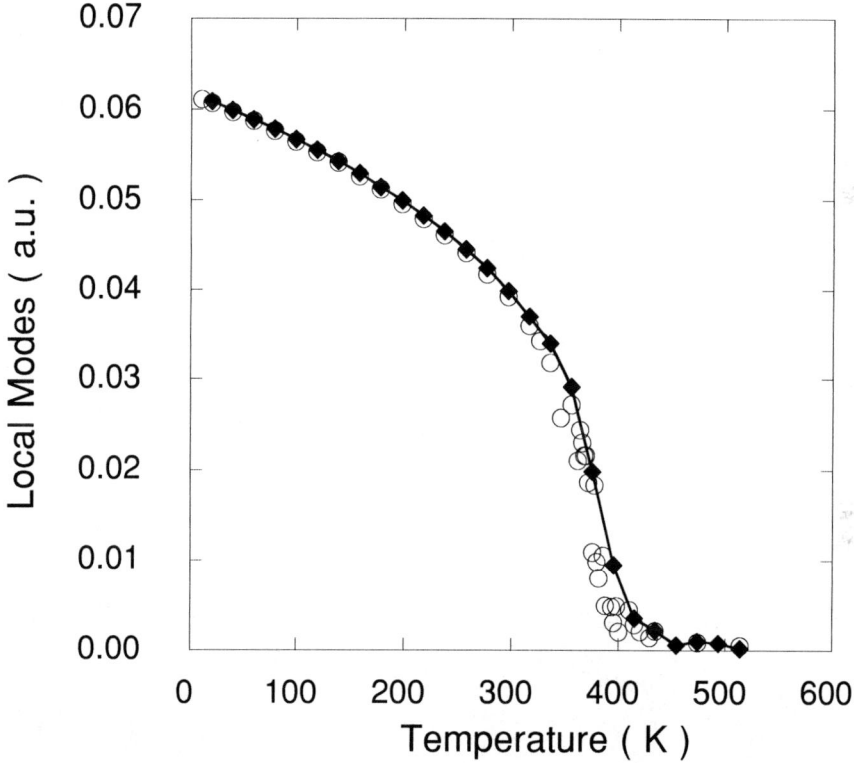

FIGURE 1. Supercell average cartesian coordinate $<u_x> = <u_y> = <u_z>$ of the local mode vectors in disordered PSN (open symbols) and in the presently studied sinusoidally modulated PSN material (filled symbols) as a function of temperature. The temperature has been rescaled so that the theoretical Curie temperature matches its experimental value in disordered PSN. For clarity, only one cartesian coordinate is displayed in each system. The two other cartesian coordinates are nearly identical to the one displayed.

$u_y(k)$ and $u_z(k)$ of the average of the local mode vectors centered *in the (001) plane indexed by k* as a function of k, for the presently studied sinusoidal modulation and at two different temperatures: 500 K, for which the phase is paraelectric and cubic, and 20 K for which the resulting structural phase is ferroelectric and rhombohedral (see Fig. 1). The cartesian coordinates $<u_x>$, $<u_y>$ and $<u_z>$ of the entire supercell average of the local-mode vectors can be derived by averaging $u_x(k)$, $u_y(k)$ and $u_z(k)$ over the different k indexes.

One can notice that $u_x(k)$ and $u_y(k)$ are equal to each other in any (001) B-plane, and are only slightly dependent on the index k at both temperature. In the paraelectric phase, $u_x(k) = u_y(k) \simeq 0$, while $u_x(k) = u_y(k) \neq 0$ in the ferroelectric phase. The most noticeable feature of Fig. 2 is that $u_z(k)$ *sinusoidally varies around its average value* $<u_z>$ as a function of k. The planes indexed by

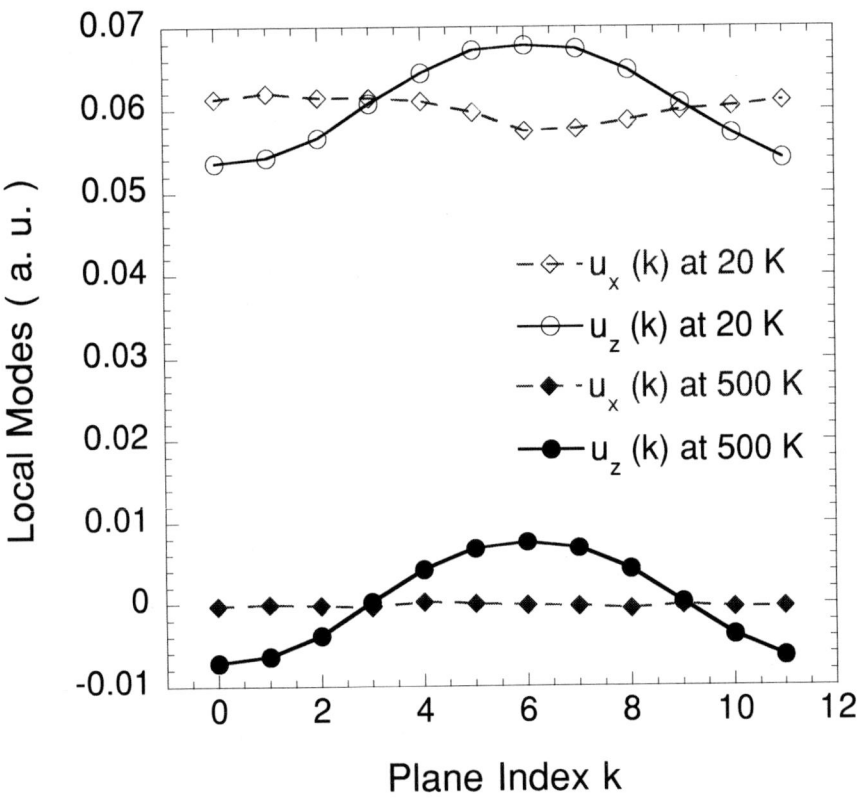

FIGURE 2. Cartesian coordinates $u_x(k)$ and $u_z(k)$ of the average of the local-mode vectors centered in the plane indexed by k as a function of k, in the presently studied sinusoidal modulation. Two rescaled temperature are considered: 500 K (filled symbols) and 20 K (open symbols). $u_y(k)$ is equal to $u_x(k)$ and is not shown for clarity.

$k=4$, 5, 6, 7 and 8 all have a $u_z(k)$ larger than $u_x(k)$ and $u_y(k)$. This indicates that the local polarizations of these planes is along the [001] direction in the paraelectric phase, while they are situated between the [111] and the [001] pseudo-cubic directions in the rhombohedral ferroelectric phase. Conversely, the planes indexed by $k=0$, 1, 2, 10 and 11 have a $u_z(k)$ which is *smaller* than $u_x(k)$ and $u_y(k)$. The local polarizations in these planes are thus along the [00$\bar{1}$] direction at high temperature, and are lying between the [111] and [110] crystallographic directions at smaller temperature. Finally, the planes for which $k=3$ or 9 are found to have a local mode $(u_x(k), u_y(k), u_z(k))$ identical to $(<u_x>, <u_y>, <u_z>)$, revealing that the polarization centered in these planes is null in the paraelectric phase, while it is parallel to the [111] direction at smaller temperature. Fig. 2 thus confirms the

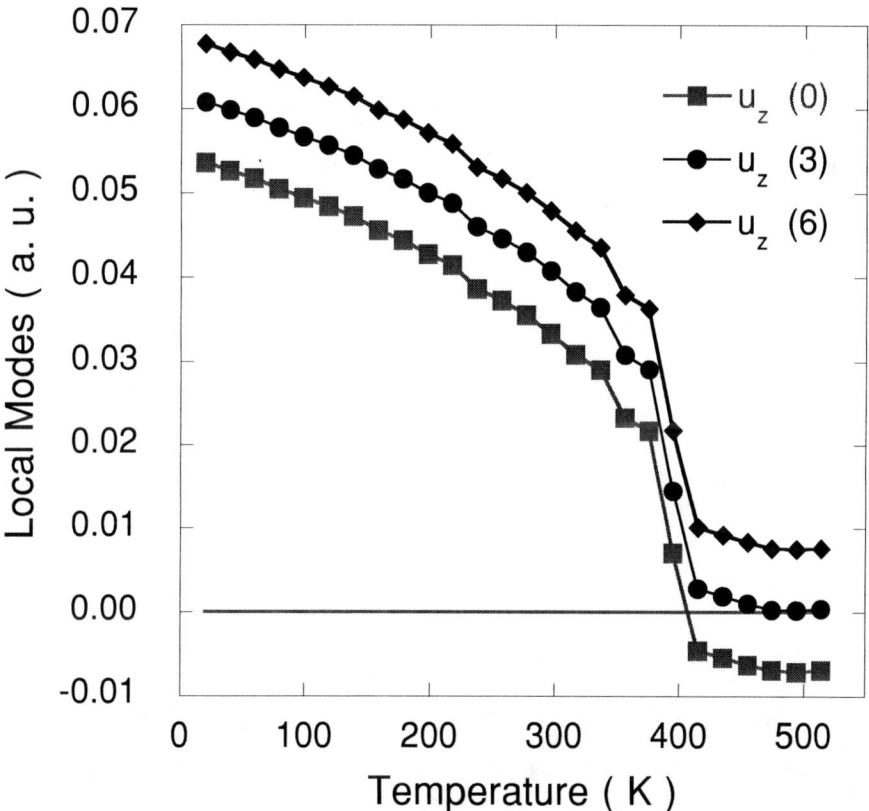

FIGURE 3. Cartesian coordinate $u_z(k)$ of the average of the local-mode vectors centered in the planes indexed by $k=0$, 3 and 6 as a function of temperature, in the presently studied sinusoidal modulation. The temperature has been rescaled so that the theoretical Curie temperature matches its experimental value in disordered PSN. The horizontal solid line indicates the zero value.

assumption of Ref. [8], i.e. that a polarization gradient along the modulated direction occurs in compositionally modulated alloys. In the paraelectric phase, this polarization gradient results in (non-null) local polarizations that can lie along the [001] or [00$\bar{1}$] directions. On the other hand, in the ferroelectric phase, the local polarizations continuously and sinusoidally rotate around the [111] direction inside the compositionally modulated PSN sample.

To get a broader perspective of the local effects of composition modulation, Fig. 3 displays the *temperature-behavior* of the z-cartesian coordinate of the local mode vectors centered in the different (001)-B planes. In fact, we "only" selected in this figure the three planes corresponding to $k=0$, 3 and 6, due to their pronounced character. That is, as indicated in Fig. 2, no displacement of $u_z(k)$ from the entire supercell average $<u_z>$ for $k=3$, and a minimum (respectively, maximum) value of $u_z(k)$ for the plane indexed by $k=0$ (respectively, $k=6$). Comparing Fig. 3 with Fig. 1 indicates that the $u_z(k)$ in these different planes all suddenly increase at the same temperature, namely at the paraelectric cubic-to- ferroelectric rhombohedral transition temperature. In fact, we also found (not shown here) that the two other cartesian coordinates $u_y(k)$ and $u_y(k)$ also exhibit a sudden jump at this temperature for any value of k. This indicate that the local polarizations associated with the different (001)-B planes *all* undergo a transition at the Curie temperature.

DISCUSSIONS

In order to understand the unusual features displayed in Figs 2 and 3, a mean-field picture was adopted, in which the local mode vectors in the (001) B-plane indexed by k are all equal to their average $(u_x(k), u_y(k), u_z(k))$ value and in which each atomic site in this plane is associated with a compositionally averaged alloy parameter:

$$\sigma(k) = [\ +1 \times \nu(k)\ )] + [-1 \times (1 - \nu(k))\ ] \tag{5}$$

It is then straightforward to demonstrate that Eq (3) can be rewritten as:

$$E_{\text{loc,m}} = -N_{xy} \sum_k Z^* u_z(k) \epsilon_{CM}(k) \tag{6}$$

where N_{xy} is the number of atomic sites in each (001) B-plane, Z^* is the Born effective charge, and $\epsilon_{CM}(k)$ is given by:

$$\epsilon_{CM}(k) = (2Q_1 + \frac{8Q_2}{\sqrt{2}} + \frac{8Q_3}{\sqrt{3}}) \frac{[\nu(k+1) - \nu(k-1)]}{Z^*} \tag{7}$$
$$= \frac{0.2}{Z^*}(4Q_1 + \frac{16Q_2}{\sqrt{2}} + \frac{16Q_3}{\sqrt{3}})cos(\frac{2\pi k}{12})\ sin(\frac{2\pi}{12})$$

Equations (6) and (7) indicate that $E_{\text{loc,m}}$ is simply the interactive energy of the dipole moments centered on the (001) B-planes with an electric field ϵ_{CM}. This electric-field is oriented along the direction of the composition modulation (i.e., along the z-axis) and takes different values —denoted $\epsilon_{CM}(k)$ — in the different (001) B-planes. Eq. (7) further reveals that $\epsilon_{\mathbf{CM}}$ is an electric-field with an *average null value*, since the integration of $\epsilon_{CM}(k)$ over the entire supercell is exactly equal to zero. Eq. (7) also tells us that $\epsilon_{CM}(k)$ simply depends on the modulation and its parameters via the difference of composition between the $k+1$ and $k-1$ planes. For the presently studied modulation, $\epsilon_{CM}(k)$ thus sinusoidally varies with the plane index k. Furthermore, we numerically find that Q_1, Q_2 and Q_3 all have negative signs. The behavior of the local modes displayed in Figs 2 and 3 can be understood by looking at the sign and the magnitude of $\epsilon_{CM}(k)$. A null value of $\epsilon_{CM}(k)$ yields a local mode $(u_x(k), u_y(k), u_z(k))$ equal to the supercell average mode $(<u_x>, <u_y>, <u_z>)$ (see Figs 2 and 3 and Eq (7) for $k=3$ and 9). A positive (respectively, negative) value of $\epsilon_{CM}(k)$ leads to a $u_z(k)$ larger (respectively, smaller) than $<u_z>$, and thus generates a local polarization which is located along the [001] direction (respectively, along the [00$\bar{1}$] direction) at high temperature, and between the total polar direction and the [001] direction (respectively, the [00$\bar{1}$] direction) in the ferroelectric phase. Increasing the magnitude of $\epsilon_{CM}(k)$ leads to an increase of the magnitude of $u_z(k)$, and thus results in a local polarization, which has a larger magnitude in the paraelectric phase and which is further away from the total polar direction in the ferroelectric phase (see Figs 2 and 3 and Eq (7) for $k=0$ or 6).

We now investigate the physical mechanism(s) responsible for the existence of the internal electric fields $\epsilon_{CM}(k)$. For that, it is relevant to assign an ionic charge of $\sigma(k)e$ to all the B-atoms sitting in the plane indexed by k. When "only" including the interactions up to the third neighbors shells, it is straightforward to realize that the *electrostatic* electric-field deriving from these plane-dependent charges is given by:

$$\epsilon_{ES}(k) = -(1 + \frac{2}{\sqrt{2}} + \frac{4}{3\sqrt{3}})\frac{[\sigma(k+1) - \sigma(k-1)]}{\epsilon_0 a^2} \tag{8}$$

where ϵ_0 and a are the dielectric constant and the cubic lattice constant of PSN, respectively.

Inserting Eq (4) and (5) into Eq (8) yields:

$$\epsilon_{ES}(k) = -(2 + \frac{4}{\sqrt{2}} + \frac{8}{3\sqrt{3}})\frac{[\nu(k+1) - \nu(k-1)]}{\epsilon_0 a^2} \tag{9}$$

$$= -\frac{0.2\,(4 + \frac{8}{\sqrt{2}} + \frac{16}{3\sqrt{3}})}{\epsilon_0 a^2} cos(\frac{2\pi\,k}{12}) sin(\frac{2\pi}{12})$$

Eq (7) and (9) are remarkably similar, providing that

$$(4Q_1 + \frac{16Q_2}{\sqrt{2}} + \frac{16Q_3}{\sqrt{3}}) = -\frac{Z^*(4 + \frac{8}{\sqrt{2}} + \frac{16}{3\sqrt{3}})}{\epsilon_0 a^2} \tag{10}$$

We numerically find that the above equality is satisfied but for an unrealistic dielectric constant ϵ_0 of 13.7. In fact, adopting a (reasonable) value around 7 for the dielectric constant of PSN results in a $\epsilon_{CM}(k)$ — derived from Eq (7) — about 2 times smaller than the electrostatic static electric field $\epsilon_{ES}(k)$ derived from Eq (9). All these findings thus indicate that, in addition to the different valence charges between Sc and Nb atoms, other chemical effects distinguishing these two kinds of atoms (such as size differences) contribute to $\epsilon_{CM}(k)$.

CONCLUSIONS

In summary, we have used the first-principles derived computational scheme of Ref. [6,13] to study finite-temperature structural properties of a PSN solid solution that is sinusoidally modulated along the [001] direction. We find that the modulated structure is (1) associated with its own internal sinusoidally-dependent electric field, which is oriented along the direction of the modulation; (2) this electric field leads to a spatially-dependent direction of the local modes in the different (001) B-planes at any temperature; (3) this electric field partially results from the difference in valence between Sc and Nb atoms. We thus expect that similar unusual properties will also occur in other heterovalent alloys.

ACKNOWLEDGMENTS

This work was supported by Arkansas Science and Technology Authority Grant N99-B-21, Office of Naval Research Grant N00014-00-1-0542 and National Science Foundation Grant DMR-9983678. We wish to thank A. García, J. Iniguez and D. Vanderbilt for very useful discussions.

REFERENCES

1. K. Abe and S. Komatsu, *J. Appl. Phys.* **77**, 6461 (1995).
2. S.-E. Park and T.R. Shrout, *J. Appl. Phys.*, **82**, 1804 (1997).
3. R.F. Service, *Science* **275**, 1878 (1997).
4. K. Uchino, *Piezoelectric actuators and ultrasonic motors*, (Kluwer Academic Publishers, Boston) (1996).
5. B. Noheda, D.E. Cox, G. Shirane, J.A. Gonzalo, L.E. Cross, and S-E. Park, *Appl. Phys. Lett.* **74**, 2059 (1999).
6. L. Bellaiche, A. García and D. Vanderbilt, *Phys. Rev. Lett.* **84**, 5427 (2000).
7. N.W. Schubring, J.V. Mantese, A.L. Micheli, A.B. Catalan and R.J. Lopez, *Phys. Rev. Lett.* **68**, 1778 (1992).
8. J.V. Mantese, N.W. Schubring, A.L. Micheli and A.B. Catalan, *Appl. Phys. Lett.* **67**, 721 (1995).
9. M. Brazier, M. McElfresh and S. Mansour, *Appl. Phys. Lett.* **72**, 1121 (1998).

10. M.S. Mohammed, G.W. Auner, R. Naik, J.V. Mantese, N.W. Schubring, A.L. Micheli and A.B. Catalan, *J. Appl. Phys.* **84**, 3322 (1998).
11. N. Sai, B. Meyer and D. Vanderbilt, *Phys. Rev. Lett.* **84**, 5636 (2000).
12. T. Mattila, L. Bellaiche, L.-W. Wang and A. Zunger, *Appl. Phys. Lett.* **72**, 2144 (1998).
13. R. Hemphill, L. Bellaiche, A. García and D. Vanderbilt, *Appl. Phys. Lett.* **77**, 3642 (2000).
14. W. Zhong, D. Vanderbilt and K.M. Rabe, *Phys. Rev. Lett.* **73**, 1861 (1994); *Phys. Rev. B* **52**, 6301 (1995).
15. L. Bellaiche and D. Vanderbilt, *Phys. Rev. B* **61**, 7877 (2000).
16. P. Hohenberg and W. Kohn, *Phys. Rev.* **136**, B864 (1964); W. Kohn and L.J. Sham, *ibid.* **140**, A1133 (1965).
17. D. Vanderbilt, *Phys. Rev. B* **41**, 7892 (1990).
18. L. Bellaiche, A. García and D. Vanderbilt, *Proceedings of the 2000 Aspen Winter Conference on Fundamental Physics of Ferroelectrics*, R.E. Cohen, ed. (AIP, Woodbury, New York, 2000), p. 79.
19. F. Chu, I.M. Reaney and N. Setter, *J. Appl. Phys.*, **77**, 1671 (1995).
20. C. Malibert, B. Dkhil, J.M. Kiat, D. Durand, J.F. Berar and A. Spasojevic-de Bire, *J. Phys.: Condens. Matter*, **9**, 7485 (1997).
21. A. García and D. Vanderbilt, in *First-Principles Calculations for Ferroelectrics: Fifth Williamsburg Workshop*, R.E. Cohen, ed. (AIP, Woodbury, New York, 1998), p. 53; *Appl. Phys. Lett.* **72**, 2981 (1998).

Domain Patterns, Texture and Macroscopic Electro-mechanical Behavior of Ferroelectrics

Kaushik Bhattacharya and JiangYu Li

Division of Engineering and Applied Science
Mail Stop 104-44, California Institute of Technology, Pasadena, CA 91125

Abstract. This paper examines the domain patterns and its relation to the macroscopic electromechanical behavior of ferroelectric solids using a theory based on homogenization and energy minimization. The domain patterns in different crystalline systems are classified, the spontaneous strain and polarization for single crystals and polycrystals are characterized, and the optimal texture of polycrystals for high-strain actuation is identified. The results also reveal why it is easy to pole PZT at compositions close to the 'morphotropic phase boundary'.

INTRODUCTION

Ferroelectric solids are widely used in transducers applications based on their piezoelectricity. The strains they display, however, are quite small, and there have been many attempts to find methods to increase them. These efforts received a significant boost when Shrout and Park [1] demonstrated strains as large as 1.7% through electrostriction of a single-crystal relaxor ferroelectric PMn-Pt. There have also been significant advances towards a fundamental understanding of the electromechanical properties of ferroelectrics. In particular, there has been much progress at the quantum mechanical level in both hard and relaxor ferroelectrics, and at the cluster level in relaxors.

This paper reports on some recent theoretical and experimental work on hard ferroelectric materials. These materials form complicated domain patterns, which can be changed through the application of an external electric field and mechanical stress. The change in domain pattern, however, may involve a significant change in polarization and strain, and lead to significant mismatch between grains in polycrystals. Therefore, inter-granular constraints severely restrict possible changes in domain patterns in polycrystals. Even in single crystals, the changes can lead to mismatch with the loading device and may consequently be suppressed. Therefore one requires very large forces and fields for switching domain patterns, and such switching is often accompanied by mechanical and electrical damage. Therefore, the use of hard ferroelectrics is often limited to those regimes where the domains are stationary. There are exceptions as during poling or in more recent memory devices.

The domain patterns also pose a challenge to a complete theoretical understanding of the macroscopic behavior of hard ferroelectrics, as they occur at a length and time scale which is inaccessible to both the classical macroscopic theories and the modern ab initio methods and effective Hamiltonians. In other words, they are sufficiently complicated and influential to be ignored, yet their interactions are sufficiently complex that studying a few unit cells or one wall does not reveal the entirety of their behavior. This paper reports on an attempt to understand these intermediate scales. It begins with a classical and phenomenological Landau-Ginzburg-Devonshire type theory, shows that this theory reveals important aspects of domain patterns, and tries to systematically average their effect to larger scales. It reports on the work of Shu and Bhattacharya [2] and Burscu, Ravichandran and Bhattacharya [3] on single crystals and then summarizes the work of Li and Bhattacharya [4] on polycrystals. This line of research parallels recent work in the study of the shape-memory effect in martensitic materials [5-8], and magnetostriction in the recently developed ferromagnetic shape-memory alloys [9,10]. In particular, this work adapts the constrained theory of magnetostriction of DeSimone and James [10] to ferroelectrics and extends it to polycrystalline media following Bhattacharya and Kohn [6].

THEORETICAL FRAMEWORK

Electroelastic Energy

Consider a ferroelectric crystal Ω subject to an applied traction $\mathbf{t}_0$ on part of its boundary $\partial_2\Omega$ and external applied electric field $\mathbf{E}_0$. The displacement $\mathbf{u}$ and polarization $\mathbf{p}$ of the ferroelectric are those that minimize the potential energy

$$\int_\Omega \frac{1}{2}\nabla\mathbf{p}\cdot\mathbf{A}\nabla\mathbf{p}+W(\mathbf{x},\mathbf{e}[\mathbf{u}],\mathbf{p})-\mathbf{E}_0\cdot\mathbf{p}dx - \int_{\partial_2\Omega}\mathbf{t}_0\cdot\mathbf{u}dS + \int_{R^3}\frac{\varepsilon_0}{2}|\nabla\phi|^2\,dx. \qquad(1)$$

Above, $\mathbf{A}$ is a positive-definite matrix so that the first term above penalizes sharp changes in the polarization, and may be regarded as the energetic cost of forming domain walls. The second term W is the stored energy density (Landau energy density) which depends on the state variables or order parameters, strain $\mathbf{e}$ and polarization $\mathbf{p}$, and also explicitly on position $\mathbf{x}$ in polycrystals and heterogeneous media. W encodes all the crystallographic and texture information, and is discussed in more detail below. The third term is the potential associated with the applied electric field, while the fourth term is the potential energy associated with the applied mechanical load. The final term is the electrostatic field energy that is generated by the polarization distribution. For any polarization distribution, the electrostatic potential ϕ is determined by solving Maxwell equation in all space subject to appropriate boundary conditions. Thus, this last term is a non-local. This potential is essentially one used in the classical Landau-Ginzburg-Devonshire theory, except the electrostatic contribution to energy is accounted for with some care.

Ferroelectric crystals are non-polar above the Curie temperature with a preferred lattice structure, but are spontaneously polarized below the Curie temperature with a spontaneous strain. Further, because of a change in symmetry, they can be spontaneously polarized and strained in one of K crystallographically equivalent variants. Thus, if $\mathbf{e}^{(i)}$ is the spontaneous strain and $\mathbf{p}^{(i)}$ is the spontaneous polarization of the ith variant ($i = 1, ..., K$), then the stored energy W is minimum (zero without loss of generality) on the $\mathbf{Z} = \bigcup_{i=1}^{K}\{(\mathbf{e}^{(i)}, \mathbf{p}^{(i)})\}$ and grows away from it. In other words it has a multi-well structure with wells $\mathbf{Z}$. Finally note that in a polycrystalline medium, if $\mathbf{R}(\mathbf{x})$ denotes the rotation that maps the crystallographic frame to a reference frame, then $W(\mathbf{x},\mathbf{e},\mathbf{p}) = W_0(\mathbf{R}(\mathbf{x})\mathbf{e}\mathbf{R}^T(\mathbf{x}), \mathbf{R}(\mathbf{x})\mathbf{p})$ where W_0 is the energy density of a reference crystal.

Since W has a multi-well structure, minimization of the potential energy in Eq. (1) leads to domain patterns or regions of almost constant strain and polarization close to the spontaneous values separated by domain walls. However, the solutions are rather complicated [11] as one has to resolve the length scales determined by the domain wall energy (square-root of smallest eigen-value of $\mathbf{A}$), and yet one has multiple domains in the setting of current interest where the specimen Ω is large compared to the domain wall thickness. In this setting, however, the domain wall energy has a negligible effect on the macroscopic behavior and may thus be dropped [12]. This leads to an ill-posed problem as the minimizers develop oscillations at a very fine scale, however there has been significant recent progress in studying such problems. Therefore $\mathbf{A}=0$ in what follows.

Homogenization

Now consider a ferroelectric polycrystal with numerous small grains. Each grain in turn forms numerous domains as shown in Fig. 1. The functional (1) (with $\mathbf{A}=0$) describes all the details of the domain pattern in each grain, and thus is rather difficult to understand. It would be very useful to obtain a functional that directly captures the *effective* macroscopic behavior of all the domains and grains without describing all the details. The theory of Γ-convergence introduced by De Giorgi [13,14] provides us a framework to derive such an effective or homogenized theory.

Suppose the texture of polycrystal is periodic with period $\varepsilon > 0$, i.e., suppose that $W(\mathbf{x},\mathbf{e},\mathbf{p})$ be a periodic function of $\mathbf{x}$ with period ε. Then, as $\varepsilon \to 0$, the effective behavior of the polycrystal is described by the functional

$$\int_\Omega \frac{1}{2}\nabla\mathbf{p}\cdot\mathbf{A}\nabla\mathbf{p} + \overline{W}(\mathbf{e}[\mathbf{u}],\mathbf{p}) - \mathbf{E}_0\cdot\mathbf{p}\,d\mathbf{x} - \int_{\partial_2\Omega}\mathbf{t}_0\cdot\mathbf{u}\,dS + \int_{R^3}\frac{\varepsilon_0}{2}|\nabla\phi|^2\,d\mathbf{x} \qquad (2)$$

in the sense that the minimizers of the functional (2) approximate in an appropriate sense [4] those of functional (1). Notice that the spatially heterogeneous energy density W in (1) replaced with a spatially homogeneous energy density $\overline{W}$ in (2). $\overline{W}$ is the *effective energy density* of the polycrystal, and can be obtained by solving the following variational problem on an unit cell:

$$\overline{W}(\mathbf{e},\mathbf{p}) = \inf_k \inf_{\mathbf{u}' \in H^1_{Per}(kY)} \inf_{\mathbf{p}' \in L^2(kY)} \frac{1}{|kQ|}\int_{kQ}\left\{W(\mathbf{y},\mathbf{e}+\nabla\mathbf{u}',\mathbf{p}+\mathbf{p}') + \frac{\varepsilon_0}{2}|\nabla\phi'|^2\right\}d\mathbf{y}. \qquad (3)$$

It turns out that in this formula (3) above, one may replace W with a relaxed or mesoscale energy $\hat{W}$ which may be defined for each $\mathbf{y}$ by holding it constant in the formula (3). $\hat{W}$ is the *effective energy density* of the single crystal.

These ideas are illustrated in Fig. 1. The energy density W describes the behavior at the smallest length-scale. It has a multi-well structure as discussed earlier. This leads to domains, and $\hat{W}(\mathbf{x},\mathbf{e},\mathbf{p})$ is the energy density of the grain at $\mathbf{x}$ after it has formed a domain pattern with average strain $\mathbf{e}$ and average polarization $\mathbf{p}$. Note that this energy is zero on a set $\mathbf{Z}^S$ which is larger than the set $\mathbf{Z}$. $\mathbf{Z}^S$ is the set of all possible *average spontaneous strains and polarizations* that a single crystal can have by forming domain patterns. But $\hat{W}$ and $\mathbf{Z}^S$ can vary from grain to grain. The collective behavior of the polycrystal is described by the energy density $\overline{W}$. $\overline{W}(\mathbf{e},\mathbf{p})$ is the energy density of a polycrystal with grains and domain patterns when the average strain is $\mathbf{e}$ and the average polarization is $\mathbf{p}$. Notice that it is zero on the set $\mathbf{Z}^P$ which describes the set of all possible average spontaneous strains and polarization of the polycrystal.

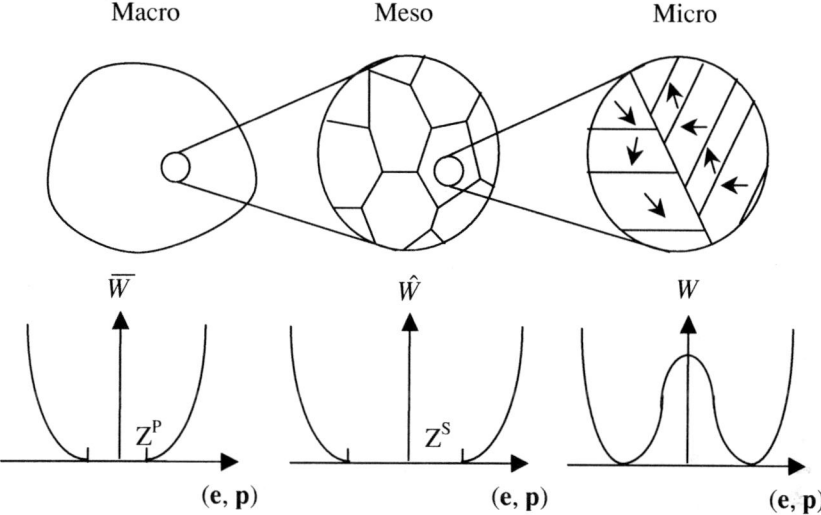

Figure 1. The multi-well structure of stored energy density W, the effective energy density of a single crystal $\hat{W}$, and the effective energy density of a polycrystal $\overline{W}$.

SINGLE CYRSTALS

Domain Patterns

The multi-well structure of the stored energy density W gives rise to minimizers that consist of domains of differently oriented polarization separated by domain walls. In order to form a domain wall between variants i and j, i.e., to find an interface separating regions of strains and polarization $(\mathbf{e}^{(i)}, \mathbf{p}^{(i)})$ and $(\mathbf{e}^{(j)}, \mathbf{p}^{(j)})$, in an energy

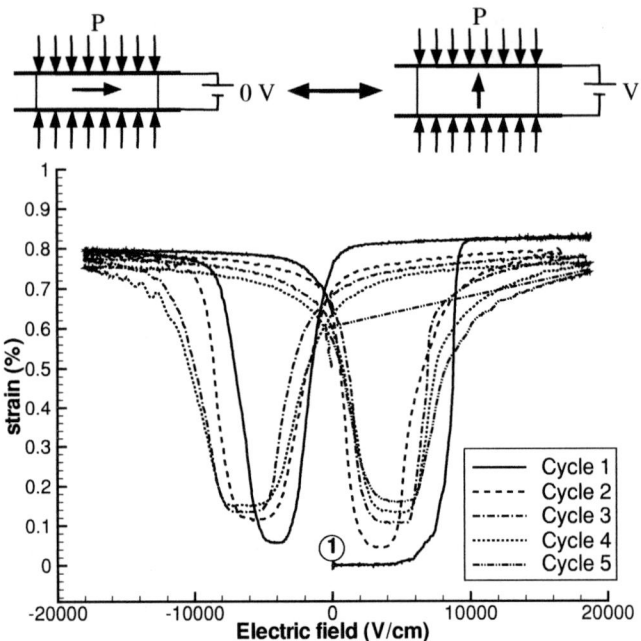

Figure 2. A single crystal of Barium Titanate with <001> orientation subjected to constant compressive stress and cyclic electric field displays very large electrostriction [3].

minimizing manner, it is necessary and sufficient to satisfy two compatibility conditions [2]:

$$\mathbf{e}^{(j)} - \mathbf{e}^{(k)} = \frac{1}{2}(\mathbf{a} \otimes \mathbf{n} + \mathbf{n} \otimes \mathbf{a}), \qquad (\mathbf{p}^{(j)} - \mathbf{p}^{(k)}) \cdot \mathbf{n} = 0, \qquad (4)$$

where $\mathbf{n}$ is the normal to the interface, $\mathbf{a}$ is the shear across the domain wall and $\mathbf{a} \otimes \mathbf{n}$ denotes the matrix with components $a_i n_j$. The first is the mechanical compatibility condition which assures the mechanical integrity of the interface, and the second the electrical compatibility conditions which assures that the interface is uncharged and thus energy minimizing. Given $(\mathbf{e}^{(j)}, \mathbf{p}^{(j)})$ and $(\mathbf{e}^{(i)}, \mathbf{p}^{(i)})$, we need to solve these equations for vectors $\mathbf{a}$ and $\mathbf{n}$. At first glance it appears impossible to solve these equations simultaneously: the first equation has at most two solutions for the vectors $\mathbf{a}$ and $\mathbf{n}$, and there is no a priori reason that these values of $\mathbf{n}$ will satisfy the second. It turns out however, that if the variants are related by two-fold symmetry, i.e.,

$$\mathbf{e}^{(j)} = \mathbf{R}\mathbf{e}^{(k)}\mathbf{R}^T, \qquad \mathbf{p}^{(j)} = \mathbf{R}\mathbf{p}^{(k)}, \qquad (5)$$

for some 180° rotation $\mathbf{R}$, then it is indeed possible to simultaneously solve the two equations in (4) [2,10].

Shu and Bhattacharya [2] studied the domains walls in materials that are cubic above Curie temperature and either <001>$_c$ polarized tetragonal, <111>$_c$ polarized rhombohedral, <110>$_c$ polarized orthorhombic, or <a11>$_c$ polarized monoclinic. The only domain walls in the tetragonal phase are 180° and 90° domain walls, and the 90° domain walls have a structure similar to that of compound twins with a rational {110}$_c$ interface and a rational <110>$_c$ shear direction. The only possible domain walls in the orthorhombic phase are 180° domain walls, 90° domain walls having a structure like that of compound twins with a rational {100}$_c$ interface, 120° domain walls having a structure like that of type I twins with a rational {110}$_c$ interface and 60° domain walls having a structure like that of type II twins with an irrational normal. The only possible domain walls in the rhombohedral phase are the 180° domain walls, and the 70° or 109° domain walls with a structure similar to that of compound twins. They also studied more complex patterns involving layers within layers.

Finally, they used the theory to motivate a configuration to obtain large electrostriction using tetragonal materials. This was subsequently confirmed experimentally by Burcsu et al. [3] as shown in Fig. 2. A single crystal in the shape of a flat plate of a tetragonal material with <001>$_c$ orientation with electrodes on top and bottom faces is subjected to a constant compressive load and a cyclic voltage. The polarization then cycles between two states, resulting in a strain close to the c/a ratio of the material. The experimental set-up allows in-situ observation and confirms that the switching from one state to another is mediated by domain walls sweeping across the specimen [15].

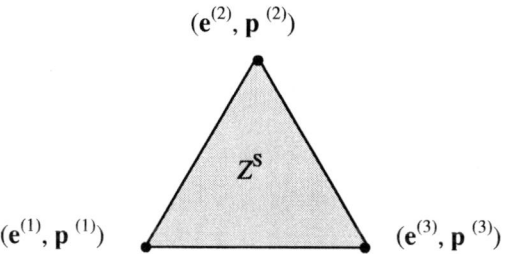

Figure 3. The set of average spontaneous strain and polarization of a single crystal.

Average Spontaneous Strain and Polarization of a Single Crystal

As the single crystal forms domain patterns, its overall behavior is described by the energy density $\hat{W}$ and the possible average spontaneous strain and polarization is given by the set $\mathbf{Z}^S$. The following result allows the characterization of this set. Suppose the ferroelectric material has $K = 2n$ variants, and suppose further that each pair of variants is compatible (each pair satisfies Eq. (4) for some vectors $\mathbf{a}, \mathbf{n}$). Then, $\mathbf{Z}^S$ is given by all possible averages of spontaneous strains and polarizations [4,10]:

$$\mathbf{Z}^S = \left\{ (\mathbf{e},\mathbf{p}) : \mathbf{e} = \sum_{i=1}^n \lambda_i \mathbf{e}^{(i)}, \ \mathbf{p} = \sum_{i=1}^n \lambda_i (2f_i - 1)\mathbf{p}^{(i)}, \ \lambda_i \geq 0, \sum_{i=1}^n \lambda_i = 1, \ 0 \leq f_i \leq 1 \right\}. \quad (6)$$

In other words, if each pair of variants is compatible, then compatibility poses no restriction, and it is possible to find energy minimizing and compatible domain pattern for any arbitrary average. This is shown schematically in Fig. 3.

Tetragonal Phase

The spontaneous strains and polarization of tetragonal ferroelectric are of the form
$$\mathbf{e}^{(1)} = \text{diag}[\beta, \alpha, \alpha], \quad \mathbf{p}^{(1)} = [p, 0, 0], \tag{7}$$
and their symmetry-related variants. The total number of variants is 6, and each pair is compatible so that one can use Eq. (6). One can readily see that the average spontaneous strains of single crystal have fixed trace $2\alpha + \beta$ and zero off-diagonal elements. Consequently these strains lie in a 2-dimensional subspace of the 5-dimensional linear space of deviatoric (shear-like) strains.

Rhombohedral Phase

Here one has eight variants, and again all pairs are compatible so that Eq. (6) holds. The average spontaneous strains lie in a 3-dimensional subspace of the 5-dimensional linear space of deviatoric strains.

Tetragonal-Rhombohedral Phase Co-Existence

Suppose one has a material where both the tetragonal and rhombohedral phases can coexist. Then one has both sets of wells. While the tetragonal variants are all pair-wise compatible amongst themselves, and the rhombohedral variants are all pair-wise compatible amongst themselves, a tetragonal variant is compatible with a rhombohedral variant if and only if the spontaneous strains and polarization satisfy a very special relationship. Thus, it is not possible, in general, to apply Eq. (6). It is possible however to estimate the set $\mathbf{Z}^S$ from inside, and use this inner (conservative) estimate to show that $\mathbf{Z}^S$ spans the 5-dimensional deviatoric strain space and 3-dimensional polarization space [4]. Thus, crystals with Tetragonal-Rhombohedral phase co-existence have a full dimensional set $\mathbf{Z}^S$ in contrast to either Tetragonal or Rhombohedral alone. This will have important consequences later.

Monoclinic Phase

In a monoclinic phase, not all pairs of variants are compatible in general and it is not possible to use Eq. (6). It is possible however to estimate the set $\mathbf{Z}^S$ from inside, and use this inner (conservative) estimate to show that $\mathbf{Z}^S$ spans the 5-dimensional deviatoric strain space and 3-dimensional polarization space [4]. Thus, crystals with a monoclinic phase also has a full dimensional set $\mathbf{Z}^S$. This will have important consequences later.

POLYCRYSTALS

A polycrystal is a collection of perfectly bonded single crystal with large number of distinct orientations. Its overall behavior is described by the energy density $\overline{W}$, and possible values of average spontaneous strain and polarization is given by the set $\mathbf{Z}^P$.

$\mathbf{Z}^P$ is the set of all strains and polarizations that can be obtained as macroscopic averages of locally varying strain and polarization fields which is stress and depolarization field free. In other words, they are accommodated within each grain by rearrangement of ferroelectric variants, and satisfy the compatibility condition at the grain boundary. Therefore the size of the set $\mathbf{Z}^P$ is an estimate of the ease with which a ferroelectric polycrystal may be poled, and also the strains that one can expect through domain switching.

Recall that the energy density $\overline{W}$ is defined through a difficult variational problem, and thus very difficult to evaluate explicitly. The set $\mathbf{Z}^P$ is also similarly difficult to evaluate. Therefore it is necessary to understand their behavior using bounds.

Taylor Bound

One can obtain a very simple upper bound on the energy $\overline{W}$, and inner bound on the set $\mathbf{Z}^P$ by assuming that the strain and polarization is constant in each grain. This is referred to as the Taylor bound in analogy to plasticity. This bound has a very simple geometric interpretation as shown in Fig. 4. The Taylor bound $\mathbf{Z}^T$ is simply the intersection of all possible sets $\mathbf{Z}^S(\mathbf{x})$ corresponding to the different grains as $\mathbf{x}$ varies over the entire crystal.

$$\mathbf{Z}^P \supseteq \mathbf{Z}^T = \bigcap_{\mathbf{x} \in \Omega} \mathbf{Z}^S(\mathbf{x}) = \{(\mathbf{e}, \mathbf{p}) \mid (\mathbf{R}(\mathbf{x})\mathbf{e}\mathbf{R}^T(\mathbf{x}), \mathbf{R}(\mathbf{x})\mathbf{p}) \in \mathbf{Z}^S(\mathbf{x}), \forall \mathbf{x} \in \Omega\}. \quad (8)$$

As we can see, the calculation of Taylor bound depends on the polycrystal texture. It turns out that this simple bound is a surprisingly good indicator of the actual behavior of the material.

Figure 4. Taylor bound as the intersection of the set of average spontaneous strain and polarizations associated with the different grains.

Tetragonal Phase

Recall that for a tetragonal ferroelectric, the set $\mathbf{Z}^S$ contains strains with only diagonal elements. Thus, the intersection of these sets corresponding to different orientation is as shown on the left in Fig. 4, and leads to a set $\mathbf{Z}^T$ with a single point unless the orientations of the grains are limited to having a common <001> axis. In other words, if a polycrystal does not have a strong <001>$_c$ texture, the intergranular constrains will be so strong as to prevent any macroscopic average spontaneous strain or polarization. Such a polycrystal would be very difficult to pole, and will show no macroscopic strain through domain switching. Therefore, it is necessary to have a polycrystal with a $<001>_c$ texture if one were to pole this material, or to use it for large strains through domain switching.

Rhombohedral Phase

The rhombohedral phase is very similar to the tetragonal. Once again, a general polycrystal is difficult to pole, and would show no macroscopic strain through domain switching. It is necessary to have a polycrystal with $<111>_c$ fiber texture if one were to pole this material, or to use it for large strains through domain switching.

Tetragonal-Rhombohedral Phase Co-Existence

Recall that when the tetragonal and rhombohedral phases co-exist, then the set of average spontaneous strains and polarizations has full dimension. So the calculation of the Taylor bound is as shown on the right of Fig. 4. The polycrystal, irrespective of texture, has a significant set of macroscopic average strains and polarizations. Thus, this would be a material which would be easy to pole. Further domain switching could be affected in these materials with limited inter-granular constrains.

Monoclinic Phase

Recall that for a monoclinic phase, as in the case just above, the set of average spontaneous strains and polarizations has full dimension. So the calculation of the Taylor bound is as shown on the right of Fig. 4. The polycrystal, irrespective of texture, has a significant set of macroscopic average strains and polarizations. Thus, this would be a material which would be easy to pole. Further domain switching could be affected in these materials with limited inter-granular constrains.

Implications

The preceding analysis shows that for a material which is cubic above Curie temperature, and $<001>_c$ polarized tetragonal below it, it is necessary to have polycrystals with a $<001>_c$ fiber texture in order to be able to pole it or to affect any domain switching. Similarly a $<111>_c$ polarized rhombohedral material requires $<111>_c$ fiber texture. In contrast, in materials with tetragonal-rhombohedral phase co-existence, or in materials with a monoclinic phase, one has some average spontaneous strain and polarization irrespective of texture.

The commonly use piezoelectric material PZT ($Pb(Zr_xTi_{1-x})O_3$) is a solid solution of Lead Titanate and Lead Zirconate. It is rhombohedral in the Zirconium rich compositions, and tetragonal in the Titanium rich compositions. It has a morphotropic phase boundary at around $x = 0.52$ where it was believed that the tetragonal and rhombohedral phases coexist but it has recently been learnt that the material is monoclinic at this composition [16,17]. It has also long been known that PZT is very difficult to pole in the Titanium rich tetragonal compositions, and the Zirconium rich rhombohedral compositions, but relatively easy to pole at the morphotropic boundary.

The preceding analysis provides an explanation, as the set $\mathbf{Z}^T$ is trivial in the tetragonal and the rhombohedral states, but large when one has tetragonal-rhombohedral coexistence or a monoclinic phase. In other words, the intergranular constrains were extremely high and prevented any macroscopic average spontaneous strain in the tetragonal or rhombohedral compositions, but not significant to prevent macroscopic average spontaneous strain and polarization at the morphotropic phase boundary.

CONCLUSIONS

This paper has summarized a line of research aimed at understanding domain patterns, and its relation to the macroscopic electro-mechanical properties. The domain patterns in different crystalline systems have been classified, large electrostrictive strains through domain switching has been demonstrated, the average spontaneous strain and polarization for single crystals and polycrystals have been characterized, and the optimal texture of polycrystals for high-strain actuation have been identified. The results also explain why it is easy to pole PZT at compositions close to the morphotropic phase boundary.

ACKNOWLEDGMENTS

The authors gratefully acknowledge the financial support of the Army Research Office through DAAD 19-99-1-0319.

REFERENCES

1. Shrout, T. R. and Park, S. E., *J. Appl. Phys.* **82**, 1804-1811 (1997).
2. Shu, Y. C. and Bhattacharya, K., *Phil. Mag. A*, submitted (2000).
3. Burcsu, E., Ravichandran, G, and Bhattacharya, K, *Appl. Phys. Lett.* **77**, 1698-1700 (2000).
4. Li, J. Y. and Bhattacharya, K., in preparation (2001).
5. Ball, J. M. and James, R. D., *Phil. Trans. Roy. Soc. Lond. A* **338**, 389-450 (1992).
6. Bhattacharya, K. and Kohn R. V., *Arch. Rat. Mech. Anal.* **139**, 99-180 (1997).
7. James, R. D. and Hane, K. F., *Acta Mater.* **48**, 197-222 (2000).
8. Bhattacharya, K., *Theory of Martensitic Microstructure and the Shape-memory Effect*, draft monograph, 2001 (http://mechmat.caltech.edu/~bhatta/publications/publications.html).
9. James, R. D. and Kinderlehrer, D., *Phil. Mag. B* **68**, 237-274 (1993).
10. DeSimone, A and James, R. D., *J. Mech. Phys. Solids*, to appear (2001).
11. Cao, W. W. and Cross, L. E., *Phys. Rev. B* **44**, 5-12 (1991).
12. DeSimone, A., *Arch. Rat. Mech. Anal.* **125**, 99-143 (1993).
13. De Giorgi, E., *Rend. Matematica* **8**, 277-294 (1975).
14. Dal Maso, G., *An Introduction to Γ-Convergence*, Boston, Birkhauser, 1993.
15. Burcsu, E., Ph.D. Thesis, California Institute of Technology, in preparation (2001).
16. Noheda, B., Cox, D.E., Shirane, G., Gonzalo, J. A., Cross, L. E., and Park, S. E., *Appl. Phys. Lett.* **74**, 2059-2061 (1999).
17. Noheda, B., Gonzalo, J. A., Cross, L. E., Guo, R., Park, S. E., Cox, D. E., and Shirane, G., *Phys. Rev. B* **61**, 8687-8695 (2000).

Prediction of the $[Na_{1/2}Bi_{1/2}]TiO_3$ Ground State

B.P. Burton, and E. Cockayne

National Institute of Standards and Technology Gaithersburg, MD 20899;
benjamin.burton@nist.gov, cockayne@nist.gov

Abstract. The Vienna Ab-initio Simulation Package (VASP) was used to perform fully relaxed, planewave pseudopotential calculations of formation energies (ΔE_{VASP}) for a large number of ordered supercells in the perovskite based system $NaTiO_3 - BiTiO_3$, including 38 supercells with the $[Na_{1/2}Bi_{1/2}]TiO_3$ (NBT) composition. The ΔE_{VASP} were used to fit a cluster expansion Hamiltonian to verify that the no other superstructure is predicted to have lower energy than the lowest ΔE_{VASP}, which is a 40 atom supercell with space group symmetry $P11m$. Its chemical ordering is characterized by pseudocubic doubling of the cell constants, with alternating [100] rows of Na and Bi atoms in (hk0) planes plus alternating [010] rows in (hk$\frac{1}{2}$) planes. Chemical ordering alone reduces space group symmetry from cubic, $Pm\bar{3}m$, to tetragonal, $P4_2/mmc$, and octahedral tilting ($a^-a^-c^+$ system) further reduces it to monoclinic, $P11m$.

INTRODUCTION

Sodium bismuth titanate, $[Na_{1/2}Bi_{1/2}]TiO_3$ (NBT), has attracted much attention as a lead-free piezoceramic [1-14] with properties intermediate between those of $Pb(Zr_{1-x}Ti_x)O_3$-based materials and Pb-based relaxor ferroelectrics such as $Pb(Mg_{1/3}Nb_{2/3})O_3$. Many studies concentrated on temperature induced phase transitions (rhombohedral? → 530K → tetragonal; tetragonal → 783 - 813K → cubic) in samples of NBT, NBT-$BaTiO_3$, or NBT-$PbTiO_3$ [6,11,12]; subject to the implicit assumption that A-site cations were disordered [12-14]. Some studies [4,5,7,9-11] however, invoked chemical long-range-order (LRO) or short-range-order (SRO) to explain diffraction data [7,9-11] or Raman spectra [4,5]. In all of these latter studies [4,5,7,9,10] it was assumed that cation ordering was of NaCl-type, $Fm\bar{3}m$, (Figure 1a) although Chiang et al. [11] proposed a $Pm\bar{3}m$ structure as a possible ordering for a sample with composition $[Na_{7.5/16}Bi_{7.5/16}Ba_{1/16}]TiO_3$ (Figure 1b).

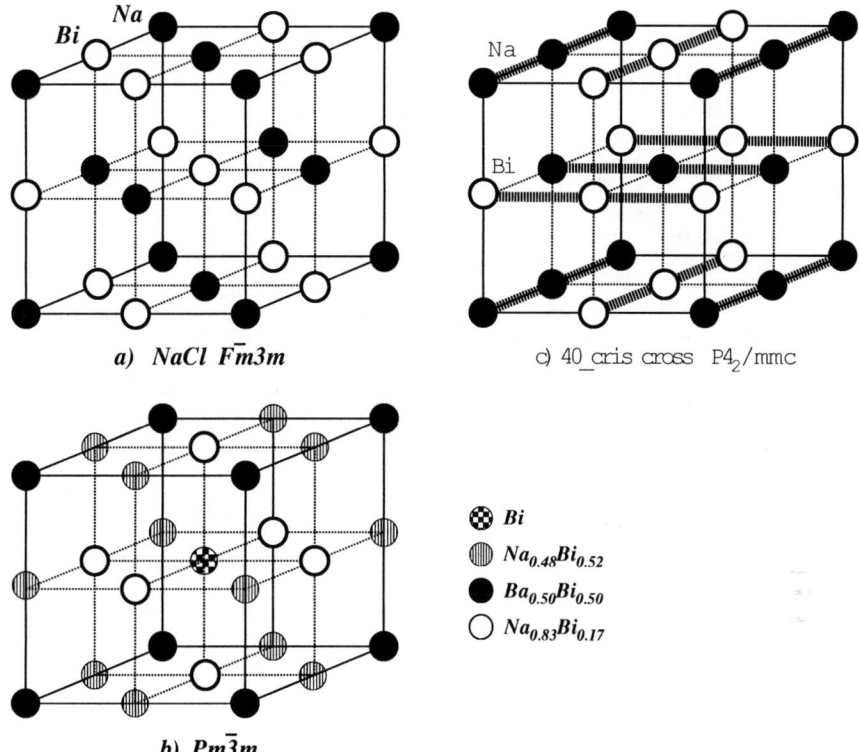

FIGURE 1. A-site patterns of chemical ordering: a) NaCl-type; b) A $Pm\bar{3}m$ partially ordered configuration proposed by Chiang et al. [11]; c) The predicted ground-state configuration 40_{CC}, $P4_2/mmc$.

It is reasonable to expect that the 1:1 mixture of Na^{1+} and Bi^{3+} in $[Na_{1/2}Bi_{1/2}]TiO_3$ would exhibit NaCl-type ordering, because NaCl is the ground-state (GS) structure for equal numbers of plus and minus charges on a simple cubic array of sites; and in NBT, Bi^{3+} and Na^{1+} have effective charges of +1 and -1 respectively, because the average A-site charge is +2. Also, the low-temperature ferroelectric phase has generally been described as rhombohedral [1,2,10] with $a^-a^-a^-$ octahedral tilting [18] which is, for example, the ferroelectric state for NaCl-ordered perovskites such as $Pb(Sc_{1/2}Ta_{1/2})O_3$ [15] in which ordering occurs on the B-sites. Soukhojak et al. [12], however, observed only $(h+\frac{1}{2}, k+\frac{1}{2}, 0)$ type superlattice reflections, not $(h+\frac{1}{2}, k+\frac{1}{2}, l+\frac{1}{2})$, and reported $a^0a^0c^+$ tilting which implies space group symmetry that is tetragonal or lower; and they found no evidence for NaCl-type LRO. The calculations discussed below clearly rule out NaCl-type LRO in the NBT GS, and predict instead that the GS is the 40_{CC} structure; a 40 atom supercell with "crisscross" (CC) rows of Na^{1+} and Bi^{3+} cations perpendicular to [001] (Figure 1c).

TOTAL ENERGY CALCULATIONS

Total energies were calculated for 65 perovskite based superstructures in the system $NaTiO_3 - BiTiO_3$, including 38 supercells with the NBT composition; these results are plotted as formation energies, relative to mechanical mixtures of $(1-X) \circ NaTiO_3 + X \circ BiTiO_3$, in Figures 2a and b. All calculations were performed with the Vienna *ab initio* simulation program (VASP) [16] using ultrasoft Vanderbilt [17] type plane-wave pseudopotentials with a local density approximation for exchange and correlation energies. Electronic degrees of freedom were optimized with a conjugate gradient algorithm, and both cell constant and ionic positions were fully relaxed. Valence electron configurations for the pseudopotentials are: Na $3s^1$; Bi $6s^26p^3$; Ti $4s^23d^2$; O $2p^6$. An energy cutoff of 395.7 eV was used, in the "high precision" option which guarantees that *absolute* energies are converged to within a few meV (a few tenths of kJ/mol; mol = ABO_3). To promote cancellation of errors, all of the calculations for low-energy structures were performed with equivalent K-point meshes: 8x8x8 within the Brillouin zone for ABO_3 pseudo primitive unit cells.

As seen in (Figure 2a) there is approximately linear variation of $\Delta E_{VASP}(X)$ between the calculated GS (40_{CC}) and the fictive end members $NaTiO_3$ and $BiTiO_3$. This trend reflects linear variation of the concentrations of electrons that are either: forced into the conduction band ($X<\frac{1}{2}$); or depleted from the valence band ($X>\frac{1}{2}$), as functions of composition. At the $X = \frac{1}{2}$, $[Na_{1/2}Bi_{1/2}]TiO_3$, composition (Figure 2b), it is clear that many superstructures have lower ΔE_{VASP} than the NaCl-type. Therefore, if cation ordering is present in the NBT samples used for experiments, it is clearly not LRO, or SRO, of the NaCl-type.

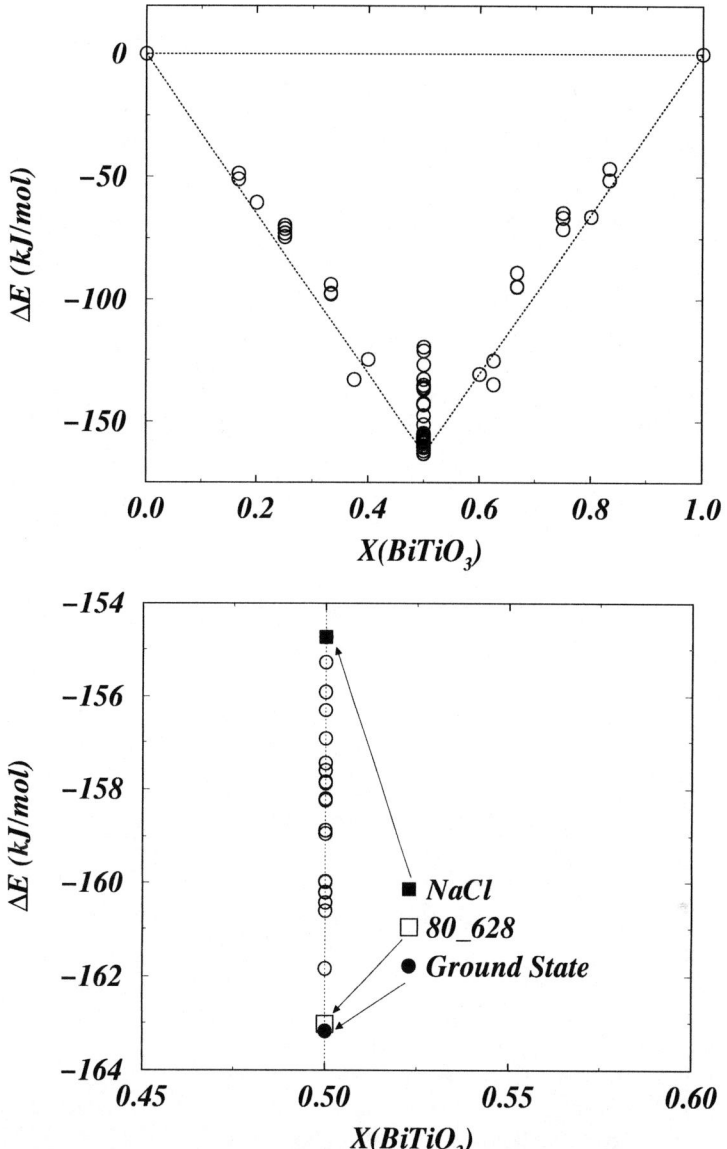

FIGURE 2. a) Formation energies per ABO_3 mol for supercells in the system $NaTiO_3 - BiTiO_3$. b) Formation energies near the 40_{CC} GS structure which is the predicted ground-state. Also marked are the formation energies for NaCl-type ordering (●) and the 80_{628} structure (□).

GROUND STATE SEARCH

The 40_{CC} structure is *not only* the lowest energy configuration that was tried, GS_{VASP}, it is also the predicted GS that one obtains by fitting a cluster expansion (CE) Hamiltonian [19,20] to the set $\{\Delta E_{VASP}\}$, GS_{CE}. The fitting was done as follows:

- Fit a CE to the set of fomnation energies $\{\Delta E_{VASP}\}$.
- Use the CE to predict a new GS_{CE}; by performing a brute force GS search on a 2x2x4 supercell (16 A-sites).
- If GS_{CE} is not an element of the set $\{\Delta E_{VASP}\}$, then calculate ΔE_{VASP} for the predicted GS.
- If GS_{CE} is the same as GS_{VASP} calculate ΔE_{VASP} for the lowest predicted excited states that are not elements of $\{\Delta E_{VASP}\}$.
- The last step was repeated three times, and did not lead to a new predicted GS; did not predict $GS_{CE} < GS_{VASP}$.

As noted above, chemical ordering in the 40_{CC} structure (Figure 1c) reduces its space group symmetry from $Pm\bar{3}m$ to $P4_2/mmc$. Octahedral tilting [18] in the $a^-a^-c^+$ system (Figures 3a-c) further reduces space group symmetry to the acentric monoclinic group $P11m$ (c-axis unique).

$[0, 1/2, 0]$ AND $[1/2, 0, 0]$ STACKING FAULTS

Disregarding octahedral tilting, the introduction of $[0, 1/2, 0]$, or $[1/2, 0, 0]$, stacking faults in every other (h,k,0) layer of the 40_{CC} GS structure leads to the 80_{628} structure (Figures 4a-d and 2b); called 80_{628} because there are 80 atoms in the supercell, and it is the 628'th 1:1 configuration in a brute force enumeration for the 2x2x4 supercell. Performing the VASP calculation, and allowing atomic positions to relax, leads to $a^+a^+c^-$ octahedral tilting in the orthorhombic space group $Pmm2$. The energy difference between 80_{628} and 40_{CC} is tiny (0.15 kJ/mol) [1] and 80_{628} is the lowest excited state shown in Figure 2b. This very small difference is expected because 80_{628} has identical (chemical) correlations [19] for all clusters within the pseudoprimitive unit cube. Such a small energy difference implies that cation ordering in 80_{628} is highly susceptible to $[0, 1/2, 0]$ and $[1/2, 0, 0]$ stacking faults. Note that in the relaxed structure a- and b-axes are symmetrically distinct, so one might expect different energies for $[0, 1/2, 0]$ and $[1/2, 0, 0]$ stacking faults; however, in terms of chemical ordering alone, $[0, 1/2, 0]$ and $[1/2, 0, 0]$ stacking faults are equivalent. It is only the tetragonal to monoclinic symmetry breaking associated with octahedral tilting that allows symmetrically distinct a- and b-axis relaxations.

---

[1] 0.15 kJ/mol is about three times the precision of VASP calculations.

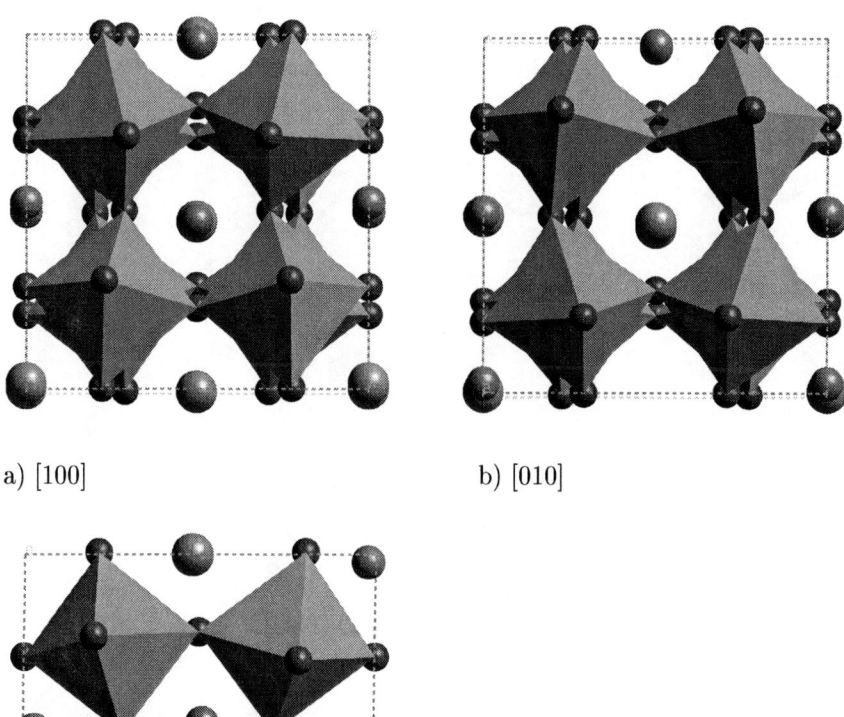

a) [100] b) [010]

c) [001]

FIGURE 3. The monoclinic 40_{CC} structure, space group $P11m$ (c-axis unique), $a^-a^-c^+$ tilt system, projected on a) [001]; b) [010]; c) [100].

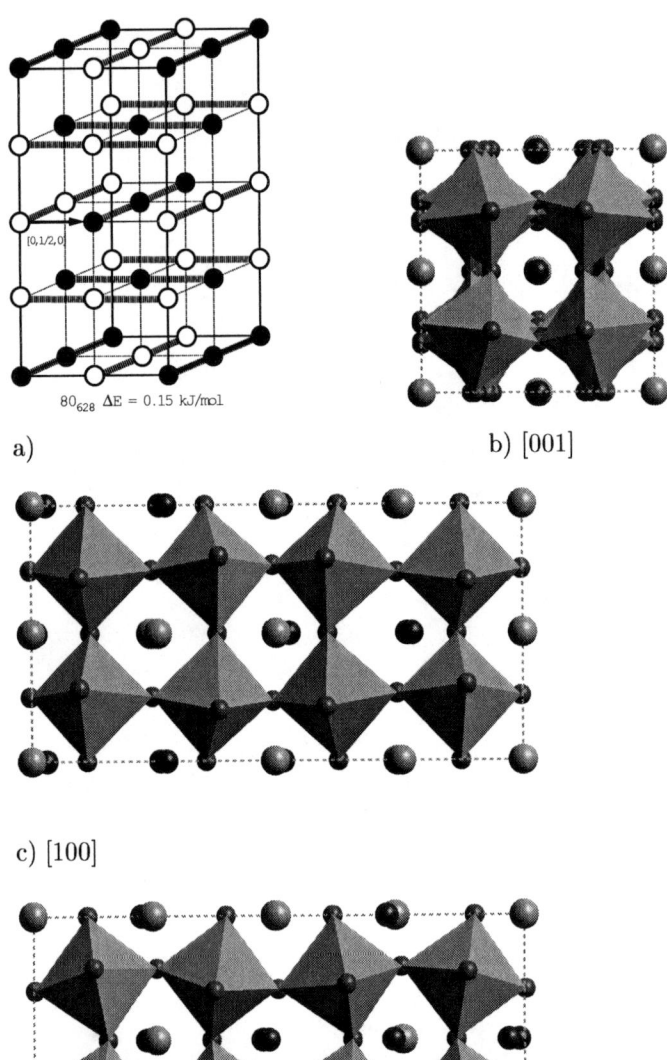

a)

b) [001]

c) [100]

d) [010]

FIGURE 4. The orthorhombic 80_{628} structure, space group $Pmm2$, tilt system $a^+a^+c^-$, which is related to the 40_{CC} structure by a $[0,1/2,0]$ stacking fault; a) Chemical ordering configuration; $\Delta E = 1.5$ kJ/mol relative to the 40_{CC} GS. Projections on: b) [001]; c) [100], d) [010].

DISCUSSION

The VASP results (Figure 2b) clearly rule out NaCl-type ordering of Na^{1+} and Bi^{3+} on A-sites, and the electron diffraction results of Soukhojak et al. [12] suggest that their samples exhibited no NaCl-type LRO, or SRO; their samples were cooled from 1350-800°C at 5°C per hr, and TEM hot-stage measurements were made between 20°C and 600°C. The predicted 40_{CC} GS with $a^-a^-c^+$ octahedral tilting is consistent with the diffraction data of Soukhojak et al. [12], but the extremely low energy difference between 40_{CC} and 80_{628} structures suggests that real samples probably have substantial disorder. Disordered samples are permissive of all tilting systems; so, the essential unanswered question is, does 40_{CC}-related LRO, or more likely SRO, influence observed octahedral tilting? Answering this question requires a reexamination of the diffraction data to see if 40_{CC}-related SRO is detectable.

CONCLUSIONS

If A-site cation ordering occurs in NBT it is clearly not of the NaCl-type. The predicted GS is the 40_{CC} structure (Figures 1c, 3a-c). Because the stacking fault energy that relates the 40_{CC} and 80_{628} structures is so small, one expects the 40_{CC} structure to be highly susceptible to disorder. Therefore, if cation ordering has a significant effect on the observed properties of NBT it is most probably SRO related to the 40_{CC} structure.

ACKNOWLEDGEMENT

The authors thank Igor Levin for many helpful discussions about octahedral tilting.

REFERENCES

1. Zvigzds, J.A., Kapostinis, P.P. and Kruzina, T.V. *Ferroelectrics* **40**, 75 (1982).
2. Vakhrushev, S.B., Kvyatkovskii, B.E., Malysheva, R.S., Okuneva, N.M., Plachenova, E.L. and Syrnikov, P.P. *Sov. Phys. Crystallogr.* **34**, 89 (1989).
3. Suchanicz, J. and Ptak, W.S. *Ferroelectrics Lett.* **12**, 71 (1990).
4. Siny, I.G., Smirnova, T.A., and Kruzina, T.V. *Ferroelectrics* **124**, 207 (1991).
5. Siny, I.G., Smirnova, T.A., and Kruzina, T.V. *Sov. Phys. Solid State* **33(1)**, 61 (1991).
6. Takenaka, T., Maruyama, K-I. and Sakata, K. *Japan. J. Appl. Phys.* **30**, 2236 (1991).
7. Park, S-E. Chung, S-J., Kim, I-T. and Hong, K.S. *J. Am. Ceram. Soc.* **77**, 2641 (1994).
8. Suchanicz, J. and Kwapuliński, J. *Ferroelectrics* **165**, 249 (1995).
9. Park, S-E. Chung, S-J., Kim I-T. *J. Am. Ceram. Soc.* **79**, 1290 (1996).
10. Park, S-E., and Hong, K.S. *J. Appl. Phys.* **79**, 383 (1996).
11. Chiang, Y-M., Farrey, G.W., and Soukhojak, A.N. *Appl. Phys. Lett.* **73**, 3683 (1998).
12. Soukhojak, A.N., Wang, G.W., Farrey, G.W., and Chiang, Y-M. *J. Phys. Chem. Solids* **61**, 301 (2000).
13. Jones, G.O. and Thomas, P.A. *Acta Cryst.* **B56**, 426 (2000).
14. Lushnikov S.G., Gvasaliya S.N., Siny, I.G., Sashin I.L., Schmidt V.H. and Uesu Y. *Solid State Comm.* **116**, 41 (2000).
15. Groves, P. *J. Phys. C: Solid State Phys.* **18** L1073 (1985).
16. Kresse, G. and Hafner, J. *Phys. Rev. B* **47**, 558 (1993); Kresse, G. Thesis, Technische Universität Wien 1993; *Phys. Rev. B* **49**, 14 251 (1994). Kresse, G. and Furthmüller, J. *Comput. Mat. Sci.* **6**, 15 (1996); *Phys. Rev.* **54**, 11169 (1996); cf. http://cms.mpi.univie.ac.at/vasp/
17. Vanderbilt, D. *Phys. Rev.* **B41** 7892, (1990).
18. Glazer, A.M. *Acta Cryst* **B28**, 3384 (1972), and *Acta Cryst* **A31**, 756 (1975).
19. Sanchez, J.M., Ducastelle, F. and Gratias, D. *Physica* **128A**, 334 (1984).
20. McCormack, R.P., Burton, B.P. *Comp. Mater. Sci.* **8**, 153 (1997).

Unique Quantum Stress Fields

Christopher L. Rogers and Andrew M. Rappe

*Department of Chemistry and Laboratory for Research on the Structure of Matter,
University of Pennsylvania, Philadelphia, PA 19104-6323.*

Abstract. We have recently developed a geometric formulation of the stress field for an interacting quantum system within the local density approximation (LDA) of density functional theory (DFT). We obtain a stress field which is invariant with respect to choice of energy density. In this paper, we explicitly demonstrate this uniqueness by deriving the stress field for different energy densities. We also explain why particular energy densities give expressions for the stress field that are more tractable than others, thereby lending themselves more easily to first-principles calculations.

Understanding a material's energetic response to strain is fundamentally coupled to understanding the physics of a broad range of phenomena, from surface reconstructions to piezoelectricity. For example, it has been demonstrated that knowledge of the spatial distribution of microscopic stress via first-principles calculations can help explain the onset of macroscopic polarization in a material [1]. Therefore it is important to understand the nature of stress at the microscopic level.

Formally, the electronic contribution to the microscopic stress must be included via quantum mechanics. There have been numerous methods developed for computing the quantum stress. (See Ref. [2] and references contained within.) The stress field, $\sigma_{\alpha\beta}(\mathbf{x})$, is a rank-two tensor field usually taken to be symmetric (torque-free). The divergence of $\sigma_{\alpha\beta}(\mathbf{x})$ must equal the force field $F^\alpha(\mathbf{x})$ of the system:

$$F^\alpha = \nabla_\beta \sigma^{\alpha\beta}. \tag{1}$$

(Note that the Einstein summation convention for repeated indices is used throughout the paper.) It is well-known that Eq. 1 does not uniquely define the stress field (regardless of whether a system is quantum-mechanical or classical) since one can add any tensor whose divergence is zero to $\sigma_{\alpha\beta}(\mathbf{x})$ and still recover the same force field [3].

The volume-averaged or total stress $T_{\alpha\beta}$ has been related to the energetic response of the system to uniform strain and to the integral over the stress field:

$$T^{\alpha\beta} = \frac{\partial E}{\partial e_{\alpha\beta}} \tag{2}$$

$$= \int \sigma^{\alpha\beta}(\mathbf{x})d^3x, \qquad (3)$$

where $e_{\alpha\beta}$ is a uniform scaling (strain) of the system. Nielsen and Martin showed that $T_{\alpha\beta}$ is a unique well-defined quantity which can be computed efficiently [2]. However Eq. 3 does not unambiguously define the stress field. One can write the total energy E as an integral over an energy density $\mathcal{E}(\mathbf{x})$. Therefore

$$T^{\alpha\beta} = \int d^3x \frac{\partial \mathcal{E}(\mathbf{x})}{\partial e_{\alpha\beta}}. \qquad (4)$$

Comparing Eq. 4 and Eq. 3 could lead to a definition of the stress field as [4,5]

$$\sigma^{\alpha\beta}(\mathbf{x}) = \frac{\partial \mathcal{E}(\mathbf{x})}{\partial e_{\alpha\beta}}. \qquad (5)$$

However, the total energy can also be expressed many other ways:

$$\int d^3x\, \mathcal{E} + \int d^3x \nabla_\alpha W^\alpha. \qquad (6)$$

If variations of W^α with respect to the dynamical variables (e.g. single-particle wavefunctions) vanish on the boundary of the system, then the minimum of E remains invariant with respect to choice of energy density. However, the energy density is changed by the presence of W, making the expression of Eq. 5 and Ref. 4 dependent on the choice of energy density. Choosing a form of the energy density is referred to as specifying an energy density "gauge", and hence the stress field of Eq. 5 is said to be not gauge invariant.

We have recently developed a formulation of the quantum stress field which is invariant to choice of energy density, and have used it to successfully calculate the stress fields in periodic systems using LDA-DFT [6]. We define the stress field as

$$\sigma^{\alpha\beta}(\mathbf{x}) = \frac{\delta E}{\delta \epsilon_{\alpha\beta}(\mathbf{x})}, \qquad (7)$$

where $\epsilon_{\alpha\beta}(\mathbf{x})$ is the strain tensor field. This definition can be derived via a virtual work theorem [3]. We then use a relationship known in continuum mechanics that equates $\epsilon_{\alpha\beta}$ with the Riemannian metric tensor $g_{\alpha\beta}(\mathbf{x})$ [7,8]. Therefore the stress field is

$$\sigma_{\alpha\beta} = -\frac{2}{\sqrt{g}} \frac{\delta E}{\delta g^{\alpha\beta}}, \qquad (8)$$

where g and $g^{\alpha\beta}$ are the determinant and inverse of $g_{\alpha\beta}$, respectively. In order to calculate $\sigma_{\alpha\beta}$, we first express the constrained energy functional for LDA-DFT [9,10] in a covariant manner via the principle of minimal coupling, and then compute the functional derivative using a procedure well-known in general relativity theory [6,11]. It should be mentioned that this prescription for obtaining $\sigma_{\alpha\beta}$ is not limited to LDA-DFT but can be generalized to derive the stress field for other DFT functionals as well as explicit many-body systems of quantum particles.

We now demonstrate the invariance of our expression for the stress field with respect to choice of energy density. In DFT there are two terms in the total energy expression which are commonly subjected to energy density gauge transformations. The first is the single-particle kinetic energy E_k which can be written as an integral over the so-called asymmetric kinetic energy density:

$$E_k = \int -\frac{1}{2} \sum_i \phi_i^* \partial_\alpha \left(\sqrt{g} g^{\alpha\beta} \partial_\beta \phi_i \right) d^3x, \tag{9}$$

where the $\{\phi_i\}$ are the single-particle wavefunctions which obey the Kohn-Sham equations generalized for curvilinear coordinates [6]. Therefore we are considering the total energy to be at a minimum. We consider here for brevity insulator occupation numbers. The variation of Eq. 9 can be written as

$$\delta E_k = \delta \left\{ \frac{1}{2} \int \sum_i \sqrt{g} \partial_\alpha \phi_i^* g^{\alpha\beta} \partial_\beta \phi_i d^3x \right\} - \delta \left\{ \frac{1}{2} \int \sum_i \partial_\alpha \left(\sqrt{g} \phi_i^* g^{\alpha\beta} \partial_\beta \phi_i \right) d^3x \right\}$$

$$= \delta \left\{ \frac{1}{2} \int \sum_i \sqrt{g} \partial_\alpha \phi_i^* g^{\alpha\beta} \partial_\beta \phi_i d^3x \right\} - \delta \left\{ \frac{1}{2} \oint \sum_i \sqrt{g} \phi_i^* g^{\alpha\beta} \partial_\beta \phi_i dS_\alpha \right\}, \tag{10}$$

where we have used the covariant version of the divergence theorem in the second line. All variations in $\{\phi_i\}$ vanish on the boundary of the system, and we require $\delta g^{\alpha\beta} = 0$ on the boundary. Therefore the surface term in Eq. 10 is zero. As a result the functional derivative required for evaluating Eq. 8 is

$$\frac{\delta E_k}{\delta g^{\alpha\beta}(\mathbf{y})} = \frac{1}{2} \sqrt{g} \sum_i \partial_\alpha \phi_i^* \partial_\beta \phi_i + \frac{\partial \sqrt{g}}{\partial g^{\alpha\beta}} \left(\frac{1}{2} \sum_i \partial_\alpha \phi_i^* g^{\alpha\beta} \partial_\beta \phi_i \right)$$
$$+ \frac{1}{2} \sum_i \int d^3x \sqrt{g} \partial_\gamma \phi_i^* g^{\gamma\kappa} \frac{\delta (\partial_\kappa \phi_i)}{\delta g^{\alpha\beta}(\mathbf{y})}$$
$$+ \frac{1}{2} \sum_i \int d^3x \sqrt{g} \partial_\kappa \phi_i g^{\gamma\kappa} \frac{\delta (\partial_\gamma \phi_i^*)}{\delta g^{\alpha\beta}(\mathbf{y})}. \tag{11}$$

Note that this result is identical to what we would obtain if we wrote E_k as an integral over the symmetric kinetic energy density:

$$E_k = \frac{1}{2} \sum_i \int \sqrt{g} \partial_\alpha \phi_i^* g^{\alpha\beta} \partial_\beta \phi_i d^3x, \tag{12}$$

and took the variation of this expression directly. Therefore the stress field is invariant with respect to choice of kinetic energy density. We mention that terms involving the variation of the wavefunctions with respect to metric vanish when the variation of the total ground-state energy is considered, since the wavefunctions obey the Kohn-Sham equations [6]. In other words, requiring $\delta E/\delta \phi_i^* = 0$ implies

$$\frac{\delta E}{\delta g^{\alpha\beta}(\mathbf{y})} = \frac{\partial E}{\partial g^{\alpha\beta}(\mathbf{y})} + \sum_i \int d^3x \frac{\delta E}{\delta \phi_i^*} \frac{\delta \phi_i^*}{\delta g^{\alpha\beta}(\mathbf{y})} + \sum_i \int d^3x \frac{\delta E}{\delta \phi_i} \frac{\delta \phi_i}{\delta g^{\alpha\beta}(\mathbf{y})}$$
$$= \frac{\partial E}{\partial g^{\alpha\beta}(\mathbf{y})}. \tag{13}$$

The Coulomb term describing the electrostatic electron-electron, electron-ion, and ion-ion interactions in the total energy functional is also commonly subjected to energy density gauge transformations. (This is entirely separate from transformations of the $U(1)$ electromagnetic gauge.) We can define E_{Coulomb} as

$$E_{\text{Coulomb}} = \frac{1}{2} \int \sqrt{g} \rho V d^3x, \tag{14}$$

with

$$\rho(\mathbf{x}) = \sum_i \frac{Z_i}{\sqrt{g}} \delta(\mathbf{x} - \mathbf{R}_i) - n(\mathbf{x}), \tag{15}$$

where Z_i is the charge of the i-th ion located at position $\mathbf{R}_i$, and n is the electronic charge density equal to $\sum_i \phi_i^* \phi_i$. The potential V can be computed from ρ via the Poisson equation:

$$\frac{1}{\sqrt{g}} \partial_\alpha \left(\sqrt{g} g^{\alpha\beta} \partial_\beta V \right) = -4\pi \rho. \tag{16}$$

The variation of E_{Coulomb} is

$$\delta E_{\text{Coulomb}} = \delta \left\{ -\frac{1}{8\pi} \int \partial_\alpha \left(\sqrt{g} g^{\alpha\beta} \partial_\beta V \right) V d^3x \right\}$$
$$= \delta \left\{ \frac{1}{8\pi} \int \sqrt{g} \, \partial_\alpha V g^{\alpha\beta} \partial_\beta V d^3x \right\} - \delta \left\{ \frac{1}{8\pi} \int \partial_\alpha \left(V \sqrt{g} g^{\alpha\beta} \partial_\beta V \right) d^3x \right\}$$
$$= \delta \left\{ \frac{1}{8\pi} \int \sqrt{g} \, \partial_\alpha V g^{\alpha\beta} \partial_\beta V d^3x \right\} - \delta \left\{ \frac{1}{8\pi} \oint V \sqrt{g} g^{\alpha\beta} \partial_\beta V dS_\alpha \right\}. \tag{17}$$

Since the variations of the potential and the metric vanish on the boundary, the surface term in Eq. 17 is zero. Therefore the functional derivative of E_{Coulomb} with respect to metric is

$$\frac{\delta E_{\text{Coulomb}}}{\delta g^{\alpha\beta}(\mathbf{y})} = \frac{1}{8\pi} \sqrt{g} \mathcal{F}_\alpha \mathcal{F}_\beta + \frac{1}{8\pi} \frac{\partial \sqrt{g}}{\partial g^{\alpha\beta}} \mathcal{F}^\gamma \mathcal{F}_\gamma$$
$$+ \frac{1}{4\pi} \int \sqrt{g} g^{\gamma\kappa} \mathcal{F}_\kappa \frac{\delta \mathcal{F}_\gamma}{\delta g^{\alpha\beta}(\mathbf{y})} d^3x, \tag{18}$$

where the electric field $\mathcal{F}_\alpha = -\partial_\alpha V$. We would obtain this same result if we initially expressed E_{Coulomb} as an integral over the Maxwell energy density:

$$E_{\text{Coulomb}} = \frac{1}{8\pi} \int \sqrt{g} \mathcal{F}_\alpha g^{\alpha\beta} \mathcal{F}_\beta d^3x. \qquad (19)$$

This demonstrates that the stress field is invariant with respect to choice of electrostatic energy density.

The non-local term involving the variation of the electric field with respect to metric in Eq. 18 is unwieldy and is in general difficult to compute. To eliminate this term it is advantageous to choose an energy density that is the Lagrangian density in electromagnetism. Therefore we write E_{Coulomb} as

$$E_{\text{Coulomb}} = \int \sqrt{g} \left(\rho V - \frac{1}{8\pi} \mathcal{F}^\gamma \mathcal{F}_\gamma \right) d^3x. \qquad (20)$$

Now when E_{Coulomb} is varied, the term containing δV will vanish due to the Poisson equation. Variation of Eq. 20 also gives a term relating to the variation of ρ with respect to metric:

$$\int \sqrt{g} \, V \frac{\delta \rho}{\delta g^{\alpha\beta}(\mathbf{y})} d^3x. \qquad (21)$$

As explained in the kinetic energy section, this variation (and those of other energy terms) will be multiplied by $\delta E/\delta \phi_i^*(\mathbf{x})$. The Kohn-Sham equation insures that variations of E with respect to ϕ_i^* vanish, and the ionic charges are fixed, thereby removing the need to evaluate Eq. 21.

In conclusion, we have demonstrated explicitly that our formulation for the quantum stress field within DFT is invariant with respect to choice of energy density. Therefore the stress field is a well-defined object that can be computed via first-principles to help gain a microscopic understanding of stress-mediated phenomena in complex materials.

The authors wish to acknowledge R. M. Martin and D. H. Vanderbilt for helpful discussions. This work was supported by the Office of Naval Research under grant number N-00014-00-1-0372 and the Air Force Office of Scientific Research, Air Force Materiel Command, USAF, under grant number F49620-00-1-0170.

REFERENCES

1. N.J. Ramer, E.J. Mele, and A.M. Rappe, Ferroelectrics **206-207**, 31 (1998).
2. O.H. Nielsen, and R.M. Martin, Phys. Rev. B. **32**, 3780 (1985).
3. L.D. Landau and E.M. Lifshitz, *Theory of Elasticity*, 3rd ed. (Butterworth-Heinemann, Oxford, 1986), pp. 1-7.
4. A. Filippetti and V. Fiorentini, Phys. Rev. B. **61**, 8433 (2000).
5. N. Chetty and R.M. Martin, Phys. Rev. B. **45**, 6074 (1992).
6. C.L. Rogers and A.M. Rappe, Phys. Rev. Lett. (submitted 2000) cond-mat/0006274.
7. Y.C. Fung, *Foundations of Solid Mechanics*, (Prentice-Hall, Englewood Cliffs, 1965), pp. 90-92.

8. L. Mistura, J. Chem. Phys. **83**, 3633 (1985); Inter. J. Thermophys. **8**, 397 (1987).
9. P. Hohenberg and W. Kohn, Phys. Rev. **136**, B864 (1964).
10. W. Kohn and L. J. Sham, Phys. Rev. **140**, A1133 (1965).
11. L.D. Landau and E.M. Lifshitz, *The Classical Theory of Fields*, 4th ed. (Pergamon, Oxford, 1975), pp. 270-273.

Dynamics of Relaxors

Robert Blinc, Raša Pirc, Vid Bobnar, and Alan Gregorovič

"Jožef Stefan" Institute, Jamova 39, 1000 Ljubljana, Slovenia

Abstract. A dynamic spherical random bond-random field (SRBRF) model based on the coupled polar cluster picture is formulated. Assuming stochastic flips of the cluster polarization, the Langevin equations of motion are written down, from which the linear and nonlinear dynamic dielectric permittivities are derived. A coupled SRBRF pseudo-spin phonon Hamiltonian is presented, which models the effects of pressure on the relative stability of the ferroelectric, relaxor-, and paraelectric- like phases and thus the temperature-pressure phase diagram of relaxor systems.

I INTRODUCTION

Relaxors represent a new low-temperature state of polar dielectrics, which can be regarded as an intermediate state between dipolar glasses and normal ferroelectrics [1,2]. Some of the concepts developed for dipolar glasses, such as the Edwards-Anderson (EA) order parameter, are applicable to relaxors as well, as recently shown for $PbMg_{1/3}Nb_{2/3}O_3$ (PMN) [3], $PbSc_{1/2}Ta_{1/2}O_3$ (PST) [4], $Pb_{1-x}La_x(Zr_yTi_{1-y})_{1-x/4}O_3$ (PLZT) [5], and $Sr_xBa_{1-x}Nb_2O_6$ (SBN) [6]. In contrast to dipolar glasses, where elementary dipole moments exist on the atomic scale, the relaxor state is characterized by the presence of nanosized polar clusters of variable sizes. This picture constitutes the basis of the superparaelectric model [1] and of the more recent reorientable polar cluster model of relaxors [2,7]. By including explicitly the long-range frustrated intercluster interactions of a spin glass type into this picture, one arrives at the so-called spherical random-bond—random-field (SRBRF) model of relaxor ferroelectrics [8], which is capable of describing the static behavior of relaxors, such as the line shape of quadrupole perturbed NMR in PMN [3] and PST [4], and the sharp increase of the quasistatic third-order nonlinear dielectric constant [3,5].

The unusually large value of the static linear dielectric permittivity can also be explained within the framework of the SRBRF model if one assumes that the mean value of the random coupling J_0 is very close to—but slightly smaller than—its r.m.s. variance J, whereas in dipolar glasses the latter is usually dominant. By including an *ad hoc* electric field dependence of J_0 into the model, one can furthermore describe the transition from the relaxor to an inhomogeneous ferroelectric state for fields E exceeding a critical value E_c [9].

While the static SRBRF model describes a relaxor system in thermodynamic equilibrium, there are a number of phenomena suggesting that relaxors, in particular their low temperature state, are dominated by nonequilibrium effects. Typical examples are the difference between the field-cooled and zero-field cooled static dielectric constant, and the occurrence of strong frequency dispersion in both the linear and nonlinear dielectric permittivity at low temperatures. It is clear that these properties can only be discussed within a dynamic model. In the present paper, we introduce a dynamic model, which is an extension of the SRBRF model. Following Vugmeister and Rabitz [2,7] we assume that polar clusters can reorient with a characteristic relaxation time τ and write down the corresponding Langevin equations of motion, which are based on the static SRBRF Hamiltonian. These equations explicitly contain the frustrated interactions between the polar clusters and thus allow us to study the effects of these interactions on both the equilibrium and nonequilibrium properties.

II DYNAMIC SRBRF MODEL

In general, the polarization of i-th polar cluster, $i = 1, 2, ..., N$, is a three component $(n = 3)$ vector $\vec{S}_i = (S_{ix}, S_{iy}, S_{iz})$, its length being restricted solely by the spherical condition $\sum_i (\vec{S}_i)^2 = 3N$. In the present work we will discuss the simpler uniaxial $(n = 1)$ case $-\sqrt{N} < S_i < +\sqrt{N}$, where S_i is subject to the spherical condition

$$\sum_{i=1}^{N} S_i^2 = N . \tag{1}$$

The SRBRF model Hamiltonian is thus

$$\mathcal{H}_S = -\frac{1}{2}\sum_{ij} J_{ij} S_i S_j - \sum_i h_i S_i - gE \sum_i S_i , \tag{2}$$

where J_{ij} are the randomly frustrated intercluster interactions, h_i local random electric fields [10], E an applied uniform electric field, and g the appropriate dipole moment [8]. As usual, J_{ij} is assumed to be infinitely ranged and distributed according to Gaussian statistics with average value J_0/N and cumulant variance J^2/N. The Gaussian random fields h_i are characterized by the random average

$$[h_i h_j]_{av} = \Delta \delta_{ij} . \tag{3}$$

The uniaxial SRBRF model (2) has potential applicability to uniaxial relaxors such as $Sr_{1-x}Ba_xNb_2O_6$ (SBN). The present results can be, however, generalized to the isotropic $n = 3$ case as long as there is no mixing of the x, y, z components [8].

The Langevin equations of motion for the variables $S_i(t)$ are written as

$$\tau_i \frac{\partial S_i(t)}{\partial t} = -\frac{\partial(\beta\mathcal{H})}{\partial S_i} - 2z(t)S_i(t) + \xi_i(t) \,. \tag{4}$$

Here τ_i is the characteristic relaxation time for the reorientation of polar clusters. In the general case, τ_i depends on the size of the cluster as well as on temperature

$$\tau_i = \tau_0 \exp[f_1(S_i^2)f_2(T)] \tag{5}$$

In the first approximation one may assume that τ_i is site independent, however, some variation of τ_i across the system should in principle not be excluded, resulting in a distribution of relaxation times [2,7]. The function $z(t)$ plays the role of a Langrange multiplier enforcing the spherical condition (1) at all times.

The stochastic Langevin forces $\xi_i(t)$ ensure the proper equilibrium distribution and are determined by their ensemble averages

$$\langle \xi_i(t)\xi_j(t') \rangle_{av} = 2\tau \delta_{ij} \delta(t-t') \,. \tag{6}$$

Equations (4) with $\tau_i = \tau$ can be solved exactly by introducing a representation of eigenstates and eigenvectors of the random matrix J_{ij}. The details of this calculation will be presented in a separate publication. In the asymptotic limit $t \gg t_c \approx 2JT\tau/\Delta$ the system reaches a steady state and the function $z(t)$ tends towards its static value z, which is determined by the equation

$$z - r + \frac{\beta^2 \Delta}{2}\frac{z}{r} = \beta^2(J^2 + \Delta) \,, \tag{7}$$

where $r \equiv \sqrt{z^2 - \beta^2 J^2}$. The result for the complex linear dielectric susceptibility is given by the following expression (for simplicity we set $g = 1$):

$$\chi(\omega) = \frac{z - i\omega\tau/2 - \sqrt{(z-i\omega\tau/2)^2 - \beta^2 J^2} - \beta J_0}{\beta(J_0^2 + J^2) - 2J_0(z - i\omega\tau/2)} \,. \tag{8}$$

This result is valid in the phase without long-range order. The $\omega \to 0$ limit of Eq. (8) reproduces exactly the static susceptibility obtained earlier by the replica method.

The third-order nonlinear susceptibility [11] is similarly obtained as

$$\chi_3(\omega) = -\frac{1}{6}\frac{\partial^3 P_{3\omega}}{\partial(\frac{1}{2}E_0)^3}\bigg|_{E=0} = \frac{(\beta^2/2)\chi_1^{[1]}(\omega)[\chi_1(\omega) - \chi_1(3\omega)]}{\chi_1(0)_0 - \chi_1(\omega)_0 + \beta^2\Delta\left[\chi_1^{[1]}(0)_0 + \frac{\chi_1(0)_0 - \chi_1(2\omega)_0}{2i\omega\tau}\right]} \,. \tag{9}$$

Here $\chi_1^{[1]}(\omega) \equiv \partial\chi_1(\omega)/(\partial(i\omega\tau))$, while $\chi_1(0)_0$, $\chi_1^{[1]}(0)_0$, etc. represent the corresponding quantities evaluated for $J_0 = 0$.

The behavior of $\chi_1(\omega)$ and $\chi_3(\omega)$ depends strongly on the temperature dependence of the relaxation time τ. It has been suggested that in relaxors τ obeys the Vogel-Fulcher (VF) law $\tau = \tau_0 \exp[U/(T - T_0)]$, where U is the barrier height (in

K) and T_0 the VF temperature. Since $\tau \to \infty$ as $T \to T_0$, the reorientation of polar clusters is frozen out at $T \leq T_0$. Thus one may assume T_0 to be a measure of the freezing temperature T_f.

In Fig. 1 we show the real and imaginary parts of $\chi_1(T,\omega)$ and $\chi_3(T,\omega)$ for several representative values of frequency. We use a single VF relaxation time with $U = 5.5J$ and $T_0 = J$. The values of the remaining parameters are: $J_0 = 0.9J$ and $\Delta/J^2 = 0.001$. The calculated response agrees qualitatively with the observed dielectric relaxation in PMN only at high temperatures $T \gg T_f$.

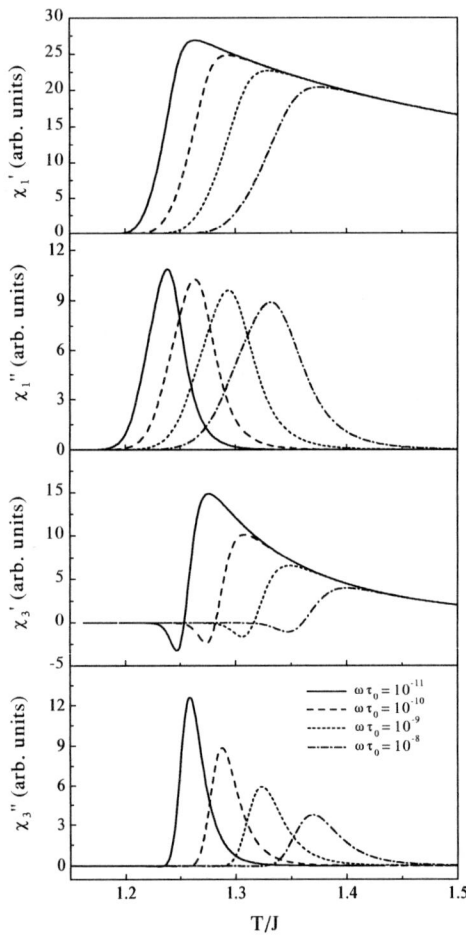

FIGURE 1. Real and imaginary parts of the linear and third-order nonlinear susceptibilities in the case of a single Vogel-Fulcher (VF) type relaxation time as functions of temperature for several values of frequency. Note that the response is strictly zero below VF temperature $T_0 = J$.

In accordance with the concept of dynamic heterogeneity in relaxors we now introduce the averaged responses $\bar{\chi}_1(\omega)$ and $\bar{\chi}_3(\omega)$, where the bar denotes an average over the distribution of relaxation times

$$\bar{\chi}(\omega) = \int_{\tau_{MIN}}^{\tau_{MAX}} \frac{d\tau}{\tau} g(\ln \tau) \chi(\omega, \tau). \tag{10}$$

This can be accomplished by performing an average over the distribution of VF temperatures $w(T_0)$, where T_0 is allowed to vary in the range $0 < T_0 < J$ with $T_f = T_0^{max} = J$

$$\bar{\chi}(\omega) = \int_0^{T_f} w(T_0) \chi(\omega, T_0) \, dT_0. \tag{11}$$

By choosing a simple linear distribution function $w(T_0) = 2(T_0 - J)/J^2$ we obtain nonzero values of $\bar{\chi}_1(\omega)$ and $\bar{\chi}_3(\omega)$ both above and below T_f, in qualitative agreement with experiments (Fig. 2).

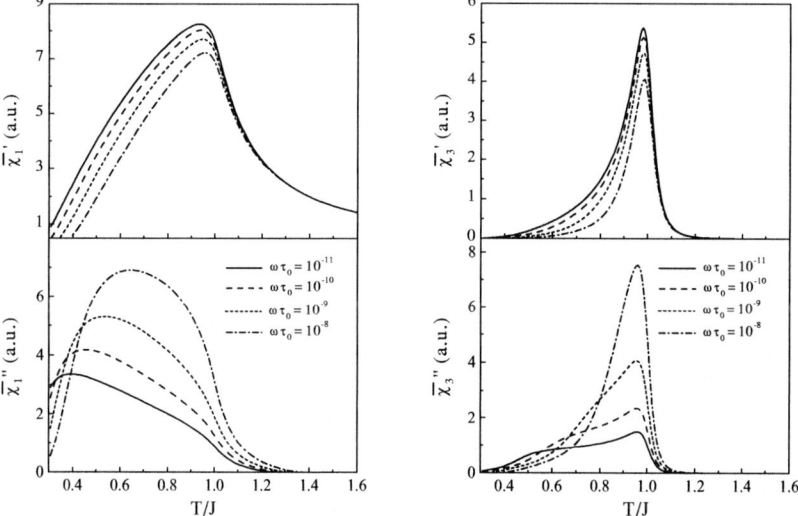

FIGURE 2. Calculated temperature dependence of the real and imaginary parts of the linear and third order nonlinear susceptibilities averaged over a linear distribution of T_0, using the parameter values indicated in the text.

An alternative approach is based on the master equation for the reorientation of cluster polarization assuming a VF relaxation time, where the barrier heights U are distributed according to a Gaussian probability function [2,7]. Such an approach was found to be applicable to PMN and PST in the region $T > T_0$. A phenomenological model of the dynamic nonlinear response was developed in Ref. [12].

III   NMR AND THE SRBRF MODEL

It should be also noted that in the fast motion regime where the mean correlation time of the order parameter field is much shorter than the inverse Larmor frequency in the local field, $\tau \ll \Delta\omega^{-1}$, the Edwards-Anderson glass order parameter

$$q_{EA} = \lim_{t \to \infty} \lim_{N \to \infty} \langle \langle S_i(0) S_i(t) \rangle_{t'} \rangle_{dis} = \frac{1}{N} \sum_i \langle S_i \rangle^2 \qquad (12)$$

can be determined from the quadrupole or chemical perturbed inhomogeneous NMR line shape of $f(\nu)$

$$M_2 = \int f(\nu)(\nu - \bar{\nu})^2 \, d\nu = \alpha^2 q_{EA} \qquad (13)$$

if the relation between the NMR frequency ν_i and the local polarization $\vec{p}_i = \langle \vec{S}_i \rangle$ is linear

$$\nu_i = \nu_0 + \vec{\alpha} \cdot \vec{p}_i. \qquad (14)$$

If the relation between the local NMR frequency ν_i and the local polarization is quadratic q_{EA} is determined from the first moment of $f(\nu)$. The results are summarized in Fig. 3.

IV   THE COUPLED PSEUDOSPIN-PHONON MODEL AND THE P-T PHASE DIAGRAM

The coupled SRBRF-phonon Hamiltonian can be written as

$$\mathcal{H} = \mathcal{H}_S + \mathcal{H}_L + \mathcal{H}_{SL}, \qquad (15)$$

where $\mathcal{H}_S$ is the bare pseudospin Hamiltonian given in Eq. (2), $\mathcal{H}_L$ represents the lattice Hamiltonian

$$\mathcal{H}_L = \frac{1}{2} \sum_{\vec{k}p} [\omega_{\vec{k}p}^2 Q_{\vec{k}p} Q_{-\vec{k}p} + P_{\vec{k}p} P_{-\vec{k}p}], \qquad (16)$$

and $\mathcal{H}_{SL}$ the pseudospin-polar phonon coupling

$$\mathcal{H}_{SL} = \sum_{\vec{k}p} Q_{-\vec{k}p} \vec{\gamma}_{\vec{k}p} \cdot \vec{S}_{\vec{k}}. \qquad (17)$$

Here we introduced the normal coordinates and momenta $Q_{\vec{k}p}$ and $P_{\vec{k}p}$, respectively, of polar optic phonons with branch index p, wave vector $\vec{k}$, and frequency $\omega_{\vec{k}p}$. Further, $\gamma_{\vec{k}p}$ represents the pseudospin-polar phonon coupling constant and

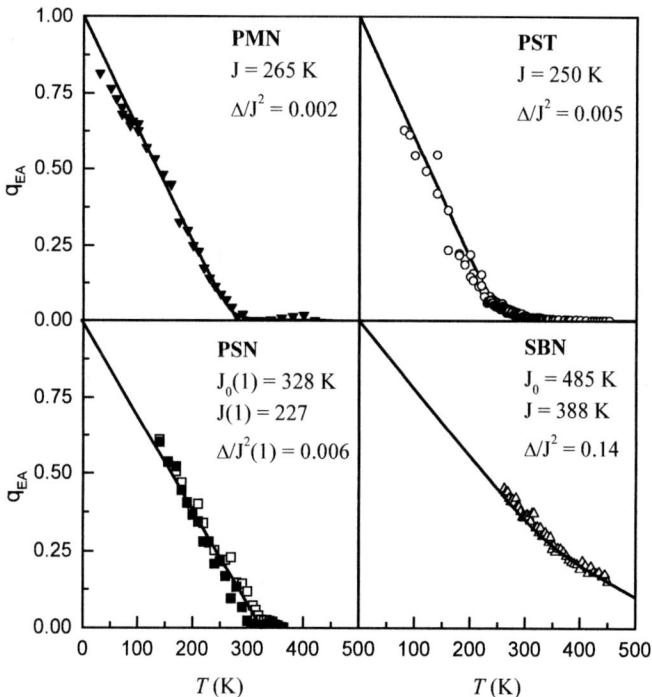

FIGURE 3. Temperature dependence of the Edwards-Anderson order parameter q_{EA} as determined from the ^{93}Nb NMR inhomogeneous line shape for the following relaxor systems: PMN, PST, PSN, and SBN.

$\vec{S}_{\vec{k}} = N^{-1/2} \sum_i \vec{S}_i \exp(i\vec{k} \cdot \vec{R}_i)$ is the Fourier transform of the order parameter field. The pseudospin-acoustic phonon coupling does not contribute to the intercluster interactions in view of the presumed cubic symmetry of the non-polar matrix.

One can derive an effective intercluster coupling by introducing displaced phonon coordinates, which are determined by the equilibrium condition $\partial \mathcal{H}/\partial Q_{-\vec{k}p} = 0$. After this transformation the terms linear in the phonon coordinates in $\mathcal{H}$ disappear and J_{ij} in Eq. (2) is replaced by an effective coupling $(J_{ij})_{\mu\nu}$. The random averaging makes the matrix of the effective coupling constants $\mathbf{J}_{ij}$ diagonal and isotropic. The parameters J_0 and J now represent the new effective parameters of the SRBRF model on a deformable lattice. They are explicitly given by

$$J_0 = J_0^{(b)} + \sum_p \frac{|\vec{\gamma}_{0p}|^2}{\omega_{0p}^2} - \frac{1}{N} \sum_{\vec{k}p} \frac{|\vec{\gamma}_{\vec{k}p}|^2}{\omega_{\vec{k}p}^2} , \qquad (18)$$

and

$$J^2 = (J^{(b)})^2 + \frac{1}{N}\sum_{\vec{k}} \left[\sum_p \frac{|\vec{\gamma}_{\vec{k}p}|^2}{\omega_{\vec{k}p}^2}\right]^2 - \left[\frac{1}{N}\sum_{\vec{k}p} \frac{|\vec{\gamma}_{\vec{k}p}|^2}{\omega_{\vec{k}p}^2}\right]^2. \tag{19}$$

where $J_0^{(b)}$ and $J^{(b)}$ stand for the unrenormalized coupling constants of $\mathcal{H}_S$ om Eq. (2).

To estimate the variation of J_0 and J with pressure we consider Eqs. (18-19) in which $\omega_{\vec{k}p} = \omega_{\vec{k}p}(P)$. The main contribution to $J_0(P)$ is given by the $\vec{k} = 0$ term in Eq. (18) involving the sum over all polar optic branches p. Since in perovskites the squares of the phonon frequencies at the zone center ω_{0p}^2 linearly increase with pressure, this term will result to leading order in a linear decrease of $J_0(P)$ with increasing pressure. The next term, which has a negative sign, involves the average over the entire phonon dispersion branch and is thus less important. This is so since the phonon frequency $\omega_{\vec{k}p}$ at $\vec{k} \to 0$ is lower than the average over the p-th branch, as indeed observed in the case in PMN. Thus we obtain

$$J_0(P) = J_0(0) - \alpha_1 P + \cdots . \tag{20}$$

In contrast to J_0, the phonon contribution to J^2 is determined by the fluctuation of the averages over the optic phonon branches. For a flat spectrum this contribution would be precisely zero. In general, the optic branches have some dispersion, but we may expect that the pressure dependence of $J(P)$ is weaker than that of $J_0(P)$.

For $J_0 < \sqrt{J^2 + \Delta}$ long-range order cannot exist and the system is in a SG phase at all temperatures. If $\Delta = 0$ as in magnetic spin glasses, a transition from a high temperature paraelectric to a SG phase occurs at $T_f = J/k$. For $\Delta \neq 0$ and $\Delta \ll J^2$ the sharp transition disappears, but the nonlinear susceptibility shows a maximum at $T_f \approx \sqrt{J^2 + \Delta}$.

For $J_0 > \sqrt{J^2 + \Delta}$ long-range order is possible and a phase transition to an inhomogeneous ferroelectric (I-FE) phase occurs below the critical temperature

$$T_c = J_0\left(1 - \frac{\Delta}{J_0^2 - J^2}\right)/k . \tag{21}$$

Since J_0 depends on pressure P according to Eq. (20), T_c is also a function of P. Thus the relative stability of the ferroelectric, relaxor, and paraelectric phases is strongly affected by the application of hydrostatic pressure.

If the system is an inhomogeneous ferroelectric at $P = 0$, and $T_c(P)$ is given by Eq. (21) in which $J_0 = J_0(P)$, then a critical pressure P_c may exist beyond which ferroelectric order is not possible at any temperature, and a ferroelectric-to-relaxor crossover occurs. The critical pressure is given by the condition $J_0(P_c) = \sqrt{J^2 + \Delta}$. The calculated pressure dependence of T_c in the entire range $0 < P < P_c$ is shown as a solid line separating the ferroelectric from the ergodic relaxor phase on the $T - P$ phase diagram presented in Fig. 4. The solid triangles represent the experimental points obtained by Samara in PLZT 6/65/35 [13], whereas the asterisk denotes the critical pressure P_c [13]. The horizontal line corresponding to the relation $J_0(P = P_c) = \sqrt{J^2 + \Delta}$ separates the ferroelectric phase from the nonergodic relaxor phase where the longest relaxation time diverges. The vertical dotted line corresponding to $T = T_f$ is determined by the peak in the static nonlinear dielectric permittivity.

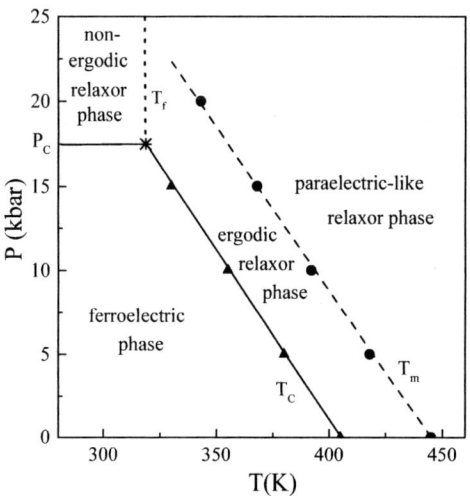

FIGURE 4. Temperature-pressure phase diagram of the relaxor ferroelectric PLZT 6/65/35. The solid and dashed lines are evaluated from the static and dynamic coupled SRBRF-phonon model, whereas the solid triangles and solid circles are experimental points obtained by Samara [13]. The vertical dotted line corresponds to the freezing temperature T_f where the static third order non-linear susceptibility $\chi_3(0)$ has a peak. T_m designates the temperature of the maximum in the real part of the linear dielectric susceptibility $\chi'(\omega)$.

Pressure strongly affects not only the static but also the dynamic properties of relaxors. At a fixed frequency ω the temperature dependence of the real part of the dielectric susceptibility $\chi'(\omega)$ shows a broad pressure dependent maximum at $T = T_m$. For $T > T_m$ the system is characterized by a weak dispersion corresponding to a paraelectric-like relaxor phase. For $T_f < T < T_m$ the system shows a strong frequency dispersion corresponding to an ergodic relaxor phase. For $T < T_f$ and $J_0 < J$, however, the system is in a nonergodic relaxor phase, where the longest relaxation time diverges. With increasing ω, the maximum value of $\chi'(\omega)$ decreases, while T_m moves toward higher temperatures. Actually, in PLZT 6/65/35 it has been found that under hydrostatic pressure P the maximum temperature $T_m(P)$ moves towards lower values with increasing P in a manner similar to $T_c(P)$.

REFERENCES

1. Cross, L. E., *Ferroelectrics* **76**, 241 (1987); **151**, 305 (1994).
2. Vugmeister, B. E. and Rabitz, H., *Phys. Rev. B* **57**, 7581 (1998).
3. Blinc, R., Dolinšek, J., Gregorovič, A., Zalar, B., Filipič, C., Kutnjak, Z., Levstik, A., and Pirc, R., *Phys. Rev. Lett.* **83**, 424 (1999).
4. Blinc, R., Gregorovič, A., Zalar, B., Pirc, R., and Lushnikov, S.G., *Phys. Rev. B* **61**, 253 (2000).
5. Bobnar, V., Kutnjak, Z., Pirc, R., Blinc, R., and Levstik, A., *Phys. Rev. Lett.* **84**, 5892 (2000).
6. Dec, J., Kleemann, W., Woike, T., and Pankrath, R., *Eur. Phys. J. B* **14**, 627 (2000).
7. Vugmeister, B. E., and Rabitz, H., *Phys. Rev. B* **61**, 14448 (2000).
8. Pirc, R., and Blinc, R., *Phys. Rev. B* **60**, 13470 (1999).
9. Bobnar, V., Kutnjak, Z., Pirc, R., and Levstik, A., *Phys. Rev. B* **60**, 6420, (1999).
10. Westphal, V., Kleemann, W., and Glinchuk, M. D., *Phys. Rev. Lett.* **68**, 847 (1992).
11. R. Pirc, R. Blinc, V. Bobnar, *Phys. Rev. B* **63**, 054203 (2001).
12. Glazounov, A.E., and Tagantsev, A.K., *Phys. Rev. Lett.* **85**, 2192 (2000).
13. Samara, G.A., *Ferroelectrics* **117**, 347 (1991); *Phys. Rev. Lett.* **77**, 314 (1996).

Inside Dielectrics: Microscopic and Macroscopic Polarization

P. Umari,[1] A. Dal Corso,[2] and R. Resta[3,4]

[1] *IRRMA–Institut Romand de Recherche Numérique en Physique des Matériaux, CH–1015 Lausanne, Switzerland*
[2] *INFM–Scuola Internazionale Superiore di Studi Avanzati, Via Beirut 4, I-34014 Trieste, Italy*
[3] *INFM–Dipartimento di Fisica Teorica, Università di Trieste, I-34014 Trieste, Italy*
[4] *Max-Planck-Institut für Festkörperforschung, D-70506 Stuttgart, Germany.*

Abstract. We address the very basic issue of what happens, at a microscopic level, inside a polarized dielectric. We show that the complete information about electronic polarization is embedded in the microscopic polarization $\mathbf{P}^{(\mathrm{ind})}(\mathbf{r})$. Previous studies in the literature have addressed the induced electronic charge density (alias the divergence of our vector field) where the most relevant information is obliterated. The physical meaning of $\mathbf{P}^{(\mathrm{ind})}(\mathbf{r})$ is best understood by imagining that the applied field is adiabatically switched on in time: $\mathbf{P}^{(\mathrm{ind})}(\mathbf{r})$ is then proportional to the microscopic transient current flowing through the sample while the field is switched on and the dielectric is polarized. We provide a quantum-mechanical expression for $\mathbf{P}^{(\mathrm{ind})}(\mathbf{r})$, and we present first-principle results for two case studies: Si and NaBr. In the case of Si, the (unperturbed) valence charge defines a continuous network of bonds. When a field is switched on, most of the polarization current $\mathbf{P}^{(\mathrm{ind})}(\mathbf{r})$ flows within narrow channels along the bonds, and percolates across the material. Although less dominant, a similar feature occurs even in NaBr. Both materials turn out to be far from the Clausius-Mossotti limit, where the transient current does not cross the cell boundaries. In ferroelectric oxides, which have a mixed ionic/covalent character, the role of percolating transient currents is expected to be dominant.

INTRODUCTION

Any textbook picture of a polarized dielectric is invariably based on the venerable Clausius-Mossotti (CM) model [1]: the polarization charge of a condensed system is regarded as the superposition of localized contributions, each providing an electric dipole. In a crystalline system the CM macroscopic polarization is *defined* as the sum of the dipole moments in a given cell, divided by the cell volume. An extreme CM view of a simple ionic crystal, having the NaCl structure, is sketched in Fig. 1. The essential point behind the CM view is that the distribution of the induced charge is resolved into contributions which can be ascribed to identifiable "polarization centers". In the sketch of Fig. 1 these are the anions, while in the

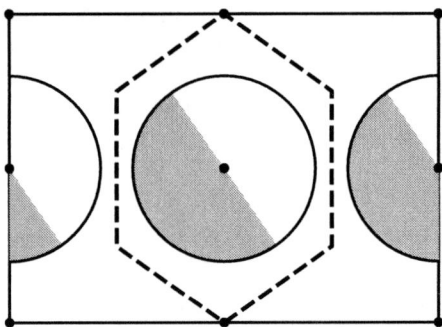

FIGURE 1. A polarized ionic crystal, having the NaCl structure, within an extreme Clausius-Mossotti model. We qualitatively sketch the polarization charge in the $(1\bar{1}0)$ plane linearly induced by a constant field **E** in the [111] direction. Only the anions are assumed to be polarizable and shaded areas indicate regions of negative charge. The dots at the upper and lower borders indicate cations, and the other dots anions; the boundary of a Wigner-Seitz cell, centered at the anion, is also shown (dashed line).

most general case they may be atoms, molecules, or even bonds. This partitioning of the polarization charge is obvious in Fig. 1, where the individual localized contributions are drawn as nonoverlapping: but what about real materials? This is precisely the case in point: the electronic polarization charge in a crystal has a periodic continuous distribution and its resolution into localized contributions is highly nonunique. Such nonuniqueness makes problematic even to define what macroscopic polarization is.

A definitive solution to the long-standing [2] problem of macroscopic polarization in condensed matter came only in relatively recent times. It is now pretty clear that—contrary to a widespread incorrect belief—macroscopic polarization has nothing to do with the periodic charge distribution of the polarized crystal: the former is essentially a property of the *phase* of the electronic wavefunction, while the latter is a property of its *modulus*. The modern theory [3–6], based on the Berry-phase concept [7,8], has been successfully applied to a large number of physical problems. The theory provides a well defined path from microscopics (the wavefunction) to macroscopics (polarization), but doesn't say anything about what microscopically happens, *in real space*, inside a polarized dielectric. The aim of the present work is to address the latter issue. Above, we have pointed out the unavoidable drawbacks of any CM-based viewpoint. In the following of this paper, we are going to present a completely novel viewpoint, introducing a real-space quantity which contains the complete information about microscopic and macroscopic polarization in a polarized dielectric. Needless to say, the present findings are totally consistent with the concepts of the modern theory of polarization. Throughout this work, we will address the electronic polarization alone, at clamped nuclei.

We will illustrate the novel concept by means of two paradigmatic case studies:

a ionic crystal (NaBr), and a covalent one (Si). All calculations are performed in a first-principles framework, using density-functional theory [9], Kleinman-Bylander norm-conserving pseudopotentials [10], a plane-wave expansion of the Kohn-Sham (KS) orbitals [11], and the local-density approximation with the Ceperley-Alder functional [12].

I INDUCED MICROSCOPIC CHARGE

We start investigating how the analogue of Fig. 1 looks like for real materials. The microscopic periodic charge $\rho^{(\text{ind})}(\mathbf{r})$ induced by a constant field has been previously studied in the literature for a few materials [13,14]: we display $\rho^{(\text{ind})}(\mathbf{r})$ for our two case studies in the form of contour plots in Fig. 2. These have been drawn using the density, *i.e.* the square modulus, of the perturbed KS orbitals, as provided by density-functional perturbation theory (DFPT) [15]. Notice that any *phase* information in the orbitals is obliterated in $\rho^{(\text{ind})}(\mathbf{r})$: ergo any information about macroscopic polarization is irretrievably lost. Notwithstanding, the left panel in Fig. 2 shows a rather well defined anionic polarization dipole, essentially contained in the Wigner-Seitz (WS) cell: this hints at the validity, apparently to a good approximation, of the schematic CM picture of Fig. 1. When we actually compute the magnitude of the dipole, however, it turns out to be by far too small. In fact its value (29.1 a.u.), divided by the cell volume, accounts for a little more than one half of the actual macroscopic polarization induced by a unit field, equal to $(4\pi)^{-1}(\varepsilon_\infty - 1)$ [16]. In the covalent material a CM model does not make any sense [17]: the induced charge is delocalized throughout the cell and nonoverlapping

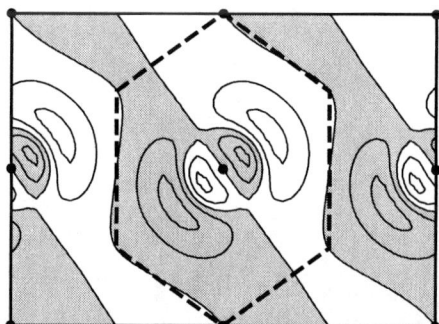 

FIGURE 2. Induced (pseudo)charge density $\rho^{(\text{ind})}(\mathbf{r})$ in the $(1\bar{1}0)$ plane linearly induced by a constant field **E** in the [111] direction. The field has unit magnitude (in a.u.) and the contours are separated by 30 charge units per cell. Shaded areas indicate regions of negative charge; dots indicate atomic positions. Left panel: NaBr, where the dots at the upper and lower borders indicate cations (and the other dots anions); the boundary of a Wigner-Seitz cell, centered at the anion, is also shown. Right panel: Si

regions can hardly be defined: this is realized by inspection of Fig. 2, right panel.

In computing the anionic dipole in NaBr, we have performed the spatial integration just by "cutting" the periodic charge at the WS-cell boundary: in other words we have assumed, rather arbitrarily, that the localized distributions around each anion are *nonoverlapping*. In fact, this is the key point: the correct value of the polarization can indeed be recovered from a CM-like picture, but only if the local dipolar distributions are allowed to overlap. Unfortunately the decomposition of the periodic charge distributions of Fig. 2 into localized overlapping contributions is highly nonunique: one could thus get *any* value for the macroscopic polarization!

The modern theory of polarization [3,4] rescues us in dictating which partitioning (into overlapping localized contributions) does provide the right polarization value: according to the theory, each contribution must be evaluated using the charge of the polarized Wannier functions [18]. But the construction of a Wannier function requires, not surprisingly, the phase of the (Bloch) orbitals: their modulus is not enough. We are not going to pursue any further the charge-partitioning issue; instead, we will define and investigate a different microscopic quantity, which— at variance with the induced charge—carries the *complete* information about the response of a dielectric to a polarizing electric field.

II INDUCED MICROSCOPIC POLARIZATION: THEORY

The central quantity of this work is the induced microscopic polarization $\mathbf{P}^{(\text{ind})}(\mathbf{r})$, which in a crystalline system is a lattice-periodical vector field. The cell average of $\mathbf{P}^{(\text{ind})}(\mathbf{r})$ coincides with the induced macroscopic polarization, while the induced charge $\rho^{(\text{ind})}(\mathbf{r})$ discussed above coincides with minus the divergence of $\mathbf{P}^{(\text{ind})}(\mathbf{r})$: clearly, the divergence obliterates important information.

The microscopic polarization has dimensions of a dipole per unit volume: we don't find, however, useful to visualize it as a "dipole density", thinking in terms of a continuous distribution of dipoles. Instead, the physical meaning of $\mathbf{P}^{(\text{ind})}(\mathbf{r})$ is better understood by imagining that the applied field is adiabatically switched on in time: $\mathbf{P}^{(\text{ind})}(\mathbf{r})$ is then proportional to the microscopic current flowing through the sample while the field is switched on and the dielectric is polarized.

Suppose our system is described by the perturbed KS Hamiltonian:

$$H_\lambda = H_0 + \lambda \Delta V_{\text{KS}} = \frac{1}{2}p^2 + V_{\text{KS}} + \lambda \Delta V_{\text{KS}} , \qquad (1)$$

where atomic Hartree units are adopted throughout, and λ measures the strength of the given perturbation in dimensionless units. The induced polarization $\mathbf{P}^{(\text{ind})}(\mathbf{r})$ is defined as either the λ-derivative of the polarization, evaluated at $\lambda = 0$, or as its first-order term, evaluated at $\lambda = 1$. If the perturbation is adiabatically switched on in a time Δt, then $\lambda = \lambda(t)$, with $\lambda(0) = 0$ and $\lambda(\Delta t) = 1$. We make the explicit assumption that the magnetization is vanishing, and therefore the transient electric current $\mathbf{j}(\mathbf{r},t)$ flowing through the system is just the time derivative of the polarization [19]:

$$\mathbf{j}(\mathbf{r},t) = \mathbf{P}^{(\text{ind})}(\mathbf{r})\dot{\lambda}(t). \qquad (2)$$

Given that we are considering only the electronic contribution, this current is:

$$\mathbf{j}(\mathbf{r},t) = -\frac{1}{2}\langle\mathbf{r}|(\mathbf{v}\rho + \rho\mathbf{v})|\mathbf{r}\rangle, \qquad (3)$$

where ρ is the single-particle KS density matrix, and $\mathbf{v}$ is the velocity operator:

$$\mathbf{v} = i[H_\lambda, \mathbf{r}]. \qquad (4)$$

Since we are using nonlocal pseudopotentials, the velocity differs from the canonical momentum $\mathbf{p}$ by a well known term, first discussed in Ref. [20], and routinely implemented in DFPT [15]. The KS density matrix is:

$$\rho = 2\sum_{i \in \text{occ.}} |\psi_i(t)\rangle\langle\psi_i(t)|. \qquad (5)$$

We are interested in the adiabatic limit, where $\Delta t \to \infty$ and $\dot{\lambda} \to 0$. Following Niu and Thouless [21], the leading order time evolution of the density matrix is:

$$\rho = 2\sum_{i \in \text{occ.}} |\varphi_i\rangle\langle\varphi_i| + \delta\rho, \qquad (6)$$

$$\delta\rho \simeq 2i\dot{\lambda}\sum_{i \in \text{occ.}}\sum_{j \neq i}\left(|\varphi_j\rangle\frac{\langle\varphi_j|\partial\varphi_i/\partial\lambda\rangle}{\epsilon_j - \epsilon_i}\langle\varphi_i| - \text{Hc}\right), \qquad (7)$$

where Hc indicates the Hermitian conjugate, and all φ's are adiabatic instantaneous KS eigenstates at $\lambda = \lambda(t)$. Eventually, only their $\lambda = 0$ value is needed in order to evaluate $\mathbf{P}^{(\text{ind})}(\mathbf{r})$. We use perturbation theory to get the λ-derivative:

$$|\partial\varphi_i/\partial\lambda\rangle = \sum_{j \neq i}|\varphi_j\rangle\frac{\langle\varphi_j|\Delta V_{\text{KS}}|\varphi_i\rangle}{\epsilon_j - \epsilon_i}. \qquad (8)$$

We then express—as usual in DFPT [15]—the perturbation sums in terms of unperturbed single-particle Green's functions:

$$G(\epsilon) = (\epsilon - H_0)^{-1} = \sum_j \frac{|\varphi_j\rangle\langle\varphi_j|}{\epsilon - \epsilon_j}. \qquad (9)$$

Straightforward manipulations provide $\delta\rho$ as a sum over occupied KS orbitals only:

$$\delta\rho \simeq 2i\lambda \sum_{i \in \text{occ.}} QG(\epsilon_i)QG(\epsilon_i)Q\Delta V_{\text{KS}} |\varphi_i\rangle\langle\varphi_i| + \text{Hc.} \tag{10}$$

where Q is the projector over the unoccupied orbitals

$$Q = 1 - \sum_{i \in \text{occ.}} |\varphi_i\rangle\langle\varphi_i|. \tag{11}$$

Of the two terms in Eq. (6), only $\delta\rho$ corresponds to a nonvanishing current. Replacement into Eqs. (2) and (3) yields therefore the central result of this work as:

$$\mathbf{P}^{(\text{ind})}(\mathbf{r}) = -2i \sum_{i \in \text{occ.}} \Big(\langle\mathbf{r}| \mathbf{v}QG(\epsilon_i)QG(\epsilon_i)Q\Delta V_{\text{KS}} |\varphi_i\rangle\langle\varphi_i|\mathbf{r}\rangle$$
$$+ \langle\mathbf{r}| QG(\epsilon_i)QG(\epsilon_i)Q\Delta V_{\text{KS}} |\varphi_i\rangle\langle\varphi_i| \mathbf{v} |\mathbf{r}\rangle \Big). \tag{12}$$

In a crystalline solid the microscopic polarization linearly induced by a macroscopic field $\mathbf{E}$ is lattice periodical:

$$\mathbf{P}^{(\text{ind})}(\mathbf{r}) = \sum_{\mathbf{G}} \mathbf{P}_{\mathbf{G}}^{(\text{ind})} e^{i\mathbf{G}\cdot\mathbf{r}}, \tag{13}$$

where $\mathbf{G}$ are reciprocal-lattice vectors. We identify the index i used so far with the band index n and the Bloch vector $\mathbf{q}$ altogether. The KS eigenvalues and eigenvectors are then, respectively: $\epsilon_i \to \epsilon_n(\mathbf{q})$, $\varphi_i \to \psi_{n\mathbf{q}}$. It is expedient to define:

$$|\tilde{\psi}_{n\mathbf{q}}\rangle = QG(\epsilon_n(\mathbf{q}))QG(\epsilon_n(\mathbf{q}))Q\Delta V_{\text{KS}} |\psi_{n\mathbf{q}}\rangle. \tag{14}$$

Given the presence of a macroscopic field, the screened potential ΔV_{KS} is the sum of a macroscopic term $\mathbf{E}\cdot\mathbf{r}$ and of a microscopic (lattice-periodical) one. The former term does not cause any harm, and $\tilde{\psi}_{n\mathbf{q}}$ can be straightforwardly evaluated following Ref. [15]. The Fourier coefficients of Eq. (12) are then:

$$\mathbf{P}_{\mathbf{G}}^{(\text{ind})} = -\frac{2i}{(2\pi)^3} \sum_{n=1}^{n_b} \int d\mathbf{q}\, \langle\psi_{n\mathbf{q}}| (\mathbf{v}e^{-i\mathbf{G}\cdot\mathbf{r}} + e^{-i\mathbf{G}\cdot\mathbf{r}}\mathbf{v})|\tilde{\psi}_{n\mathbf{q}}\rangle, \tag{15}$$

where n_b is the number of occupied bands, and the integral is over the Brillouin zone.

The $\mathbf{G} = 0$ term in Eq. (13) is by definition the induced macroscopic polarization: in fact we recover from Eq. (15) the previously known [15] expression for it. Also previously known is the expression for the Fourier coefficients $\rho_{\mathbf{G}}^{(\text{ind})}$ of the induced charge: we recover such expression as minus the divergence of $\mathbf{P}^{(\text{ind})}(\mathbf{r})$:

$$\rho_{\mathbf{G}}^{(\text{ind})} = i\mathbf{G}\cdot\mathbf{P}_{\mathbf{G}}^{(\text{ind})}, \tag{16}$$

which clearly obliterates the transverse (divergence-free) part of our vector field.

III INDUCED MICROSCOPIC POLARIZATION: RESULTS

We have implemented the above formulation within first-principle DFPT for two paradigmatic case studies: a ionic crystal (NaBr) and a covalent one (Si): the flux lines of the vector fields $\mathbf{P}^{(\text{ind})}(\mathbf{r})$ for the two materials are displayed in Fig. 3.

Starting with NaBr, left panel, we notice that most of the action is at the anionic sites, while little happens at the cations. In a CM model the transient polarization current would displace charge within each cell, and would vanish at the WS cell boundary. This is *not* what happens here: although the current looks strong in the cationic region and weak on the cell boundary, by inspection of the data we find that its maximum value on the cell boundary is of the same order as the macroscopic (*i.e.* cell-averaged) value. This fact is in agreement with our previous finding: when discussing Fig. 2 we have indeed emphasized that the local dipole is by far too small and accounts for only a part of the macroscopic polarization: the remaining part is due to a current percolating through the crystal, traversing the cell boundaries, and unrelated to the local dipole.

The landscape is even more extreme in the case of Si, shown in the right panel of Fig. 3. The field $\mathbf{P}^{(\text{ind})}(\mathbf{r})$ is everywhere much stronger than for NaBr: we have drawn the two panels to the same scale and for the same value of $\mathbf{E}$, and therefore the cell averages of the two fields are in a ratio 1:7 [16]. The polarization current flows through most of the unit cell, although it is weaker in the interstitial region, and stronger elsewhere. A strong current flows through the WS cell boundary, no matter whether we choose it as atom- or bond-centered. There is an "active region", percolating all the way across the sample around the bonding network, where the transient polarization current mostly flows. In these channels (whose planar section is zig-zag like in Fig. 3) the current has uninterrupted flux lines of almost constant magnitude and direction.

We provide a complementary visualization of $\mathbf{P}^{(\text{ind})}(\mathbf{r})$ (or equivalently of the transient polarization current) in Fig. 4, by showing its component normal to a few selected planes in both materials. In NaBr (top panels) the current is stronger in planes (b) and (c), and weaker in plane (a), which crosses the cations. Although at the chosen scale nothing seems to happen in the latter plane, there is one point (on the WS cell boundary, midway between two nearest neighbor anions) where the magnitude of microscopic polarization is of the same order as the macroscopic one. In Si (lower panels) one clarly detects the channels where the polarization current mostly flows, discussed above: starting with the (d) plane, passing through the middle of the longitudinal bond, we detect a huge flux tube, which splits in three in (e), steering in order to keep clear of the core regions; these three channels then cross the transverse bonds in their middle in (f).

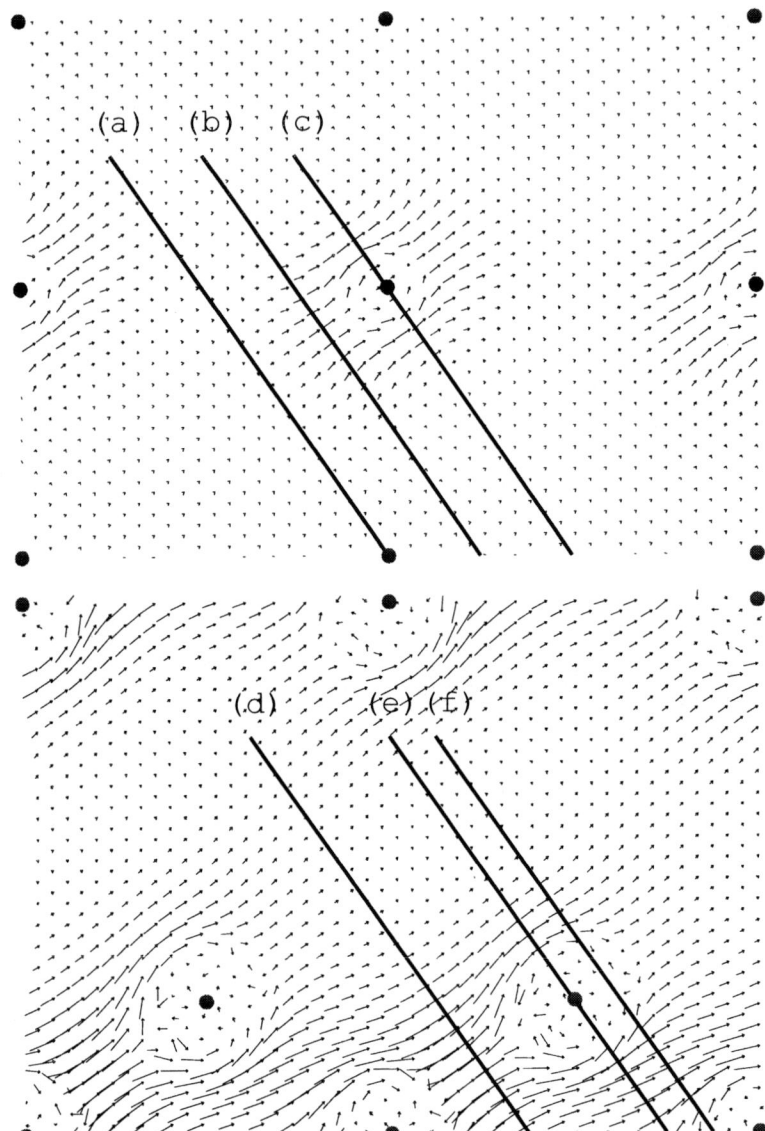

FIGURE 3. Flux lines of the microscopic polarization field $\mathbf{P}^{(\mathrm{ind})}(\mathbf{r})$ for NaBr (top panel) and Si (bottom panel). Geometries and field $\mathbf{E}$ same as in Fig. 2, only the panels have been enlarged to show more detail. The length of the arrows is proportional to the magnitude of $\mathbf{P}^{(\mathrm{ind})}(\mathbf{r})$. The oblique lines and their labels refer to the next figure

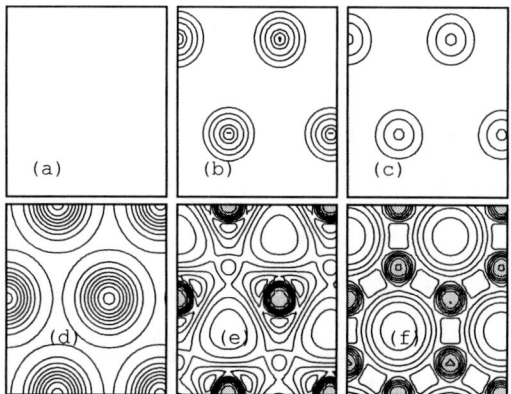

FIGURE 4. Projections of the microscopic polarization $\mathbf{P}^{(\mathrm{ind})}(\mathbf{r})$ along the direction of the field $\mathbf{E}$, shown as contour plots in planes normal to $\mathbf{E}$. See also Fig. 3 for the labeling of the planes. Contours are separated by 0.24 a.u.; shaded areas indicate regions of negative polarization. Top panels for NaBr: (a) plane through the cations; (b) plane midway between anions and cations; (c) plane through the anions. Bottom panels for Si: (d) plane through the longitudinal bonds; (e) plane through the atoms; (f) plane through the transverse bonds.

IV CONCLUSIONS AND PERSPECTIVES

The present work perpicuously shows that the polarization mechanism is little related to the formation of local dipoles, in either ionic or covalent materials. In NaBr the local dipoles account for only about one half of the macroscopic polarization, while in Si the concept itself of local dipoles is ill defined [17]. Microscopic polarization is much better described as the time integral of a transient polarization current flowing through the unit cell, in both our case-study materials. In the case of Si, the (unperturbed) valence charge defines a continuous network of bonds. When a field is switched on, most of the polarization current $\mathbf{P}^{(\mathrm{ind})}(\mathbf{r})$ flows within narrow channels along the bonds, and steers to keep clear of the core regions: the dominant feature is that the current percolates across the material. Although less dominant, a similar feature occurs even in NaBr. This is quite an unexpected result: so far, ionic crystals have been regarded as prototypical CM materials [22], while a different view emerges from the present findings.

For ferroelectrics, which are mixed ionic/covalent materials, we expect a dominant role of percolating transient currents. The present viewpoint is likely to provide in the future a thorough microscopic understanding of the main trends (*i.e.* high or low polarizability) in materials of different chemistry and/or structure.

We have addressed throughout the present work the polarization *linearly induced* by a given perturbation: what about spontaneous polarization? For a ferroelectric material, could one define a microscopic vector field $\mathbf{P}(\mathbf{r})$, such that its divergence provides the *unperturbed* charge density, while its cell average provides the *spontaneous* macroscopic polarization? The answer is yes, but the result is expected to bear little physical significance: in fact only the longitudinal (curl free) part of $\mathbf{P}(\mathbf{r})$ is uniquely defined, while its transverse (divergence free) part remains completely arbitrary, owing to fundamental reasons rooted in classical electrodynamics of continuous media [23]. Because of similar reasons, our induced polarization $\mathbf{P}^{(\text{ind})}(\mathbf{r})$ is instead uniquely defined whenever the magnetization is assumed to vanish.

ACKNOWLEDGMENTS

Partly supported by the Office of Naval Research, through grant N00014-96-1-0689.

REFERENCES

1. O. F. Mossotti, Memorie di Matematica e di Fisica della Società Italiana delle Scienze Residente in Modena, **24**, 49 (1850); R. Clausius, *Die Mechanische Behandlung der Electrica* (Vieweg, Berlin, 1879).
2. R. M. Martin, Phys. Rev. B **9**, 1998 (1974).
3. R. Resta, Ferroelectrics **136**, 51 (1992); R. D. King–Smith, and D. Vanderbilt, Phys. Rev. B **47**, D. Vanderbilt and R. D. King-Smith, *ibid.* **48**, 4442 (1993).1651 (1993); R. Resta, Europhys. Lett. **22**, 133 (1993).
4. R. Resta, Rev. Mod. Phys. **66**, 899 (1994).
5. G. Ortíz and R. M. Martin, Phys. Rev. B **43**, 14202 (1994).
6. R. Resta, Phys. Rev. Lett. **80**, 1800 (1998).
7. *Geometric Phases in Physics*, edited by A. Shapere and F. Wilczek (World Scientific, Singapore, 1989).
8. R. Resta, J. Phys.: Condens. Matter **12**, R107 (2000).
9. *Theory of the Inhomogeneous Electron Gas*, edited by S. Lundqvist and N. H. March (Plenum, New York, 1983).
10. L. Kleinman and D. M. Bylander, Phys. Rev. Lett. **48**, 1425 (1982).
11. We use a kinetic–energy cut–off of 300 Ry; the Brillouin–zone integration is performed using 10 special points for NaBr and 28 for Si (in the irreducible wedge).
12. D. M. Ceperley and B. J. Alder, Phys. Rev. Lett. **45**, 566 (1980); J. Perdew and A. Zunger, Phys. Rev. B **23**, 5048 (1981).
13. A. Baldereschi, R. Car, and E. Tosatti, Solid State Commun. **32**, 757 (1979).
14. R. Resta and A. Baldereschi, Phys. Rev. B **23**, 6615 (1981).
15. S. Baroni, P. Giannozzi, and A. Testa, Phys. Rev. Lett. **58**, 1861 (1987); P. Giannozzi, S. de Gironcoli, P. Pavone, and S. Baroni, Phys. Rev. B **43**, 7231 (1991); S. Baroni, S. de Gironcoli, A. Dal Corso, and P. Giannozzi, Rev. Mod. Phys., in press.
16. Our calculated ε_∞ values, consistent with the charge densities shown in Fig. 2, are 2.69 in NaBr and 12.9 in Si. The corresponding experimental values are 2.59 and 11.4. The calculations are performed at the theoretical equilibrium lattice constants: 11.81 and 10.14 a.u. for NaBr and Si, respectively.

17. R. Resta and K. Kunc, Phys. Rev. B **34**, 7146 (1986).
18. P. Fernàndez, A. Dal Corso, and A. Baldereschi, Phys. Rev. B **58**, R7480 (1998).
19. L. L. Hirst, Rev. Mod. Phys. **69**, 607 (1997).
20. S. Baroni and R. Resta, Phys. Rev. B **33**, 7017 (1986).
21. D. J. Thouless, Phys. Rev. B **27**, 6803 (1983); Q. Niu and D. J. Thouless, J. Phys A **17**, 2453 (1984).
22. G. D. Mahan and K. R. Subbaswamy, *Local density theory of polarizability* (Plenum, New York, 1990), Ch. 5.
23. See Sect. II C in Ref. [19].

Kinetic Monte Carlo Simulations of Crystal Growth in Ferroelectric materials

Charles Tahan[1], M. Suewattana, P. Larsen[2]
Shiwei Zhang, and H. Krakauer

Department of Physics, College of William and Mary
Williamsburg, VA 23187-8795

Abstract. We study the growth process of ferroelectric materials by kinetic Monte Carlo simulations. An ionic model with long-range Coulomb interactions is used to model the relaxor single crystals. The growth is characterized by thermodynamic processes involving adsorption and evaporation, with solid-on-solid restrictions. An algorithm is developed in order to simulate growth under such a model, for which existing formalism of the kinetic Monte Carlo algorithm is inadequate. We study the growth rates and the order structure of the grown crystals as a function of temperature, chemical composition, and growth orientation. Tests of our algorithm on NaCl gave good results. Preliminary results on growth in Ba-based heterovalent binaries showed 1:2 ordering along the [111] direction over limited scales.

I INTRODUCTION

The discovery of ultra-large coupling piezoelectrics [1] has intensified the theoretical interest in these remarkable materials. In addition to understanding equilibrium properties, where calculations by density-functional theory methods coupled with effective Hamiltonians have shown great promise [2], understanding how to grow large single crystals is clearly of key importance. In this paper, we explore theoretical approaches to study the crystal growth process of ferroelectric materials.

We use a "second-principles" procedure which simulates the growth process with kinetic Monte Carlo based on a simple model Hamiltonian. The model was proposed by Bellaiche and Vanderbilt [3] in 1998 for the perovskite alloys. It reduces the system to only the B-sites and assumes that the long-range Coulomb interactions between ions at these lattice sites are the driving mechanism for ordering. This simple electrostatic model successfully reproduced most of the B-site compositional long-range orders in equilibrium Monte Carlo simulations.

---

[1]) Present address: Dept. of Phys., University of Wisconsin-Madison Madison, WI 53706
[2]) Present address: St. John's College, Oxford OX1 3JP, United Kingdom

Using this electrostatic model for the perovskite alloys, we study the crystal growth process by kinetic Monte Carlo (KMC) simulations [4]. KMC methods have seen various applications to study growth and kinetic processes in Ising-like lattice models. In particular, a sampling algorithm introduced by Bortz, Kalos, and Lebowitz (BKL) [5] works efficiently for the short-range interactions present in these systems. For the electrostatic model here, however, an enhanced sampling algorithm, which we develop below, is needed in order to treat long-range Coulomb interactions efficiently.

In this paper, we describe our theoretical and computational approach for growth simulations of ferroelectric crystals, and present preliminary results. The rest of the paper is organized as follows. In Section II, we discuss the electrostatic model, standard KMC, our new sampling algorithm to implement KMC for this model, and additional technical issues to treat the long-range Coulomb interactions in a growth simulation with a finite-sized simulation cell. In Section III, we show results on the growth processes of rock-salt type crystals. We conclude in Section IV with some remarks on future directions.

II APPROACH

A The model

The electrostatic model of Bellaiche and Vanderbilt [3] is a simple but remarkably successful ionic model of perovskite alloys. In this model, the electrostatic energy due to point-charge ions in the ideal cubic structure is assumed to be the dominant factor for the observed B-site ordering. For an A(BB')O$_3$ compound, the total electrostatic energy can be written as

$$E_t(C) = \sum_{(l\tau,l'\tau')} \frac{Q_{l\tau}Q_{l'\tau'}}{\epsilon |\mathbf{R}_{l\tau} - \mathbf{R}_{l'\tau'}|}, \quad (1)$$

where $\mathbf{R}_{l\tau}$ is the position of the ion on site τ ($\tau = \{A, B, O_1, O_2, O_3\}$) of cell l and ϵ is the dielectric constant. Because charges at the A-sites and O-sites have fixed values, it can be shown that, up to a constant, the configurationally averaged electrostatic energy depends only on the B-site charges

$$E_B(C) = \frac{1}{\epsilon a} \sum_{(l,l')} \frac{q_l q_{l'}}{|\mathbf{l} - \mathbf{l}'|}, \quad (2)$$

where a is the cubic lattice constant, $\mathbf{R}_{lB} = l a$, and $q_l \equiv Q_{lB} - 4e$ for compounds whose A-sites are occupied by Pb or Ba.

Different classes of alloys can be conveniently described using this electrostatic model for the B-sites. For example, IV$_x$IV$'_{1-x}$ denotes a homovalent binary alloy having tetravalent B-atoms, e.g., Pb(ZrTi)O$_3$, while II$_{(1-x)/3}$IV$_x$V$_{2(1-x)/3}$ indicates a heterovalent ternary such as $(1-x)$Ba(MgNb)O$_3 + x$BaZrO$_3$.

The model of Eq. (2) leads to the following Hamiltonian for our system,

$$\mathcal{H}(C) = E_B(C) + \Delta\mu N, \tag{3}$$

where N is the total number of occupied sites, i.e., the total number of ions in the grown crystal. The second part of the Hamiltonian accounts for the chemical potential difference between the crystal and the melt. The magnitude of $\Delta\mu$ controls the "sticking" rate from the melt.

B Kinetic Monte Carlo (KMC) method

The kinetic Monte Carlo (KMC) method is one of several simulation techniques used to simulate the relaxation processes of systems away from equilibrium (e.g. growth processes). It has been applied successfully to crystal growth and surface/interface phenomena [4,6], mostly in the context of kinetic Ising models. Our growth simulation uses the general approach of KMC, although significant enhancement to the algorithm must be introduced to make it practical for the long-range electrostatic interactions.

The objective of the growth simulation is to create a model that describes the dynamics of crystal growth as stochastic processes such as adsorption, evaporation, and surface migration. Our simulations presently include the first two only, namely the adsorption and evaporation of the adatoms. The adatoms represent the B-site ions in the single crystal perovskite alloy, characterized entirely by their charge and interacting with each other through the potentials that we discuss in Section II.D. The structure of the simulation is a three-dimensional lattice where adatoms can individually and singularly occupy lattice sites.

We carry out our growth simulations in an $L \times L \times \infty$ cell on an ideal cubic lattice. Periodic boundary conditions (PBC) are imposed in the x-y plane. The z-direction is free and is the growth direction. We start our growth simulations with an $L \times L \times H_0$ substrate which serves as the growth seed. The crystal configuration C at any time is specified by the sites which are occupied and the charge q_l at each occupied site $l = (i, j, k)$.

In our simulations, we impose a solid-on-solid (SOS) restriction, which does not permit the appearances of vacancies. The SOS restriction means that we can write H as

$$\mathcal{H}(C) = E_B(C) + \Delta\mu \sum_{(i,j)} h_{ij}, \tag{4}$$

where h_{ij} is the height of the present crystal configuration at position (i, j).

We now outline the basic theoretical background for the kinetic Monte Carlo method. The goal is to simulate the time evolution of the system through a Markov chain of configurations. We define $P(C, t)$ as a time-dependent distribution of configurations, C as the current crystal configuration, and C' as a crystal configuration related to C by one time step. The transition rate from C to C' is denoted by $w(C \to C')$. The transition rate is to be chosen to simulate the physical system as realistically as possible.

We can then write down the following master equation

$$\frac{\partial P(C, t)}{\partial t} = -\sum_{C} w(C \to C')P(C, t) + \sum_{C'} w(C' \to C)P(C', t), \qquad (5)$$

where the first term on the right describes the loss because of transitions away from C, while the second term describes the gain because of transitions into C. In the equilibrium limit (as $t \to \infty$), the Boltzmann distribution

$$P_{eq} = Z^{-1} \exp\left[\frac{-\mathcal{H}(C)}{kT}\right] \qquad (6)$$

is reached. We require that detailed balance be satisfied:

$$\frac{w(C \to C')}{w(C' \to C)} = \frac{P_{eq}(C')}{P_{eq}(C)} = \exp\left[-\frac{\mathcal{H}(C') - \mathcal{H}(C)}{kT}\right]. \qquad (7)$$

The KMC technique can be viewed as a method of solving equation (5). We adopt the following choice of transition rates $w(C \to C')$

$$w_a = \exp\left(\Delta\mu/kT\right) \qquad (8)$$
$$w_e = \exp\left(-\Delta E_B(C)/kT\right), \qquad (9)$$

where w_a and w_e are the rates for adsorption and evaporation, respectively, of an adatom. In this choice, the adsorption rate is a constant, while the evaporation rate depends on the change in total potential energy in the crystal when an adatom evaporates from the surface: $\Delta E_B(C) \equiv E_B(C') - E_B(C)$.

For kinetic Ising models, the algorithm of BKL [5] allows an efficient stochastic realization of the kinetic process under the choice in Eq. (9). In this algorithm, a site (i, j) is selected randomly in each step. An event is then selected at (i, j) by Monte Carlo sampling [7] from the list of all three possible events, {*adsorption, evaporation, nothing*}. For Ising-like models, where the interaction is limited to near-neighbors, the energy difference $\Delta E_B(C)$ is completely determined by the *local* environment at site (i, j). The relative probabilities $\{P_a, P_e, P_n \equiv 1 - P_a - P_e\}$ can thus be easily obtained from the global maximum of w_e, i.e., the minimum possible $\Delta E_B(C)$ for any C.

C KMC algorithm for long-range interactions

The electrostatic model of Eq. (2) means that the energy change needed in Eq. (9) depends on the *entire* configuration C. It is therefore difficult to determine the global minimum, $\Delta E_{\min} \equiv \min[\Delta E_B(C)]$. Furthermore, even if $\Delta E_{\min}$

could be identified, the energy change $\Delta E_B(C)$ for most configurations would be much greater than $\Delta E_{\min}$, which would cause P_n to approach unity, thereby driving evaporation and adsorption probabilities to zero and rendering the algorithm ineffective.

Our new algorithm goes a step beyond the standard KMC algorithm. [5] It considers all $N = L \times L$ surface sites *simultaneously* and creates an event list which includes every possible event for every possible surface site. The algorithm proceeds as below:

(i) Generate a list, E, of all possible events per time step. There are $2N$ possible events: an evaporation or an adsorption could happen on each of the $N = L \times L$ surface sites.

(ii) Calculate the rates (w) of adsorption and evaporation for each site on the surface ($2N$ rates). Denote the total rates by W, $W = \sum_{i}^{2N} w_i$.

(iii) Normalize these $2N$ rates by W, giving probabilities, P_i, for adsorption and evaporation on sites $1, 2, \cdots, N$.

(iv) Generate a random number $r \in [0, 1)$ and choose the first event E_i such that $\sum_{k=1}^{i} P_k \geq r$. An event will always be chosen.

(v) Generate the new configuration C based on chosen event E_i.

(vi) Assign a "real time" increment $\Delta t_{\text{real}} = -1/W \ln(r')$ to this MC step, where r' is another random number on $[0, 1)$.

The added complexity comes from the need to store and update an array of surface potentials, the calculation of an event list from these surface potentials, and finding the potential event in this list. The benefit comes in that an event is guaranteed to take place with each iteration of the algorithm and that the need for $\Delta E_{\min}$ is negated completely. Evaporation/adsorption rates for all possible sites are normalized; The sum of the probabilities that an adsorption or evaporation occurs at any site is unity.

We end the discussion of the algorithm with a comment on step (vi). The issue of time in a KMC simulation is a subtle one. Often the Monte Carlo (MC) time t_{MC} is used as a measure of the real time. This is approximate. We follow the same principle, but an adjustment is necessary because of our particular sampling algorithm. In our simulation, an event is forced to happen in each step regardless of the total rates W for the configuration at hand. We therefore rescale Δt_{MC} in each step to reflect the total normalization factor W.

D The Coulomb interaction in a growth simulation

In addition to requiring a modified KMC sampling algorithm, the long-range Coulomb interactions themselves also require special treatment in a simulation with a finite-sized simulation cell. In order to calculate the evaporation rate for a surface charge at $o = (i, j, k)$, we need to compute pair-wise interactions between this charge and every other charge in the present crystal configuration. Indeed, for each charge at $l \neq o$ inside the cell, the interactions between o and all the images of l from periodic boundary condition (PBC) also need to be included. A neutralizing background is then added to ensure that the total pair-wise interaction between o and l is finite. That is

$$V_{o-l} = \frac{q_o q_l}{|l - o|} + \sum_{l'} \frac{q_o q_l}{|l' - o|} - \sum_{o'} \frac{q_o q_l}{|o' - o|} \equiv q_o q_l \, v_{o-l}, \qquad (10)$$

where l' denotes the image positions of l due to PBC, and o' denotes those of o.

In an equilibrium simulations, PBC is implemented for the simulation cell in all directions. The formula indicated in Eq. (10) is the so-called Ewald method and evaluation of V_{o-l} is straightforward [8,9]. In our growth simulation, however, PBC is only in the x-y plane, while the z-direction is the direction of growth and is free. The sums in Eq. (10) are therefore restricted to two-dimensions. The sites o and l are in general not in the same x-y plane. We place all the images in the x-y plane of l. For l' this is natural. For o' we project o to the x-y plane of l along the growth direction, and then take the images of the projected position. In other words, the charges are in 3-D but the Ewald sum is limited to 2-D. A procedure for this "fractional-dimension" situation was developed to evaluate v_{o-l}. The energy change in Eq. (9) for evaporation of the charge at o is thus given by

$$\Delta E_B = -\frac{q_o}{\epsilon a} \sum_l q_l v_{o-l}, \qquad (11)$$

where the sum is over all sites l in the crystal inside the simulation cell. We use a look-up table for v_{o-l} in our simulation to improve computational efficiency.

III RESULTS

Our results here will focus on the rock-salt structure, i.e., that of sodium chloride (NaCl). This structure is composed of alternating layers of Na^{1+} and Cl^{1-} in both the [111] and [$\bar{1}$11] directions. It typifies the crystal ordering of a wide variety of materials. Heterovalent binaries such as described by $II_{1/2}VI_{1/2}$ ($q_B = \pm 2$) or $III_{1/2}V_{1/2}$ ($q_B = \pm 1$) are good perovskite examples of rock salt type ordering. The rock-salt structure also represents the simplest crystal our ionic model can grow. It can also be modeled with Ising-like short-range interaction, for which various results are available for comparison and benchmark.

Our crystal growth simulation is a model of two parameters. The variation of the temperature of the system, T, and the chemical potential difference, $\Delta \mu$, will control the changing characteristics of the growing crystal. We choose $q_B = \pm 1$, i.e., the $II_{1/2}VI_{1/2}$ compound. Clearly, a trivial energy scale factor will allow us to convert the system to $III_{1/2}V_{1/2}$ compounds.

We define the growth rate of the crystal as the number of adatoms adsorbed divided by the total real MC time. The growth rates from the simulation, over a range of temperatures, are given in Figure 1 as a function of the chemical potential difference $\Delta\mu$. If we let $a \sim 8$ a.u. and $\epsilon \sim 10$, our temperature scale is about 1000K, i.e., $kT = 1$ in the simulation corresponds to 1000K. We use an $L = 20$ crystal matrix. Each simulation ran for 100 MC steps or 100 crystal layers (including incomplete layers), whichever came first. (An MC step is defined as L^2 attempts at the procedure outlines in Section II.C.) The selected temperatures range from $kT = 0.025$ to $kT = 2.0$. Every data point represents the average growth rate value of ten separate crystal simulation runs. The error bars, which are the normal statistical error $(\sigma/\sqrt{n})$ in MC calculations, are smaller than symbol size in most cases.

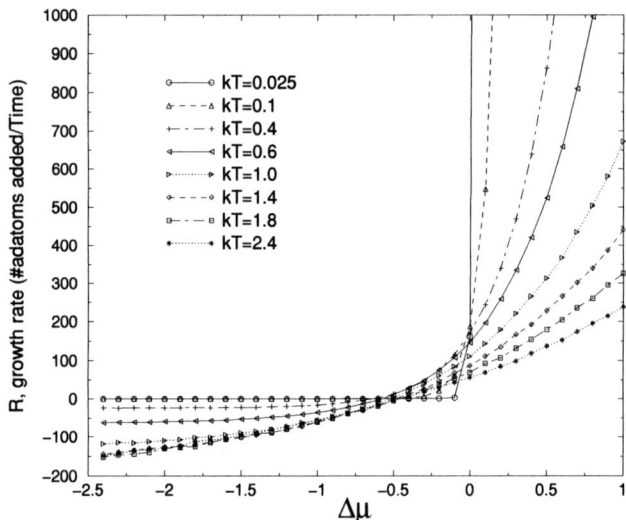

FIGURE 1. Growth rates for rock salt.

The qualitative features of these growth rate curves are consistent with the model and the expected behavior of crystal growth. As $\Delta\mu$ increases, an adatom is more likely to adsorb to the crystal surface. As kT decreases, the adsorption rate will increase but more importantly, the "selectiveness" of evaporation will increase. A lower kT will in effect increase the energy differences between competing configurations. The direct result, as growth is concerned, will be that adatoms will increasingly have more neighbors instead of less (layer-by-layer growth vs. rough) and charge similar adatoms will seem even more repulsive. For very high $\Delta\mu$, adatoms will stick anywhere, no matter the location or ionic adversity, and the growth rate will be high. Alternatively, if the temperature becomes too high, the crystal will melt, the preferred phase becomes the liquid phase and result in the negative growth.

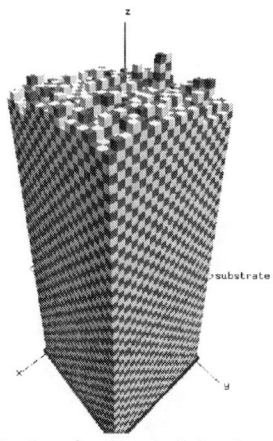

FIGURE 2. Long-Range Simulation for Rock Salt: layer-by-layer growth ($kT = 0.1$ and $\Delta\mu = -0.75$). Grey and white cubes represent the two different species. Initial substrate (seed) is indicated.

An interesting phenomena is the presence of "towers" when very rough growth occurs. At very high adsorption rates, crystal formations nucleate on the surface independent of one another and likely with several defects. As these initial formations grow, they will eventually meet and will not necessarily match up. That is, the alternation of charges will not fit the rock salt description. Since the crystal is growing so fast, there is not time to "correct for" these imperfections with evaporations and the most energy efficient alternative is the creation of individual towers of the correct crystal order. Since surface diffusion is not yet included in our simulation, it remains to be seen how robust these towers are in more realistic situations.

Simulations on heterovalent binaries of the form $II_{1/3}V_{2/3}$ are presently underway. The ground state crystal ordering of Ba-based materials is predicted (and confirmed by equilibrium MC [3]) to be $[111]_{1:2}$. In our model, $q_1 = -2$ and $q_2 = +1$, where the subscript on q is the species identifier of this two species system. In adsorption, the quantity of species one is therefore 1/3 and that of species 2 is 2/3, maintaining the charge neutrality of the B sublattice. Preliminary growth simulations in the [001] direction have shown 1:2 ordering along [111] over limited scales. We believe ordering over extended scales will be achieved as we move to direct simulations of growth along the [111] direction.

IV SUMMARY AND DISCUSSIONS

In conclusion, we have presented an approach for growth simulations of ferroelectric materials. We use an ionic, long-range model and an enhanced kinetic

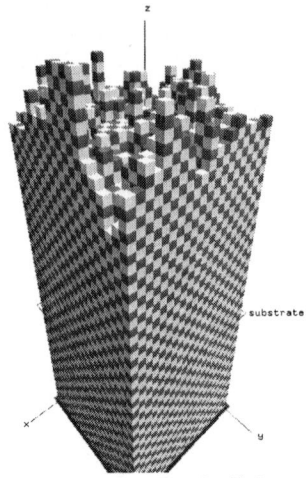

FIGURE 3. Long-Range Simulation for Rock Salt: rough growth with towers ($kT = 0.1$ and $\Delta\mu = -0.65$).

Monte Carlo formalism. Test results from our simulations on rock salt exhibit the expected growth and ordering behavior and are encouraging.

Clearly much progress would have to be made in theoretical and computational capabilities of growth simulations in order for them to contribute in a significant way to the understanding and control of the growth process of relaxor ferroelectric crystals, which has largely remained an experimental issue. Our goal here is a second-principles procedure to study growth processes which can be systematically enhanced. The two components of our approach are a model Hamiltonian and kinetic Monte Carlo. Either can be improved or even replaced eventually. For example, the electrostatic model could be modified to include charge transfer [10]. Within the framework of KMC, the effect of surface migration/diffusion could be incorporated. The impact of the solid-on-solid restriction should be explored. We are also studying growth along directions other than [100] in our simulations of heterovalent binaries.

ACKNOWLEDGMENTS

We thank T. J. Walls for his contributions in early stages of this work. This work was supported by ONR N00014-97-1-0049 and NSF DMR-9734041. SZ is a Research Corporation Cottrell Scholar.

REFERENCES

1. S.-E. Park and T.E. Shrout, J. Appl. Phys. **82**, 1804 (1997).
2. See, e.g., *Fundamental Physics of Ferroelectrics 2000*, Ed. R. E. Cohen, AIP Conf. Proceedings **536** (2000).
3. L. Bellaiche and D. Vanderbilt, Phys. Rev. Lett. **81**, 1318 (1998).
4. See, e.g., K.Binder and M. H. Kalos in *Monte Carlo Methods in Statistical Physics, 2nd Ed.*, Ed. K. Binder, Springer-Verlag, Berlin (1986).
5. A.B. Bortz, M.H. Kalos, J.L. Lebowitz, J. Comput. Phys. **17**, 10 (1975).
6. A. Levi and M. Kotrla, J. Phys.: Condensed Matter **9**, 299 (1997).
7. See, e.g., M. H. Kalos and P. A. Whitlack, *Monte Carlo Methods*. John Wiley and Sons, New York (1986).
8. B. Nijober and F. de Wette, Physica **23**, 309 (1957).
9. D. M. Ceperley, Physical Review B **18**, 3126 (1978).
10. Zhigang Wu and H. Krakauer, Phys. Rev. B, in press (2001).

First-principles study of Pb_2MgTeO_6 perovskite

Razvan Caracas and Xavier Gonze

Université Catholique de Louvain, Unité de Physico-Chimie et de Physique de Matériaux, pl. Croix du Sud 1, B-1348 Louvain-la-Neuve, Belgium

Abstract. The Pb_2MgTeO_6 perovskite is cubic at high temperature and becomes incommensurately modulated with a rhombohedral, $R\bar{3}$, average structure at lower temperatures. Both the cubic and the rhombohedral average structure have been investigated using the local density approximation (LDA) of the density functional theory. They are insulators with a 3 eV LDA gap. The electronic band structure is composed of weakly dispersive electronic bands. The peaks in the electronic density of states are assigned to the different atomic and/or molecular orbitals. The ab initio determination of the structure is in reasonable agreement with the experiment, with differences up to 2% for the lattice constants. The relevant unstable modes and their eigenvectors are analyzed.

I INTRODUCTION

Incommensurate (IC) phases occur when a periodic structure is subject to a spatially periodic modulation whose wavelength is incommensurate with respect to the basic underlying lattice. There are more than 130 inorganic IC compounds where the modulation has a displacive and/or occupational character. For most of these compounds the mechanisms governing the phase transitions leading to IC structures are unknown.

A few perovskites present an incommensurate (IC) phase: Pb_2CoWO_6 [1], Pb_2CdWO_6 [2], $PbSc_{1/2}Ta_{1/2}O_3$ and $PbSc_{1/2}Nb_{1/2}O_3$ [3] and Pb_2MgTeO_6 [4,5]. In all these compounds the incommensurability has been discovered recently. The average structures are relatively simple with just a few atoms in the unit cell. Out of all these materials the Pb_2MgTeO_6 presents also three other remarkable properties: (i) the temperature range of existence of the IC phase is very large, down to 0 K; (ii) it has not a lock-in transition (blocking of the modulation wavevector to a commensurate value, generating a commensurate structure) and (iii) the cations are well ordered.

Pb_2MgTeO_6 has a face-centered cubic lattice, $Fm\bar{3}m$, at high temperatures. This phase transforms at 194 K to an IC phase with an $R\bar{3}m$ average structure. This latter phase transforms at 142 K to another IC phase with $R\bar{3}$ symmetry of the average structure and with modulation wavevector $q = \delta a^* + \delta b^* + \delta c^*$, with δ close to $1/9$.

We determined the electronic, structural and dynamical properties of the high-temperature $Fm\bar{3}m$ cubic phase and of the low-temperature $R\bar{3}$ rhombohedral phase. We also analyzed the instabilities in the cubic phase.

The paper is organized as follows. First, details of the calculation methods are presented. Then the crystal structures of the two polymorphs are briefly discussed. Next, we present the calculated electronic properties and the theoretical determination of the crystal structure of both phases. The dynamical properties and the analysis of the instabilities will be discussed in the end.

II COMPUTATIONAL DETAILS

All the calculations were based on the local density approximation (LDA) of the DFT [6,7]. We used the ABINIT code, a common project of the Université Catholique de Louvain, Corning Incorporated and other contributors (ABINITv2.x, 1999-2000, [8]). The ABINIT software is based on pseudopotentials and planewaves. It relies on the adaptation to a fixed potential of the band-by-band conjugate gradient method [9] and on a potential-based conjugate-gradient algorithm for the determination of the self-consistent potential [10]. As usual with planewave basis sets, the numerical accuracy of the calculation can be systematically improved by increasing the cut-off kinetic energy of the planewaves. The wavefunctions describe only the valence and the conduction electrons, while the core electrons are taken into account using pseudopotentials. We used Troullier-Martins pseudopotentials [11] for Mg, Te and O and extended Teter norm-conserving pseudopotentials for Pb, the considered valence electrons for Pb, Mg, Te and O being $5d^{10}6s^26p^4$, $3s^2$, $5s^25p^4$ and $2s^22p^4$ respectively.

For the characterization of the electronic properties, a set of convergence tests has been done in order to choose correctly the grid of special k points [12] and the cut-off for planewave kinetic energy. During these tests the grid density and the cut-off energy value have been consecutively and independently increased. The variation of the total energy has been monitored. A difference of 1 mHa (1 Hartree = 27.211 eV) between two successive grids/cut-off energies has been considered indicative of sufficient convergence. A grid of 32 special k points in the full Brillouin zone (primitive 4x4x4), folding to 6 points and to 2 points in the irreducible part of the Brillouin zone (IBZ) of the rhombohedral and cubic structures respectively, and a planewave kinetic energy cut-off of 26 Hartree have been finally adopted for the calculation of the electronic properties. The determination of the crystal structure and of the dynamical properties needed separate convergence tests of the density of the grid of k points (see later).

The structural relaxation was conducted using the Broyden-Fletcher-Goldfarb-Shanno minimization (BFGS), modified to take into account the total energy as well as the gradients (as in usual BFGS), [13]. The dynamical properties are calculated using the linear response technique [14].

Several increasing k-points grids were used in order to ensure the convergence of the structural parameters. The results of each k-point grid were tested in the next denser grid and the residual stresses and the cartesian forces were monitored. A value of the stresses of less than $5.0 * 10^{-5}$ hartree/bohr^3 and a value of forces of less than 1.0^{-4} hartree/bohr were considered sufficiently accurate. Finally the same grid of k-points (Monkhorst-Pack 4x4x4) was enough to attain the desired convergence.

III CRYSTAL STRUCTURE

The structures of the two polymorphs, the cubic, Fm$\bar{3}$m high-T phase and the rhombohedral, R$\bar{3}$, low-T phase are very similar [4,5]. The high-temperature cubic phase has a double ABO$_3$ perovskite-type structure (10 atoms in the primitive cell). The metallic atoms are alternatively arranged along the [100] and [111] directions as Mg-O-Te-O-Mg and Mg-Pb-Te-Pb-Mg respectively. The O atoms are not situated at the high-symmetry positions (1/4 0 0) but are slightly displaced toward the Te atoms (0.26 0 0) [5].

The low-temperature rhombohedral phase is stable below 142K. It is incommensurate, with space group R$\bar{3}(\delta\delta\delta)$. The rhombohedral structure is obtained from the cubic one by squeezing the primitive cubic unit cell, with the rhombohedral angle $\alpha = 59.9°$. In the same time the O atoms are displaced from their positions (in the cubic lattice they lie on the Mg-O-Te-O directions) and come closer to the 3-fold diagonal rotation axis. The Pb atoms are also displaced from their initial positions (1/4 1/4 1/4) away from the above-mentioned rotation axis. From experimental data they seem to be randomly distributed between one of the three possible positions which result from this displacement. However, these displacements are very small and the incommensurability affects mainly the O atoms and not the Pb atoms. So, the Pb positions are approximated as (1/4 1/4 1/4) for both experimental structures.

IV ELECTRONIC PROPERTIES

We started our study with the determination of the groundstate properties to have a basis for treating later the dynamical properties. The analysis of the valence charge density reveals the main features of the chemical bonds. The Mg atoms are ionized, loosing their valence electrons. The Pb atoms keep their $5d$ electrons. They partly loose the $6s$ and $6p$ electrons, the remaining ones participating in weak covalent bonds with the surrounding O atoms. The Te atoms preserve some of their $5s$ and $5p$ electrons and form relativey strong covalent bonds with the six oxygens of the TeO$_6$ octahedra.

The valence electronic bands are weakly dispersive (Fig. 1). Two zones within the valence bands are readily identified, one of low energy and another one of high

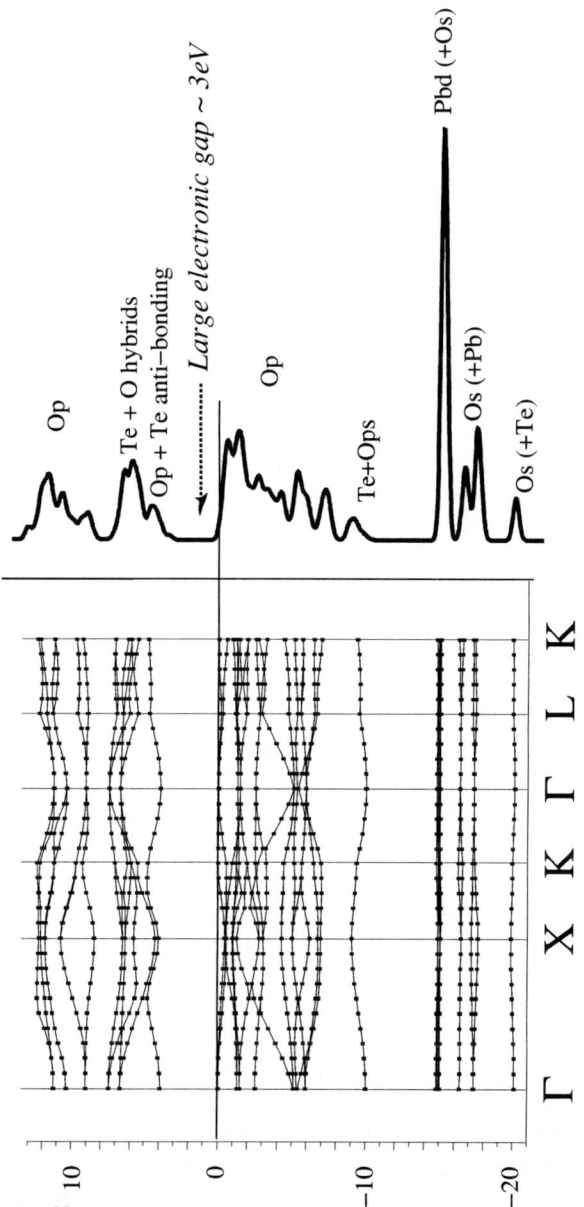

FIGURE 1. The electronic band structure and the corresponding density of states in the cubic phase of Pb_2MgTeO_6

energy. In the lower energy part, the electronic charge corresponding to the different peaks in the density-of-states (DOS) may be assigned mainly to O2s electrons, with some participation from Te and Pb electrons. The 5d electrons of the Pb atoms are clearly differentiated, generating a sharp peak in the DOS. The remaining Te electrons together with some hybrids of O2p and 2s electrons form an isolated band in the intermediate range of energy. In the higher energy part the valence bands correspond to hybrids of O2p electrons. Pb_2MgTeO_6 presents a large electronic gap of about 3 eV. The first conduction bands have an anti-bonding character. The electronic band structure and the corresponding DOS are almost identical for both the cubic, high-T and the rhombohedral, low-T phases.

V AB INITIO STRUCTURAL DETERMINATION

The theoretical results of the structural relaxation are summarized in 1 for the cubic phase and in 2 for the rhombohedral phase. The lattice constants for both the $Fm\bar{3}m$ and $R\bar{3}$ phases are slightly underestimated (-2%) with respect to the experimental values. We observe that within the experimental uncertainty, the primitive cubic and rhombohedral lattice constants are equal. Our calculations give smaller values than the experiment for both, with a tiny difference between them, less than 0.2 %. The α angle of the rhombohedral structure is slightly overestimated. This overestimation is translated in terms of a compression rather than extension of the initial cubic lattice along a diagonal three-fold axis.

The theoretical values of the interatomic bond distances in the cubic phase are smaller than the experimental values (up to 3.5-3.7 % discrepancy).

In case of the rhombohedral phase at 0 K, the experimental structure is incommensurately modulated. Consequently there is a large spread of the bond distances around the mean values (corresponding to the average structure) within some continuous intervals. In the following we will first make comparisons with the average values for the interatomic distances, and then we will see the position of our results within the continous intervals.

The calculated Mg-O bond lengths are slightly larger than the average experimental values, while the calculated Te-O bond lengths are much smaller (-6.8 %) than the experimental average values. The Pb-O bond lengths are just slightly smaller then the experimental ones. The two O-O unequivalent bond distances for the oxygens belonging to the Mg-centered octahedra are larger (+0.1 %; +6.6 %) in our calculations than the mean experimental values. In the same time the two unequivalent bonds between the oxygens belonging to the Te-centered octahedra are smaller (-4.1 %; -8.4 %) than the experimental data. Thus, our calculations tend to increase the volume of the MgO_6 octahedra while decreasing the volume of the TeO_6 octahedra, if we take the mean values as reference.

In any case we see that all our results fall close to the middle of the continous intervals of existence of the real bonds. The deviations of the theoretical results with respect to the mean bond length values are all smaller than the spread of the bond lengths in the IC structure.

VI DIELECTRIC PROPERTIES

The determination of the dielectric properties plays an important role in the calculation of the long-range interatomic forces. We obtained for the cubic phase a dielectric constant of 5.44, which corresponds to a refractive index of 2.33. The rhombohedral phase presents slightly smaller values.

We further computed the Born effective charges, defined as the change in polarization due to an atomic displacement under zero external electric field. The Pb effective charge (+3.938) is higher than the nominal charge of the Pb^{2+} ion (+2). Mg effective charge (+2.436) is somewhat larger than the nominal charge of the Mg^{2+} ion (+2). The Te effective charge (+4.486) is smaller than the nominal charge of the Te^{6+} ion (+6). The effective charge of the oxygens is expected to be anisotropic, due to the anisotropy of the environment around the oxygens. But the difference between the computed three diagonal values is extremely small (-2.462 -2.462 -2.463), and close to the nominal charge of the O^{2-} ions (-2).

For the rhombohedral structure the deviations of the effective charge with respect to the nominal charges are slightly increased. In the same time the anisotropy of the oxygen effective charge is accentuated.

These Born effective charges do not follow the trends that have been observed for the ferroelectric perovskites. We remark the absence of large anomalies of the O effective charges (common in perovkites like $BaTiO_3$, $PbTiO_3$ [15]). This might be attributed to the fact that the chains Mg-O-Te-O-Mg do not allow the flow of a polarization current. Such a feature was dominating the physics of Born effective charges for ferroelectrics perovskites. A band-by-band anlysis [16] might be an appropriate tool for the understanding of Born effective charges in Pb_2MgTeO_6.

VII DYNAMICAL PROPERTIES

We further determined the vibration modes at the Γ point for both structures. The unstable modes and their eigenvectors will be briefly discussed here, while the detailed analysis of all the modes will be reported elsewhere. We first analyze the u-modes (among which the IR-active modes), then the g-modes (among which the Raman-active modes).

There are no experimental data concerning the IR spectra. Due to symmetry considerations, there must be four T_{1u} modes and one T_{2u} mode in the cubic phase. In our calculations the cubic phase presents a first instability reflected in a TO T_{1u}-mode situated at $13.92\ i\ cm^{-1}$. This imaginary mode is a ferroelectric mode (similar to those found in $BaTiO_3$ or $PbTiO_3$). It displaces the Pb atoms in the negative sense of the [100] axis while all the other atoms are displaced in the opposite direction. The displacement of the O and Te atoms are similar, confirming thus the covalent character of the bonding between these two atoms. This mode, which may induce a rhombohedral or tetragonal distortion in the perovskite family, will induce a rhombohedral one in the case of the Pb_2MgTeO_6. In the rhombohedral phase all the u-modes have positive frequencies.

The Raman spectra have been determined experimentally, on powder samples [4]. For the high-T cubic phase our high-frequency results are in good agreement with the experimental data. We find also a second instability. This unstable Raman mode is situated at 91.19 i cm^{-1}. It has an antiferrodistortive character. If the MgO_6 octahedra rotate clockwise around the [100] direction, the TeO_6 octahedra will rotate anticlockwise. In the same time a small shift of the apex oxygens in the opposite [011] direction is superposed onto this rotation. This mode may induce a rhombohedral distortion.

For the rhombohedral phase, the comparison of our results (corresponding to the average structure), with the experimental data [4] is rather difficult to realize. Indeed, in the experiment, there are broad peaks composed of several smaller peaks superposed on a central peak due to the incommensuration. Some of the main experimental peaks might correspond to our calculated data, an identification that we are currently pursuing. In the rhombohedral phase all the calculated g-modes are stable.

Although we have only preliminary results for the phonon band structures, we can mention that for the cubic phase instabilities persist also in other high-symmetry points: W, K, L. A 2x2x2 grid of q points gives already accurate results.

In the rhombohedral phase preliminary data from a 2x2x2 grid of q points indicate the presence of a ferroelastic instability. The ferroelastic character of the instability agrees with the experimental data, although the wavevector differs in our calculations (along the Γ-X line, with a minimum close to (1/9 0 0)) and in the experiment (along the Γ-L line, with a minimum close to (1/9 1/9 1/9)).

VIII CONCLUSIONS

We showed that Pb_2MgTeO_6 is an insulator with a large band gap. The Mg and Pb atoms are ionized, while the Te and O atoms form relatively strong covalent bonds. The theoretical structural determination provided results in relatively good agreement for the cubic phase. In the case of the rhombohedral phase the comparison between our results and the experimental average data is somewhat meaningless. The experimental data have been taken at a temperature where the real structure is incommensurately modulated. So there is a continuous range of values for each bond length. With respect to these variations, all our data fall close to the middle of the experimental intervals, which is quite encouraging for the continuation of the forthcoming study of the IC phase. The analysis of the phonons in the Γ point reveals the presence of two instabilities in the cubic phase, corresponding to a ferroelectric distortion and an antiferroelectric distortion. The rhombohedral phase does not present any instability in the Γ point. Preliminary results indicate a ferroelastic instability in the rhombohedral phase.

TABLE 1. Cubic phase: Ab initio structural determination and comparison with the experiment.

Parameter	Theory	Exp. (Ref. [5])	Diff.
a_0 (Å)	7.833	7.984	-1.89 %
Ox	0.2645	0.2600	0.0045
Mg-O bond length (Å)	2.072	2.081	-0.43 %
Te-O bond length (Å)	1.844	1.911	-3.50 %
Pb-O bond length (Å)	2.608	2.709	-3.78 %
O-O bond length (Å)	2.930	2.935	-0.17 %

TABLE 2. Rhombohedral phase: Ab initio structural determination and comparison with the experiment. We present first the theoretical results, then the range of values obtained in the real incommensurate structure and their average. Next, we compare our theoretical results with the average values. The last column gives the spread in the IC structure relative to the average.

Parameter	Theory	Exp. IC [5]	Exp. Average [5]	Diff.	Spread
a_0 (Å)	5.531		5.645	-2.01 %	
α (°)	60.27		59.9	0.37 °	
Ox	0.2669	0.1965-0.3050	0.2408	0.0261	0.1085
Oy	0.7646	0.6828-0.7726	0.7384	0.0262	0.0898
Oz	0.7051	0.7356-0.7788	0.7559	-0.0508	0.0432
Pbx	0.2507	0.2320-0.2680	0.2500	0.0007	0.036
Mg-O bond length (Å)	2.084	1.910-2.172	2.029	2.71 %	12.91 %
Te-O bond length (Å)	1.848	1.833-2.106	1.983	-6.80 %	13.77 %
Pb-O bond length (Å)	2.776	2.409-3.236	2.831	-1.93 %	29.21 %
O-O bond length (Å) [a]	2.968	2.452-3.262	2.783	6.64 %	29.11 %
O-O bond length (Å) [a]	2.928		2.924	0.13 %	27.70 %
O-O bond length (Å) [b]	2.614		2.855	-8.44 %	28.37 %
O-O bond length (Å) [b]	2.612		2.724	-4.11 %	29.74 %

[a] in MgO_6 octahedra [b] in TeO_6 octahedra

ACKNOWLEDGMENTS

We thank G. Baldinozzi and Ph. Ghosez for interesting dicussions and suggestions. Part of this research has been supported by the FRFC project No. 2.4556.99

REFERENCES

1. Bonin, M., Paciorek, W., Schenk, K., and Chapuis, G., *Acta Crystal.*, **B51**, 48–54 (1995).
2. Baldinozzi, G., Calvarin, G., Sciau, P., Grebille, D., and Suard, E., *Acta Crystal.*, **B55**, 1–16 (1999).
3. Malibert, C., Baldinozzi, G., A.Bulou, and J.-M.Kiat, *Ferroelectrics*, **235**, 241–249 (1999).
4. Baldinozzi, G., Sciau, P., and Bulou, A., *J. Phys.: Condens. Matter*, **9**, 10531–10544 (1997).
5. Baldinozzi, G., Grebille, D., Sciau, P., Kiat, J.-M., Moret, J., and Berar, J.-F., *J. Phys.: Condens. Matter*, **10**, 6461–6472 (1998).
6. Hohenberg, P., and Kohn, W., *Phys. Rev. B*, **136**, 864–872 (1964).
7. Kohn, W., and Sham, L. J., *Phys. Rev. B*, **140**, 1133–1138 (1965).
8. Gonze, X., Caracas, R., Sonnet, P., Detraux, F., Ghosez, P., Noiret, I., and Schamps, J., *First-principles study of crystals exhibiting an incommensurate phase transition*, American Institute of Physics, Melville, New York, 2000, vol. 535, pp. 13–20.
9. Payne, M. C., Teter, M., Alan, D., T.A.Arias, and J.D.Joannopoulos, **64**, 1045–2097 (1992).
10. Gonze, X., *Phys. Rev. B*, **54**, 4383–4386 (1996).
11. Troullier, N., and J.L.Martin, *Phys. Rev. B*, **43**, 1993–2006 (1991).
12. Monkhorst, H., and Pack, J., *Phys. Rev. B*, **13**, 5188–5192 (1976).
13. Press, W., B.P.Flannery, S.Teukolsky, and W.T.Vetterling, Cambridge University Press, Cambridge, 1989.
14. Gonze, X., *Phys. Rev. B*, **55**, 10337 (1997).
15. Ghosez, P., Michenaud, J.-P., and Gonze, X., *Phys. Rev. B*, **58**, 6224 (1998).
16. Ghosez, P., Gonze, X., Lambin, P., and Michenaud, J.-P., *Phys. Rev. B*, **51**, 6765–6768 (1995).

Why Is There An Isotope Effect On T_c In Hydrogen-Bonded Ferroelectrics Upon Deuteration But Absent Upon Replacing ^{16}O By ^{18}O?

Annette Bussmann-Holder and Naresh Dalal*

Max-Planck-Institut für Festkörperforschung, Heisenbergstr.1, D-70569 Stuttgart, Germany
* Department of Chemistry and NHMFL, Florida State University, Tallahassee, FL32306,USA

Abstract. Isotope effects in ferroelectrics have not attracted much attention in the past, with the exception of the large effect on the transition temperature T_c in hydrogen-bonded ferroelectrics upon deuteration. Reinvestigating this class of compounds experimentally reveals that the exchange of oxygen by its isotope does merely not affect T_c of hydrogen-bonded KH_2PO_4 and squaric acid, in contrast to the recently observed isotope induced ferroelectricity in $SrTiO_3$. The new findings in hydrogen-bonded systems are in agreement with theoretical predictions of a vanishing isotope effect in the classical regime and substantially support our earlier conclusions on the coexistence of order-disorder and displacive dynamics in these compounds.

INTRODUCTION

The general understanding of ferroelectricity is mostly based on the fact, that one is dealing with purely ionic systems [1] where effects arising from the electron-phonon interaction can be neglected due to the tightly bound electronic shells. Within this framework, the ferroelectric transition temperature T_c is independent of isotopic substitutions and not much experimental or theoretical effort has been devoted to this topic in the past. The exception of the large positive isotope effect on T_c upon deuteration in hydrogen-bonded compounds has been taken as evidence that here the tunneling dynamics of the protons play the decisive role in the triggering the structural instability and order-disorder effects dominate the transition [2]. This division of ferroelectrics into hydrogen-bonded and non-hydrogen-bonded systems obviously requires that two different theoretical concepts have to be introduced to model the phase transition mechanism: the displacive soft mode concept and the order-disorder proton tunneling mechanism [3]. As already outlined in a variety of recent work, this strict division is neither possible for non-hydrogen-bonded compounds nor for the KH_2PO_4 (KDP) type systems. While in the former a competing coexistence of both features has been observed experimentally [4] and shown theoretically to arise from competing time scales [5], in the latter high precision NMR experiments detected the coexistence but having the same time scale [6]. These new findings have the important consequence that subtle electron-ion interactions are crucial for the phase transition mechanism and that charge transfer effects trigger the lattice instability. While the importance of charge transfer effects is known in perovskite type ferroelectric oxides [7, 8], it was until then speculated but never decisively detected to be dominant also in hydrogen-bonded lattices [9]. Very recent first principles electronic structure calculations [10] have indicated that rearrangements in the electronic distribution take place at T_c and are the only source of the lattice instability. We have also shown - experimentally as well as theoretically [6, 7, 11]- that these effects are common and important to both, hydrogen-bonded and non-hydrogen-bonded systems. Small charge transfer effects lead in both types of ferroelectrics to a cancellation of short-range and long range forces and induce the structural instability. Within this framework, an isotope effect on T_c has been predicted to take place in non-hydrogen-bonded compounds only in the quantum limit [12]: a fact which has recently been verified experimentally in $SrTiO_3$ [13], but to vanish in the classical regime, a regime which applies to KDP and analogues. The large effect on T_c upon deuteration in the hydrogen-bonded ferroelectrics consequently seems to be in contrast to our predictions, but – as has been shown [11] - here geometrical O...H...O bond details gain importance and give rise to the large isotope effect on T_c when the proper electron-phonon interactions are included in the modelling. The consequences of such ideas are that order-disorder and displacive dynamics coexist, and these features have been revealed, soon after the first prediciotns, by means of ultra-high-precision NMR techniques. All of these ideas point to the fact, that a division of ferroelectrics into hydrogen-bonded and non-hydrogen-bonded systems is unnecessary, but that a single model is capable to grasp the microscopic phase transition mechanism. In the following we will show - from experiment as well as theory - that there is more detailed evidence for our ideas that ferroelectricity is the consequence of small electronic rearrangements (charge transfer) which triggers the lattice instability.

THE MODEL

As already outlined in the introduction, the phase transition temperature T_c is expected to exhibit no isotope effect as long as purely ionic concepts are used [1]. As a consequence, there have been no experimental efforts to seek for this effect until very recently when ferroelectricity was induced in the quantum paraelectric $SrTiO_3$ by replacing ^{16}O by its isotope ^{18}O [13]. This rather huge effect had been predicted a few years before [12] within the framework of the nonlinear polarizability model. This model which is based on the idea, that small oxygen ion p - transition metal d charge transfer takes place in the perovskite type compounds, also predicts that the isotope effect vanishes with increasing transition temperature. The huge isotope effect in hydrogen-bonded ferroelectrics is a consequence of this charge transfer, which influences the details of the O...H...O bond geometry and consequently changes T_c. Since the small geometrical effects in this bond are unimportant to the PO_4-unit of e.g. KH_2PO_4, which carries the polarization, it is expected from the theory that no oxygen isotope effect should be observable here, since T_c is in the classical regime. Instead of modelling KDP-type compounds within a coupled pseudospin – nonlinear polarizability model, we use a coupled double well potential model, where one double well is attributed to the protons (deuterons), while the other one arises at the PO_4-cluster site and is due to strong local and nonlinear electron-phonon interaction effects. The coupling between both subsystems arises solely via electronic hybridization terms. Phenomenologically such a scenario can be best described within a lattice dynamical model where the coupled subsystems both exhibit a nonlinear polarizabilty:

$$H = T + V_2 + V_4^{(PO_4)} + V_4^{(H)}$$

$$T = \tfrac{1}{2}\sum_n [m^{(PO_4)}(\dot{Q}_{1n}^{(PO_4)})^2 + m^{(H)}(\dot{Q}_{2n}^{(H)})^2]$$

$$V_2 = \tfrac{1}{2}\sum_n [f(v_{2n}^{(H)} - v_{1n}^{(PO_4)})^2 + f(v_{2n}^{(H)} - v_{1n-1}^{(PO_4)})^2 +$$

$$f'_{(PO_4)}(Q_{1n}^{(PO_4)} - Q_{1n-1}^{(PO_4)})^2 - f'_{(H)}(Q_{2n}^{(H)} - Q_{2n-1}^{(H)})^2]$$

$$V_4^{(PO_4,H)} = \tfrac{1}{2}\sum_n [g_2^{(PO_4,H)}(Q_{in}^{(PO_4,H)} - v_{in}^{(PO_4,H)})^2$$
$$\tfrac{1}{2}g_4^{(PO_4,H)}(Q_{in}^{(PO_4,H)} - v_{in}^{(PO_4,H)})^4]$$

(1)

Here the Q, v are the site n-dependent displacement coordinates of ion i (i=PO_4, H) with conjugate momenta, f and f' are harmonic nearest and next nearest neighbour shell-shell and core-core couplings and g_2, g_4 are attractive harmonic and repulsive fourth order core-shell coupling constants. All couplings are indexed with respect to the coupling ions The Hamiltonian can be mapped onto the one used in our former work [11]: i.e. $f'_{(H)} \approx J$ and the proton local double well potential corresonds to the Morse potential. The shell-shell coupling f contains the details of the O...H...O bond geometry. Similarly two limiting cases can be considered here: i) the phase transition is driven by the protonic subsystem, $g_2^{(PO_4)} > 0$, $g_4^{(PO_4)} = 0$, and ii) the nonlinear oxygen ion polarizability triggers the transition corrsponding to

$g_2^{(H)} > 0$, $g_4^{(H)} = 0$. In the case of interest, where both double well potentials exist, the self-consistent phonon approximation can be applied which corresponds to:

$$g_2^{(PO_4,H)}(Q_{in}^{(PO_4,H)} - v_{in}^{(PO_4,H)}) + g_4^{(PO_4,H)}(Q_{in}^{(PO_4,H)} - v_{in}^{(PO_4,H)})^3 =$$
$$g_2^{(PO_4,H)} w_{in}^{(PO_4,H)} + g_4^{(PO_4,H)}(w_{in}^{(PO_4,H)})^3 =$$
$$g_2^{(PO_4,H)} w_{in}^{(PO_4,H)}[1 + \frac{3 g_4^{(PO_4,H)}}{g_2^{(PO_4,H)}} < (w_{in}^{(PO_4,0)})^2 >_T$$
$$= g_T^{(PO_4,H)} w_{in}^{(PO_4,H)} \quad (2)$$

where $<(w)^2>_T$ is the selfconsistently determined thermal average over the relative displacement w which defines the contribution to the dipole moment stemming from the protons, PO_4-clusters, respectively, and is explicitly given by:

$$<(w)^2>_T = \sum_{q,j} \frac{\hbar}{2Nm\omega_{q,j}} w_{q,j}^2 \coth\frac{\hbar\omega_{q,j}}{2kT} \quad (3)$$

where the momentum q and branch j dependent frequencies have to be calculated from the equations of motion. From the equations of motion the eigenvectors $w_{q,j}^2$ can be determined explicitly:

$$(w_{q,j}^{(PO_4)})^2 = \frac{[m^{(PO_4)}\omega^2 - 4f'_{(PO_4)}\sin^2 qa]^2[(2f'_{(H)}+g_T^{(H)})m^{(H)}\omega^2 - 2f'_{(H)} g_T^{(H)}]^2}{[m^{(PO_4)}\omega^2 - 4f'_{(PO_4)}\sin^2 qa - g_T^{(PO_4)}]^2 4[f'_{(H)} m^{(H)}\omega^2 \cos qa]^2}$$
$$\times (w_{q,j}^{(H)})^2$$

(4)

In the vicinity of T_c both $g_T^{(PO_4)}, g_T^{(H)}$ go to zero indicating that a finite polarization corresponding to $<w^{(PO_4,H)}> = \pm g_2^{(PO_4,H)}/g_4^{(PO_4,H)}$ sets in. In this limit equation 4 becomes:

$$(w_{q,j}^{(PO_4)})^2 = \frac{1}{\cos^2 qa}(w_{q,j}^{(H)})^2 \quad (5)$$

Since we are interested in the various isotope effects on T_c, equation 4 is evaluated in the limit that the polarization arising from the protons is small, which is well verified by experimental data. This admits to find an anlytical expression for T_c:

$$kT_c = \frac{|g_2^{(PO_4)}|}{3g_4^{(PO_4)}} \frac{3}{V_c \int q^2 dq [\frac{1}{2f}\cot^2 qa + \frac{1}{4f'_{(PO_4)}}\cos ec^2 qa]} = \frac{|g_2^{(PO_4)}|}{3g_4^{(PO_4)}} f_{\tilde{D}} f \qquad (6)$$

Obviously equation 6 is independent of any masses but depends on the bond geometry through f in a nonlinear way. This finding agrees perfectly with our previous analysis. Note that in the quantum limit a strong mass dependence in T_c appears [12].

In order to check the validity of the model predictions, specific heat measurements have been performed for hydrogen bonded compounds squaric acid and KH_2PO_4 replacing in both compounds ^{16}O by $^{17,18}O$ and check the effect of this replacement on the corresponding transition temperatures. Fig. 1 shows the specific heat, C_p, data for the normal (^{16}O) squaric acid (solid circles) and on its 75% $^{17,18}O$-labeled crystal (represented by squares). Only a minor shift of about 1 K can be noted in the T_c'. This small shift which is not expected from equ. 6 is here attributed to a 5% pressure effect stemming from the heavier isotope, in accordance with the previous calculations. Similar minor changes of the order of a degree or less were also observed for 18-labeled $NH_4H_2PO_4$ and RbH_2PO_4 (data not included).

Since the theoretical modelling predicts an appreciable isotope effect on T_c in the quantum limit, i.e. for T_c less than approximately 45K, it would be interesting to carry out the same type of experiments for hydrogen-bonded samples whose T_c is reduced by e.g. replacing potassium by ammonium.

CONCLUSIONS

In summary we have modelled hydrogen-bonded systems by two coupled double – well potentials, which is in very strong correspondence to our previous modelling [6, 11]. We find that the isotope effect on T_c, when other ions than the hydrogens are replaced, should be vanishing. The experimental results fully confirm the predictions. In addition the use of the combination of a displacive and an order-disorder component reconfirms that the combined character is present in this class of materials and leads to the conclusion that charge transfer and covalency are vital in understanding and modelling also H-bonded compounds.

ACKNOWLEDGEMENT

We wish to thank Kim Pierce for preparing the ^{17}O-labeled samplesand Mathew Gierce for help with the specific heat measurements.

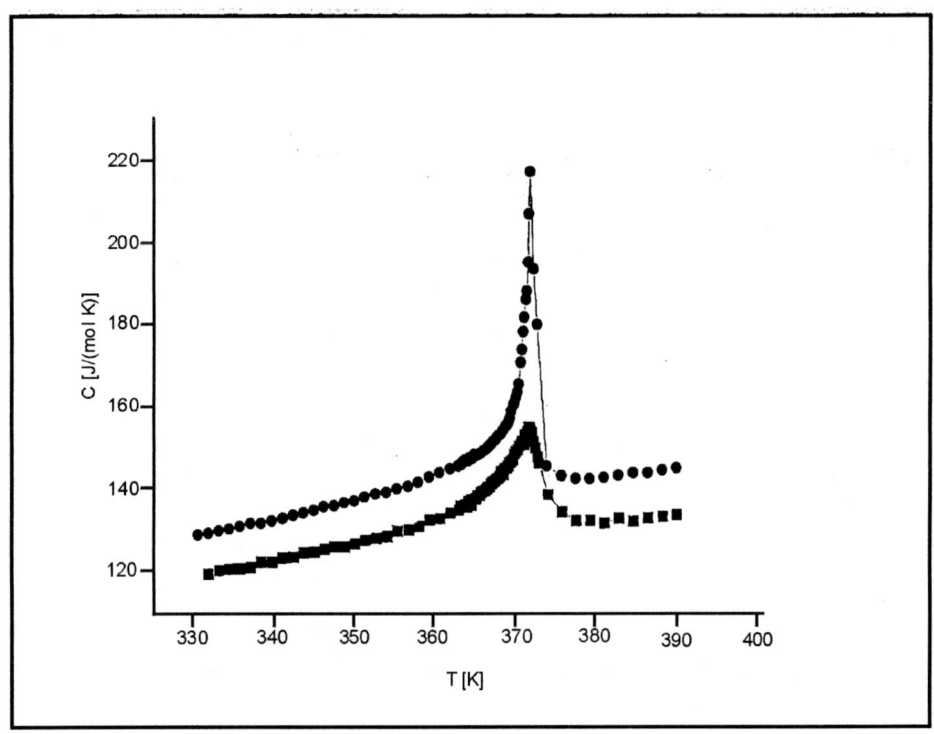

FIGURE 1. Specific heat of squaric acid as a funtion of temperature. Circles refer to data taken from a ^{16}O sample, squares are data from a ^{17}O sample.

REFERENCES

1. Lines, M. E., and Glass, A. M., *Principles and Applications of Ferroelectrics and Related Materials*, Oxford: Carendon Press, 1977.
2. Blinc, R., and Žekš, B., *Advances in Physics* **21**, 693 (1972); Blinc, R., *J. Phys. Chem. Solids* **13**, 204 (1960)
3. Blinc, R., and Žekš, B., *Soft Modes in Ferroelectrics and Antiferroelectrics*, Amsterdam, Oxford: North-Holland Publishin Company, 1974.
4. Sircon, N., Ravel, B., Yacobi, Y., Stern, E. A., Dogan, F., and Rehr, J. J., *Phys. Rev. B* **50**, 13168 (1094).
5. Stachiotti, M., Dobry, A., Migoni, R., and Bussmann-Holder, A., *Phys. Rev. B* **47**, 2473 (1993).
6. Dalal, N., Klymachyov, A., and Bussmann-Holder, A., *Phys. Rev. Lett.* **26**, 5924 (1998).
7. Migoni, R., Bilz, H., and Bäuerle, D., *Phys. Rev. Lett.* **37**, 1155 (1976); Bilz, H., Benedek, G., and Bussmann-Holder, A., *Phys. Rev. B* **35**, 4840 (1987).
8. Cohen, R. E., and Krakauer, H., *Phys. Rev. B* **42**, 6416 (1990).
9. Mehring, M., and Suwalack, D., *Phys. Rev. Lett.* **42**, 317 (1979).
10. Koval, S., Kohanov, J., Migoni, R., and Bussmann-Holder, A., IC/2000/188.

11. Bussmann-Holder, A., and Michel, K. H., *Phys. Rev. Lett.* **80**, 2173 (1998).
12. Bussmann-Holder, A., and Büttner, H., *Phys. Rev. B* **41**, 9581 (1990); Bussmann-Holder, A., Büttner, H., and Bishop, A. R., *J. Phys.: Cond. Mat.* **12**, L115 (2000).
13. Itoh, M., Wang, R., Inaguma, Y., Shan, Y. I., and Nakamura, T., *Phys. Rev. Lett.* **82**, 3540 (1999).

Model of Polar Clusters in Relaxors: Charge Transfer and Local Configuration Instability Effects

Valentin S. Vikhnin*, Robert Blinc[+], Rasa Pirc[+], and Siegmar Kapphan[#]

*A.F. Ioffe Physical Technical Institute, St. Petersburg 194021, Russia
[+]J. Stefan Insitute, POB 3000, 1001Ljubljana, Slovenia
[#]Osnabrück University, Fachbereich Physik, 49069Osnabrück, Germany

Abstract. A model of polar clusters and of chemical fluctuation clusters (CFC) in PMN-like relaxors is proposed. The model is mainly based on three phenomena: on a charge transfer state formation corresponding to localized Charge Transfer Vibronic Excitons (CTVE) induced by Mg-Nb disorder, on an appearance of electronic and hole polarons in CFC due to charge compensation, and on Local Configuration Instability (LCI) effects. The origin of co-operative phenomena, of the giant piezoelectric effect and characteristic dielectric losses, including Vogel-Fulcher law as well as peculiarities of EPR and NMR do find a explanation within this model. Another LCI based model of isocharged doped ferroelectric solid solutions (concentrated as well as diluted) is proposed and developed for examples like SBN, $Sr_{1-x}Ca_xTiO_3$ (SCT), and $K_{1-x}Li_xTa_{1-y}Nb_yO_3$ (KLTN) systems. The origin of polar clusters, their interactions, and co-operative phenomenon was considered here taking into account the LCI effect. It is shown that polaron and CTVE-trapping effects including the case of oxygen vacancy clusters play an important role for both types of relaxors discussed.

INTRODUCTION

Three characteristic types of ferroelectric relaxors are mainly considered in connection with their uncommon properties. There are PMN-like [1] concentrated solid solutions with locally non-compensated non-isocharged ion substitution, SBN-like concentrated solid solutions with isocharged ion substitution [2], and SCT-like diluted solid solutions on the basis of incipient ferroelectrics [3]. For all these ferroelectric relaxors the principal question is still open: what is the origin of polar clusters accompanied by specific co-operative dynamics with poly-dispersive dielectric response ? In recent years this problem attracted constant attention leading to substantial progress (see, for instance, [4,5]). In the present work we shall consider the possible nature of polar clusters in these strongly different situations, and shall discuss some key effects which are based on the properties and behaviour of polar clusters. For instance, we shall discuss the origin of the giant piezoelectric effect (GPEE) detected in [6] as well as EPR and NMR peculiarities in PMN-type relaxors.

The surprising appearance of new phase transitions with unexpected properties for weakly co-doped KLTN can also be explained by the polar cluster model which allows to consider this system as a "weak" relaxor.

The Model of Polar Clusters in PMN-type Relaxors

We shall assume that two principal elements take part in the active cluster formation in PMN-type relaxors. The are (i) electronic and hole polarons (Nb^{4+} and O^- centers in the PMN case) which are located in the CFC and give rise to charge compensation, and (ii) small clusters of localized and ferroelectrically correlated charge transfer vibronic excitons (CTVE) [7-10] with high cluster polarization.

(i) Initially extra-charged CFC induce polaron charge transfer on nano-scale distance with following charge compensation. As a result, the homogeneous CFC in the ordered relaxor state can be presented as a periodic succession of ordered and disordered regions. It corresponds to the following chemical and charge composition: $Pb^{2+}Nb^{4+}O^{2-}_3$ (ordered electronic polaron region), and $Pb^{2+}Mg^{2+}_{2/3}Nb^{5+}_{1/3}O^{2-}_2O^{1-}$ (disordered hole polaron region) for the PMN case. Here polarons are correlated with conservation of possibilities of strong change of reciprocal distance, that is, without formation of the well defined pair states like bi-polarons and CTVE. But the magnetic polaron-polaron exchange is enough, in average, for formation of spin-less co-operative polaron state. Only separately located polarons give EPR-signals.

(ii) The CTVE are connected with strongly correlated pairs of neighboring electronic and hole polarons [7-10], that is, with O^- - Nb^{4+} pairs in the PMN situation. Small polar clusters (PC) with CTVE origin are embedded into the CFC.

As a result, the structure of nano-scale clusters in PMN-like relaxors becomes principally unhomogeneous. It is determined by short range ordered CFC containing a set of small "defect"-type CTVE-induced polar clusters (PC) much less in size than the maternal CFC, and with average PMN-concentration of components. Here the correlated CTVE in the PC are induced by internal CFC fields controlled by the appearance of the electron-hole polaronic system within the CFC.

The polarization of these small PCs has two main parts: the CTVE order-disorder dipole centres as well as a displacive type part of strongly polarisable Pb^{2+} ions. The CTVE reorientations together with reorientations of the displacive part of the PC polarization give rise to the total PC polarization reorientation. The latter leads to the possibility of ferroelectric or glass-type ordering of the PC within CFC due to different possible conditions for PC-PC indirect interactions via the soft lattice as well as via the soft CFC degrees of freedom (quasilocal vibrations and polaronic pseudo-spins).

As a result, all CFC region can be ferroelectrically ordered under the conditions of ferroelectric order of PCs within CFC. But a CFC can be also non-polar under the conditions of glass-like order of these PCs. Note, that such order conditions are strongly dependent from CFC-interface boundary conditions which control the magnitude of CFC polar mean field. Indeed, if the CFC ferroelectric order condition, $M_1 \gg (M_2)^{1/2}$, where M_1, M_2 are the first (mean field) and the second (average square of fluctuation field) moments of polar field distribution within the CFC respectively, are satisfied we get the situation of a well-defined quasi-uniform in-cluster polarization for the CFC. Taking into account the possibility of reorientations this total cluster polarization on the one hand, and "cluster polarization – cluster polarization" interaction via long wave TO phonons [11] on the other, we get here good conditions for the validity of the Spherical Random Bond Random Field (SRBRF) model [12]. In accord with estimations, we get a fulfilling of the above mentioned inequality for a high enough concentration of small CTVE-clusters (PC) within CFC, or/and under the action of the external electric field. This is the case for PMN-like relaxors according to our estimations. But the CFC remain non-polar in the case of the opposite sign for this inequality, under the conditions of nano-scale glass-like behavior with distributed CFC-parameters according to their real distributions.

Note, that electronic and hole polarons under consideration take part in charge transfer via internal hopping motion which leads to a wide dispersion of dielectric losses, and to motion averaging of NMR spectra which had been recently observed in model relaxors PMN and PST [13]. Moreover, indirect "polaron pseudo-spin-polaron pseudo-spin" interaction via phonon and quasilocal vibration fields leads, similar to [14], to the correlation of different polaronic jumps with the appearance of co-operative order-disorder type soft relaxator behavior. Such soft relaxators obey a critical dependence of the relaxation time $\tau_{soft} \sim (T-T_0)^{-1}$, T_0 is the critical temperature of polaron pseudo-spin ordering. We can connect this critical temperature with the freezing temperature, $T_0 \approx T_f$. We shall show later that this assumption leads to the possibility of the derivation of a Vogel-Fulcher law from the model proposed. The main mechanism of polaron pseudo-spin formation here is the appearance of a polaron multi-well potential near the effective attractive Nb^{5+}-center as a core in the PMN lattice. This type pseudo-spin centers are appearing in the electronic polaron CFC region when such a core is located on the concentration fluctuation in this CFC, or within intermediate regions with average PMN composition.

Note also, that the reorientations of PC-polarization are connected with single CTVE successive reorientations due to near resonance cross-relaxation in the CTVE-clusters which are more slow than the soft reorientations of the polaronic "cloud" within CFC (except of the very vicinity of freezing temperature). Polaronic "cloud" reorientations are following in on adiabatic manner behind these slow PC reorientations. As a result, the slow PC reorientations control the dynamic behavior of co-operative response for the total relaxor system (which can be successfully described by a dynamical SRBRF model [12]). The corresponding slow relaxation rate for the reorientations of a single polar cluster connected with CFC obeys a critical dependence which is described by the Vogel-Fulcher law.

Indeed, the rate of CTVE- cross-relaxation is proportional to the overlap integral S for the vibrational wave functions in the fourth power. But the S-value is determined by $S = \exp[-V_b/2\hbar\omega]$ for the ground vibration states, $\hbar\omega$ is the vibration frequency of CTVE-state. Here the key factor is the magnitude of CTVE potential barrier, V_b. The main contribution to the latter is determined by CTVE-induced polarization of the soft pseudo-spin polaron subsystem not so far from the freezing temperature. Taking into account that the critical temperature for the ordering of the latter subsystem equals to the freezing temperature, we obtain that the dominating critical contribution to the above discussed potential barrier is proportional to $\sim (T-T_f)^{-1}$. As a result, the slow relaxation rate for cross-relaxation reorientations of polar CTVE-cluster has a Vogel-Fulcher critical dependence, $\tau^{-1} \sim \exp[-V(T-T_f)^{-1}]$, where V is the barrier parameter.

Last not least, we deal with Jahn-Teller (JT) Nb-ion related electronic polarons. Indeed, there is a quasi-degenerated (in the random relaxor field) triplet $4d^1$ state linearly interacting with trigonal and, especially important, tetragonal JT-distortions of the cluster. We shall consider the mechanism of the PC polarization increasing under the action of LCI on the tetragonal JT-vibration for electronic polarons. The starting point will be an initially weak linear JT-effect within the harmonic approximation of the lattice vibrations description. This case corresponds to a very soft JT tetragonal vibration with rather high local susceptibility. As a result, a weak addition tetragonal field can strongly change the magnitude of the resulting tetragonal JT-distortion.

Indeed, let us consider the influence of third order anharmonicity of the cross-type with local potential $U_{anh.} = V_3 Q_\theta Q^2_u$, where Q_u is a fast relaxating local vibration variable, Q_θ is a slow variable tetragonal JT-distortion of the NbO_6 cluster with trapped electronic polaron. Such type temporal hierarchy allows to replace the square of the fast dynamical variable by its averaged with temperature value, $<Q^2_u>_T$. As a result, we obtain the effective tetragonal field acting on the tetragonal JT-variable which is described by an effective potential $U^*_{anh.} = V_3 Q_\theta <Q^2_u>_T$. Such type perturbation induces a strong Q_θ value change in the actual case of the soft JT-vibration. In addition, this Q_θ value change will rise with temperature exponentially up to temperatures where the higher anharmonicities with opposite temperature dependence of the induced Q_θ value become more important.

This type of LCI effect in the JT-subsystem can be important for the modulation of the equilibrium charge transfer value by a temperature induced strong change of the tetragonal JT distortion. For instance, we have an important suppression of the equilibrium Q_θ value for $V_3 < 0$ case. As a result of such type decreasing of the JT-tetragonal distortion, we have to deal here with a corresponding increasing of the oxygen-niobium charge transfer value due to a corresponding decreasing of the harmonic potential magnitude for charge transfer fluctuations. Namely, the latter undisturbed harmonic term here will be replaced by a new expression with important contribution of above mentioned cross-type third order anharmonicity. This contribution will rise with temperature at least in the low temperature region and increases the equilibrium charge transfer: $\alpha q^2 \rightarrow (\alpha + \alpha_1 Q_\theta) q^2$, where q is a oxygen – niobium charge transfer magnitude, and the new vibronic parameter α_1 is proportional to $\alpha_1 \sim V_3 <Q^2_u>_T / \alpha^{-1}$.

Such an effect of the niobium-oxygen charge transfer increasing with temperature being induced by the JT-instability increasing of the charge transfer harmonic fluctuation can lead to pronounce consequences in the magnitude of the PC equilibrium polarization, P_{PC}. Namely, it will be increasing with temperature under the conditions discussed above as a result of direct $\sim(q\, P_{PC})$ interaction [9]. But with high enough temperature a decreasing of P_{PC} due to the opposite influence of the next, higher order cross anharmonicity can occur. The resulting bell-like temperature dependence of the cluster polarization density was really detected [5].

Note that this model explains simultaneously the bell-like temperature dependence of the anomalously high g-factor shift from $g \approx 2$ up to $g \approx 1.5$ for the EPR signal (S=1/2) in PMN [15] approximately in the same temperature region where the peak for α-spots (polar cluster contribution) was detected in [5]. We shall immediately get within the model that such type g-factor change can be at least qualitatively predicted under the assumption that this EPR signal is induced by separately located electronic JT-polarons. Indeed, the orbital contribution to the effective g-factor of JT-polarons will increase with the q-value increasing with temperature due to the accompanied occupation of a new degenerated orbital states of the active Nb-ion.

The appearance of electronic and hole polarons in the CFC, and of CTVE in the PC leads to a specific and strong vibronic mechanism for the piezoelectric effect in PMN-type relaxors. We propose here an explanation of the GPEE [6] in relaxor ferroelectrics on this basis. Here a vibronic model of GPEE can be proposed where a strong electron-lattice interaction is responsible for the piezoelectric coupling.

A microscopic basis of the model is the co-operative JT- and pseudo-Jahn-Teller- (PJT) effect on the electronic and hole polarons, and on the CTVE in the PC embedded to the CFC. Active JT-centers (Nb^{4+} and O^- ions in the PMN) are created by charge transfer between cluster regions with different chemical composition as well as due to CTVE creation in the intermediate regions between different CFC.

The GPEE phenomenon is appearing due to the simultaneous action of two effects.

First, there is the appearance of a mean field in the CFC due to ferroelectric ordering of PC. As a result, a high value of total cluster polarization (polarization of CFC with ferroelectrically ordered PC) in zero field is appearing. Here we can replace the P_{PC} value by the corresponding mean field, $P_{PC} \rightarrow <P_{PC}>$, with taking into account strong indirect PC-PC interaction of the ferroelectric type via soft quasilocal mode of all CFC, $\omega^2_{CFC} = \lambda(T-T_f)$, where T_f is a freezing temperature. The high magnitude of such a mean field is directly connected with realization of LCI conditions for the polaron subsystem, because $<P_{PC}> \sim \omega_{CFC}^{-2}$ is fulfilled (but not so near to the freezing transition). Such a LCI is controlled by above mentioned soft quasilocal mode of CFC.

That is, GPEE occurs in a situation when the CFC is transformed to the total reorienting polar cluster induced by ferroelectric interaction of the PC (CTVE-clusters) embedded within this CFC. The LCI case gives the basis for such a behavior.

Second, the principal contribution to the GPEE is caused by a new vibronic mechanism of cluster electrostriction induced by polaronic and CTVE subsystems. Here the simultaneous action within polaron quasi-degenerate states of the quadratic co-operative PJT effect on the cluster polarization as well as of the linear PJT effect on the non-polar local distortions takes place. The same order and importance has here the simultaneous action of two linear PJT-effects on the cluster polarization together with a linear PJT-effect on the same non-polar distortions. Both these effects can be considered within perturbation theory for quasi-degenerated states of the second and of the third order respectively. Note also, that the temperature dependence of the charge transfer due to the LCI within the JT-polaron structure discussed above induces the low lying excited polaronic states corresponding to polarons of intermediate radius. These low lying states play an important role in the strengthening of the resulting cluster electrostriction. Here an interference between these types of PJT effect leads to a strong, vibronic increase of intra-cluster electrostriction.

An important role in the model plays a strong CTVE-CTVE correlation within PC. The corresponding order-disorder soft mode can be connected with the soft mode [16] with condensation at the Burns temperature of PMN, $T_{BM} \approx 630K$.

The simultaneous existence of cluster low frequency (ω_{CFC}) modes near LCI conditions together with giant cluster electrostriction (with high magnitude of the parametrers W and $\tilde{W}$ for CFC and PC regions respectively) will be responsible for GPEE in relaxors.

The total cluster potential within GREE regime can be presented in the following form

$$U_{Cl.} = \frac{1}{2}\omega_{CFC}^2 P_{CFC}^2 - J P_{CFC} <P_{PC}> - <P_{PC}> E - P_{CFC} E - \tilde{W} <PPC>^2 Q_0 - W P_{CFC}^2 Q_0 + \frac{1}{2}\varpi_0^2 Q_0^2, \quad (1)$$

J is the CFC-PC polarization interaction, ϖ_0, Q_0 and $<\delta Q_0^2>$ are elastic frequency, dynamic elastic distortion, and average square of elastic distortion induced by random field in the cluster respectively. Minimization of (1) leads to the cubic equation,

$$Q_0 = \frac{W(J<P_{PC}>+E)^2}{\varpi_0^2(\omega_{CFC}^2 - 2WQ_0)^2}, \quad (2)$$

where the small value of the denominator and high values of the W, J, and $<P_{PC}>$ parameters protect the high resulting magnitude of the equilibrium elastic distortion for the cluster. For the important real case of strong effect of intra-cluster random field, $<\delta Q_0^2> \gg Q_0$, we get the final expression for GPEE from the equation (2):

$$Q_0 = \frac{2WJ<P_{PC}>E}{\varpi_0^2(\omega_{CFC}^4 + 4W^2<\delta Q_0^2>)}, \quad (3)$$

which can give values of the piezoelectric coefficient in agreement with experiment [6] under the conditions of strong vibronic CFC-PC and CTVE-CTVE interactions, and soft CFC quasilocal mode. The role of additional doping by 8% PT in the case which was considered in [8] can be connected with an additional CFC mode softening.

Note, that an oxygen vacancy can be also effective as a charge compensator in PMN-type relaxors. Here oxygen vacancies partly trap electronic polarons to dipole-type, localized on a single Nb-ion, states. Corresponding pseudo-spin variables connected with polaron-vacancy center reorientations as well as with polaron hopping between different vacancies take part in soft quasilocal cluster mode formation. A very similar situation is appearing in the case of CTVE-trapping on the oxygen vacancies. Here CTVE reorientations due to quasi-resonance cross-relaxation between different vacancies also lead to soft cluster mode formation. Note also, that cluster dynamics will be strongly changed under reduction or oxidation treatments in this case.

Local Configuration Instability as an Origin of Relaxor-type Properties of Ferroelectric Solid Solutions SBN and SCT

Let us consider the polar cluster formation due to LCI in SBN-, and in SCT- cases.

We shall assume that Sr^{2+} ions in SBN, and Ca^{2+} ion in SCT are weak off-center ions whose off-center displacements are small enough and have the same order of magnitude of the corresponding quantum fluctuations for the case of the absence of chemical fluctuations in the nearest neighbouring coordination spheres. As a result, Sr^{2+} ion in SBN has no electric dipole moment (on-center case for the quantum mechanical state) for the situation of average composition of the SBN components in spite of the existence of a weak multi-well potential. But such an ion has a strong anharmonic contribution to its vibrational potential in this case. As a result, the weak off-center ion can be under conditions of the LCI which is induced by the chemical fluctuation field, or which is appearing with temperature, or which is induced by a Coulomb impurity fields. The principal dynamical property of Sr^{2+} ions in SBN, and of Ca^{2+} ions in SCT is common: there is proximity to the LCI point. The latter allows to consider the SCT-case within the same model as the SBN-case.

Chemical fluctuations induce an important variation of the square of the frequency of the quasilocal polar vibration of such an "isolated" Sr^{2+} ion in SBN. An increasing of the local concentration of Sr^{2+} ions leads to a corresponding increasing of the Sr^{2+} ion off-center displacements for such ions involved in the chemical fluctuation. The latter effect is controlled by a local decreasing of short range repulsive forces in Sr-rich chemical fluctuations. A rather wide distribution of off-center Sr^{2+} ion displacements takes place here, with is accompanied by a wide spectrum of off-center ion hopping rates. But dipole relaxation is appearing only for large enough Sr-rich chemical fluctuation regions where active ion off-center displacements are also large enough, and the single well localization criterion for active Sr^{2+} ions is fulfilled.

Three different type of clusters in SBN and in SCT are appearing here.

(i) There are polar clusters which appear in the regions where the average distance between off-center ions is less or of the same order an effective correlation radius in the region of chemical fluctuations. Here an indirect dipole-dipole interaction of the ferroelectric type is realized via the soft quasilocal mode of the chemical fluctuation cluster. The latter leads to polar cluster formation as a result of order-disorder behaviour. Note that the distribution of the off-center displacements due to the distribution of elastic fields of chemical fluctuations leads here to a validity of the dynamics of a random-field Ising model. Corresponding peculiarities are detected in the experiments [17]. Note also that the appearance of a system of weak off-center ions with a distribution of dipole moments gives rise to the wide distribution of dipole relaxation rates. This is the origin of a wide spectrum of dielectric losses.

Let us write for a model with simplest description the effective potential "U" for soft variable "x" of the active Sr^{2+} ion with harmonic (with K-parameter) and anharmonic contributions. The main influence, is local electrostriction (with parameter W) which is corresponding to a modulation of the frequency square of a soft polar quasilocal mode by the chemical fluctuation elastic field "e":

$$U = \frac{K}{2}x^2 + W(ex^2) + \frac{\beta}{4}x^4 \qquad (4)$$

Following from (1) the off-center Sr^{2+} ion displacement for the actual K<0, W<0 case equals to

$$x_{off} = \sqrt{-\frac{(K+2We)}{\beta}} \qquad (5)$$

where LCI is corresponding to a (K + We) = 0 equation fulfilling. But in the situation of the absence of off-center effects for separate impurities, a new, co-operative mechanism of polar cluster appearance exists. It is based on the influence of dipole co-operative fluctuations of impurities within the cluster. The mean field of such a fluctuation can change the conditions of the stability of the initial non-dipole state via the appearance of LCI of the first order. Namely, ferroelectric dipole fluctuation can lead to the renormalization of the anharmonicity of the fourth order up to a change of its sign. The LCI of the first order is appearing here within stabilization of the cluster polarization due to six order anharmonicity effect. The temperature lowering increases the cluster polarization via increasing of the mean field of the dipole fluctuation. The effective potential $U_{eff.}$ of an active ion in the above discussed approach for (K+2We) > 0 region is

$$U_{eff.} = U(x) - 2Jn\,x^2 tgh\frac{Jn\,x^2}{2kT} + \frac{\gamma}{6}x^6 \,, \qquad (6)$$

n is a number of active centers which are under mean field approximation conditions, and which take part in the polar cluster formation (cluster strength). At the high temperature approximation we immediately obtain from (6) the renormalized effective parameter of the anharmonicity of the fourth order which becomes negative after reaching of the n-threshold. Here the LCI of the first order is corresponding to the polar cluster creation, and the dependence of such LCI temperature from cluster strength "n" can be described by the following equation:

$$4\sqrt{(K\gamma/3)} = |\beta - (Jn)^2(kT)^{-1}| \qquad (7)$$

(ii) The important role in the polar cluster effect belongs to another type, near neighboring clusters which obey rather different properties in comparison with polar clusters. There are regions where the system of active ions (incipient off-center ions) is very near to the conditions of the co-operative LCI of the first order discussed above, but LCI is not realized. That is, polar cluster minimum of the cluster potential has little bit higher energy than non-polar ground state. We have to deal here with incipient co-operative LCI case, because a small polar field can induce real co-operative LCI of the first order. Such type "critical" clusters display a high susceptibility. The latter property can switch on a strong polar cluster – polar cluster indirect interaction via such critical clusters on the one hand, and is responsible for strong increasing of the correlation radius on the other hand. As a result, the mean field approach becomes valid not only for intra-cluster, but also for inter-cluster interaction. This circumstance, for instance, explains a ferroelectric phase transition taking place for the SBN case (not a transition to a glass-like state).

(iii) Last type of cluster is a paraelectric one, where the cluster is outside of LCI conditions. The role of these clusters in co-operative phenomena is connected with random elastic field formation which act on the polar cluster via local piezoeffects.

A resulting mechanism of a polar cluster formation here is the co-operative LCI with concentration threshold and temperature dependent effect for cluster polarization appearance. Note, that critical clusters amplify a percolation behaviour.

But the role of a mechanism [18] induced by small self-organized clusters with active internal degrees of freedom (order-disorder and displacive type) in phase transition phenomena should not be disregarded. The SCT-case is characterized by self-organized (i) "$Ca_{Sr} – Ca_{Sr}$" and (ii) "Ca_{Sr}-oxygen vacancy- Ca_{Ti}" clusters. Both type active Ca-ions are off-center ions, but with important Ca-Ca correlation. As a result, single Ca-ion multi-well potentials become asymmetric, and a set of cluster order-disorder states is appearing. The Ca-rich regions are also characterized by extra-softening of the polar quasilocal displacive type vibration. Such two types of degrees of freedom directly interact with the soft lattice mode and can induce a ferroelectric or glass-like phase transition [18]. The later is due to a resulting polarization percolation occuring for a system of these small self-organized clusters in a soft lattice.

Nano-cluster Model of "Weak" Relaxor $K_{1-x}Li_xTa_{1-y}Nb_yO_3$

An important role in the investigations of ferroelectric and glass-like phase transitions (PT) induced by impurities with internal degrees of freedom is played by studies of three component solid solutions consisting in both, weak and strong off-center impurities in the soft matrix. Studies of such a system $K_{1-x}Li_xTa_{1-y}Nb_yO_3$ (KLTN) with weak (Nb^{5+}) and strong (Li^+) off-center impurities interacting with the soft lattice manifests rather uncommon properties of induced PT [19-22]. In KLTN with x=0.0014 and y=0.024 (KLTN 0.14/2.4) two soft mode driven PT [21], and a third low temperature PT with strong hysteresis, memory effects, and cusp-like dielectric constant peak [19,20] were reported. The latter revealed only slightly pronounced dielectric dispersion obeying stretched exponential law behavior [20]. Two components of the soft TO_1-mode are clearly split [21] below the highest temperature PT point at which a step-like freezing of dielectric relaxation is appearing for (KLTN 0.14/2.4) [20]. The sequence of PT driven by soft mode and "reentrant" dipole glass-like phase formation at lowest temperature was suggested in [20] for the description of (KLTN 0.14/2.4) properties. But the main surprising result of the [19-

22] experiments was the appearance of the drastic influence of extremely small Li^+ impurity concentrations (even in the case of one order of magnitude less than the critical concentration threshold for initial $KTaO_3$:Li as in the case of KTLN 0.14/2.4) on the PT scenario in KLTN. Here the key aspects are the nature of the drastic influence of a small concentration of impurities on the induced PT phenomena, and the nature of new cusp-like peak of the dielectric susceptibility at low temperature.

We develop a superparaelectric (SP) model of these phenomena with taking into account of a set of cluster local instabilities whose local order parameter interacts with the soft lattice. We assume that the concentration of strong off-center Li^+ impurities is much less than the concentration of weak off-center Nb^{5+} impurities. Starting position of the model is the formation of weak off-center impurity clusters with strong off-center impurities as the cluster core. An important role belongs here to the interaction between cluster internal degrees of freedom and a soft lattice. Corresponding bi-linearly interacting variables $d_{i,Li}$, $<d>_{i,Nb}$, $P_{i,TO}$, $P_{i,Cl.}$ are the following, respectively. First, there is rather slow reorienting relaxator-like electric dipole moment of strong off-center Li^+ ion in the core of the SP cluster. Second, taking into account a strong ferroelectric type correlation between reorienting electric dipole moments of weak off-center Nb-ions within the cluster, we introduce a total (averaged) reorienting dipole moment of all Nb ions within SP cluster (fast relaxator). The third and the fourth are displacive type polarizations of soft lattice TO phonons and SP cluster respectively.

Taking into account the bi-linear interactions mentioned above, a polar cluster local instability of the first order [11] induces a first order ferroelectric PT of order-disorder type in the matrix at low temperature. The latter is appearing due to cluster-cluster indirect interaction via soft modes mentioned above. In accord with the model, the probable direction of the order parameter in the low temperature phase is along [111] (along eight possible Nb^{5+} ion off-center displacements). The latter leads to the three fold degeneracy for the [100]-type off-center positions for Li^+ impurities in the field of the order parameter. The next key point in the ordering mechanism formation at below-threshold concentrations is indirect Li^+-Li^+ interaction via highly polarizable "clouds" of weak off-center Nb-ions within the superparaelectric cluster. This mechanism is responsible for a strong strengthening of such a interaction. Here actual relatively large average Li^+-Li^+ distances due to the diluted situation for this type ions correspond to glass-type Li^+-Li^+ correlations. The glass-type Li^+-dipole ordering is appearing here in parallel way with ferroelectric ordering in the [111]-direction and co-exists with this ferroelectric ordering. Such type behaviour supports a phenomenological assumption [20] about reentrant dipole glass state formation at low temperatures, and gives the grounding for it from the side of the microscopic model. But it is not disregarded that the low temperature phase has [100] direction of the order parameter with qualitatively the same scenario of the reentrant glass effect.

The soft phonon – cluster order parameter interaction is responsible for the TO_1 soft mode splitting which appears after passing of the highest temperature SP PT of the second order. Critical increasing of the correlation radius up to the values higher than average distance between SP cluster cores (that is, between strong Li^+ defects) induces the co-operative effects in cluster polarization reorientations in the narrow region near

SP PT. The latter switch on a critical reorientation slowing down in this narrow region. Such type phenomenon can lead to abrupt freezing of dielectric relaxation in the region of the highest temperature PT in good agreement with experiment [19]. Here soft cluster – soft lattice mode mixing takes place with accompanied enrichment of selection of rules for Raman scattering polarizations (with softening of sharp polarization dependences in our case). The interaction of two in-cluster relaxators, and one soft in-cluster oscillator mode on the one hand, and soft lattice oscillator-type mode on the other, lead here to the Central Peak Phenomenon with an appearance of a PT of dynamical origin proposed by us in [23]. The possibility to explain the main experimental results [19-22] of PT studies in the KLTN supports the model proposed (see also [24] for more details) and allows to consider the KLTN as a "weak" relaxor.

REFERENCES

1. Smolenskii, G.A., Isupov, V.A., Agranovskaya, A.I., and Popov, S.N., *Sov. Phys. Solid State* **2**, 2584-2594 (1960).
2. Glass, A.M., *J. Appl. Phys.* **40**, 4699- 4713 (1969).
3. Kleemann, W., Albertini, A., Chamberlin, R.V., and Bednorz, J.G., *Europhys. Lett.* **37**, 145-152 (1997).
4. Egami, T., Teslic, S., Dmowski, Davis, P.K., Chen I.-W., and Chen, H., *J. Korean Physical Society* **32**, S935-S938 (1998).
5. Tkachuk, A., Wu, Z., Chen, H., Zschack, P., Han, P., and Colla, E., *APS March Meeting*, Minneapolis, March, 2000, Bulletin of APS, Minneapolis, 2000, Abstract K 20 6, p. 510.
6. Park, S.-E., and Shrout, Th. R., *J. Appl. Phys.* **82** (4), 1804-1811, (1997).
7. Vikhnin, V.S., *Ferroel.* **199**, 25 – 40 (1997).
8. Vikhnin, V.S., *Z. Phys. Chem.* **201**, S. 201 – 213 (1997).
9. Vikhnin, V.S., *Ferroel. Lett.*, **25**, 27 – 35 (1999).
10. Vikhnin, V.S., Eglitis, R.I., Kotomin, E.A., Kapphan, S.E., and Borstel, H., *Williamsburg Workshop – 2001*, AIP Conference Proceedings, 2001, submitted.
11. Vikhnin, V.S., Blinc, R., Pirc, R., *Ferroel.* **240**, 355-360 (2000).
12. Pirc, R., and Blinc, R., *Phys. Rev.* **B 60**, 13470-1376, (1999).
13. Blinc R., to be published (2001).
14. Blinc, R., and Zeks, B., *"Dynamics of Order-Disorder Type Ferroelectrics and Antiferroelectrics,"* in *Soft Modes in Ferroelectrics and Antiferroelectrics,* edited by North-Holland, Amsterdam, Oxford, and American Elsevier, New York, 1974, pp. 125-174.
15. Takesue, N., Fujii, Y., Koyama, K., Motokawa, M., and You, H., *Ferroel.*, in press (2001).
16. Vakhrushev, S.B., and Shapiro, S., private communication (2000).
17. Kleemann, W., Bobnar, V., Dec, J., Lehnen, P., Pankrath, R., Prosandeev, S.A., *Ferroel.*, in press.
18. Vikhnin, V.S., Eglitis, R.I., Markovin, P.A., and Borstel, G., *Phys. Stat. Sol.*, (b) **212**, 53-63 (1999).
19. Trepakov, V., Savinov, M., Kapphan, S., Vikhnin, V., Jastrabik, L., and Boatner, L., *Ferroel.* **239**, 305-312 (2000).
20. Trepakov, V., Savinov, M., Kapphan, S., Licher, J., Jastrabik, L., and Boatner, L.A., Abstract O Fr A7, 222 in *ICDIM-2000*, Johannesburg, April, 2000; *Rad. Eff. Def. Solids*, at press (2001).
21. Galinetto, P., Giolotto, E., Camagni, P., Samoggia, G., Trepakov, V., and Boatner, L.A., Abstract P Th 26, 193 in *ICDIM-2000*, Johannesburg, April, 2000; *Rad. Eff. Def. Solids*, at press (2001).
22. Giulotto, E., Galinetto, P., Camagni, P., Samoggia, G., Trepakov, V.A., Jastrabik, L., and Syrnikov, P.P., *J. Phys.: Cond. Matter* **12**, 6935-6942 (2000).
23. Vikhnin, V.S., Trepakov, V.A., and Kapphan, S., *Ferroel. Lett.* **25**, 153-160 (1999).
24. Vikhnin, V.S., Kapphan, S.E., and Trepakov, V.A., *Ferroel.*, to be published.

Pressure as a Probe of Ferroelectric Properties: Quantum Regime

G. A. Samara* and L. A. Boatner[†]

*Sandia National Laboratories, Albuquerque, NM 87185-1421
[†] Oak Ridge National Laboratory, Oak Ridge, TN 37831

Quantum fluctuations can strongly influence the low temperature response of a system near a structural phase transition. Among the manifestations of quantum fluctuations at ferroelectric phase transitions are the suppression of the transition temperature, T_c, below its classical value, the emergence of a special critical point - the quantum displacive limit where $T_c = 0\ K$, and the ultimate development of a quantum paraelectric state. To study these quantum effects as well as the crossover from the classical to the quantum regime, it is necessary to shift T_c to the appropriate low-temperature range. This is conventionally done by chemical substitution, but we find that high pressure is a "cleaner" variable. After a brief summary of earlier theoretical and experimental work, results on $KTaO_3$ and KTN crystals doped with ~ 0.05 at % Ca or Ba are presented. The addition of these two dopants introduces dipolar defects into the $KTaO_3$ lattice which strongly enhance the dielectric susceptibility of the host lattice in the quantum regime. In a Ca-doped KTN crystal with 2.3 at % Nb, pressure induces a crossover from normal ferroelectric to a relaxor state which on further increase in pressure crosses over to a quantum paraelectric phase. With Ba-doping $KTaO_3$ remains a quantum paraelectric. Because the characteristic energies are small in the quantum regime, the properties are very strongly dependent on pressure and biasing electric fields. The results are discussed in terms of the physics involved.

INTRODUCTION

The occurrence of ferroelectric (FE) transitions in solids is determined by a competition between cooperative, long-range forces which try to order the system and fluctuations which favor disruption of this order. When the transition occurs at high temperature (T), i.e., in the classical regime, thermal fluctuations are at work. These fluctuations dominate in the high T phase and there is no ordering; but, on lowering T the fluctuations decrease and eventually the ordering forces win out, and the system orders at a transition temperature, T_c. On the other hand, if the transition occurs at sufficiently low T, quantum fluctuations, or zero-point motions, come into play, and they can strongly influence the response of the system.

Starting with the early work of Barrett,[1] there have been considerable theoretical[2-6] and experimental[7-14] efforts devoted to the study of the manifestations of quantum effects on ferroelectric behavior. Among these manifestations are new critical exponents, the suppression of T_c below its classical value, and ultimately the complete suppression of T_c and the formation of a quantum paraelectric state.

As there are no known naturally occurring ferroelectrics with T_c's sufficiently low to make the full range of quantum effects discernible, experimental study of these effects necessitates that T_c be shifted down to the appropriate range by application of external fields. Ideally such fields should not change the symmetry of the high temperature phase. Hydrostatic pressure is an excellent such field and we have used it successfully earlier to study these effects in a variety of crystals.[11,16] In fact, these studies yielded the first true quantum ferroelectric (KH_2PO_4 at ~ 17 kbar) and the first phase diagram for such a crystal.[10,15,16] Another "external" field is chemical substitution which yields mixed crsytals.[8-10,13] However, this substitution introduces randomness, compositional fluctuations, and clustering and breaks translational symmetry. These effects undoubtedly change some of the interaction parameters of the system, and it is not clear that they can be neglected.

In this paper we discuss these quantum effects in ABO_3 perovskite ferroelectrics. These materials undergo ferroelectric transitions driven by soft, long-wavelength transverse optic phonons, or soft FE modes.[17] On cooling from the high temperature, high symmetry phase, the frequency of the soft mode decreases and ultimately vanishes (for a second-order transition) at T_c, thereby transforming the crystal to the low temperature FE phase. Our emphasis will be on the model system, Nb-substituted potassium tantalate, $KTa_{1-x}Nb_xO_3$ or (KTN). Although in pure $KTaO_3$ the high temperature phase is just barely stabilized by quantum fluctuations, and the crystal does not undergo a phase transition down to ~ 0 K, the substitution of Nb for Ta and Na for K yields mixed crystals, KTN and $K_{1-y}Na_yTaO_3$ (or KNaT) with broad ranges of transition temperatures.[9,11] By increasing pressure on samples of fixed composition, or alternatively by adjusting x and y at zero pressure, the transition temperatures can be lowered to ~ 0 K: for this reason, KTN and KNaT have been used as model systems to study the onset of quantum effects.[8-11] Similarly the system $Sr_{1-x}Ca_xTiO_3$ has attracted recent interest.[12,13] Among the crystals we investigated were a Ca-doped KTN crystal (0.055 at. % Ca) and a Ba-doped $KTaO_3$ crystal (0.056 at. % Ba). The presence of Ca^{2+} and Ba^{2+} ions in these samples enhances the polarizabilities of the host lattices and introduces lattice defects, but does not qualitatively change the observed quantum effects.

BACKGROUND AND THEORETICAL CONSIDERATIONS

Inherent in the soft mode concept for FE transitions is the premise that the crystal is unstable in the harmonic approximation with respect to the soft mode.[17] Specifically, the square of the harmonic frequency, ω_o^2, is presumed to be sufficiently negative (i.e., ω_o is imaginary) that this mode cannot be stabilized by zero-point anharmonicities alone. Thermal fluctuations then renormalize ω_o and make it real at finite temperatures, thereby stabilizing the lattice. The transition temperature is the temperature where the renormalization is complete. Formally, the anharmonic crystal potential can be written as a series expansion in the displacements (or normal mode coordinates), and perturbation or self-consistent treatments are then used to solve for the normal modes. Considering only quartic anharmonicities for the purposes of the discussion to follow, self-consistent treatments yield[17] for the renormalized frequency of mode j with wavevector q:

$$\omega_T^2(jq) = \omega_0^2(jq) + \sum_{\mu k} g_{j\mu}^{(4)}(qk) \times \frac{1}{2\omega(\mu k)} \coth\frac{\omega(\mu k)}{2k_B T}, \quad (1)$$

where $g_{j\mu}$ are effective fourth-order coupling constants and the summation is over all modes μ and wavevectors k At suitably high temperatures, the second term on the right-hand side of Eq. (1) can be expected to vary linearly with T so that

$$\omega_T^2(jq) = \omega_0^2(jq) + \alpha T, \quad (2)$$

where α is a positive constant. This linear T dependence is observed experimentally.

Some of the effects of quantum fluctuations on the dielectric (or soft-mode) response are shown schematically in Fig. 1. As noted above, in the high temperature classical regime the soft mode is stabilized by thermal fluctuations, i.e., the αT term in Eq. (2). These fluctuations decrease with decreasing T and ultimately the stabilization vanishes at some T_c, the classical transition temperature designated by T_c^{cl}, (curve A). The influence of quantum fluctuations is contained in Eq. (1). Specifically, when T_c dips into the regime of zero-point motion, decreasing T does not appreciably decrease the total fluctuations. Consequently, the high temperature paraelectric phase will extend below its classical limit, i.e., the transition temperature in the quantum regime, designated by T_c^q, falls below T_c^{cl} (curve B). Ultimately, at low enough temperatures, zero-point fluctuations can suppress the occurrence of the phase transition altogether (curve C).

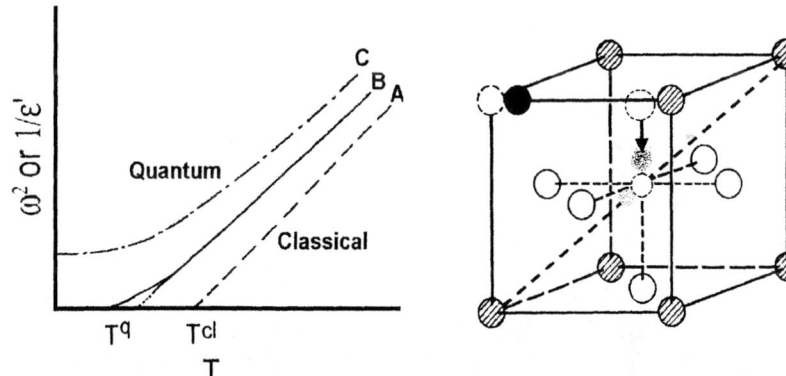

Fig. 1 The suppression of the FE phase transition temperature by quantum fluctuations.

Fig. 2 The cubic perovskite structure of KTaO$_3$ and some defects produced by Nb^{5+}, Ca^{2+} and Ba^{2+} substitutions.

One of the consequences of the suppression of the phase transition is the presence of a special critical point, namely $T_c^q = 0$ K. This point, which is referred to as the *quantum displacive limit*, is characterized by special critical exponents.[2-4]

Another consequence of the complete suppression of the phase transition is the formation of a quantum paraelectric state. This state is characterized by a high static dielectric susceptibility (or low soft-mode frequency) which is constant at low T over an extended range of temperatures. $SrTiO_3$ and $KTaO_3$ exhibit such a state at atmospheric pressure[7,18] and other systems exhibit it at high pressure.[11,19]

Historically, there have been three periods of interest in quantum effects in ferroelectrics. The first, around 1950, was motivated by the contrast between the dielectric, $\varepsilon'(T)$, response of $BaTiO_3$, which obeys the Curie-Weiss law $\varepsilon' = \varepsilon'_\infty + C/(T-T_o)$, and the $\varepsilon'(T)$ responses of $SrTiO_3$ and $KTaO_3$ which deviate markedly from such a law at low temperatures. Barrett[1] explained this deviation in terms of an extension into the quantum regime of Slater's[20] model of the ionic polarizability of $BaTiO_3$ – a model based on a mean-field classical statistical ensemble of anharmonic oscillators. Barrett derived the following expression for ε'

$$\varepsilon' = A + \frac{B}{\frac{1}{2}T_1 \coth(T_1/2T) - T_o}, \quad (3)$$

where the lowest quantum level for the oscillating ion has an energy $= kT_1$, so that for $T < T_1$, ions occupy their lowest energy states and further reduction in T does not change $\varepsilon'(T)$. Equation (3) has been widely used to fit experimental data, the constants being treated as empirical parameters. Note that Eq. (3) reduces to $\varepsilon' = A + B/\left(\frac{1}{2}T_1 - T_o\right) =$ constant as $T \to 0$ K and to $\varepsilon' = A + B/(T-T_o)$, a Curie-Weiss law, at high temperatures. Thus, T_o is the Curie-Weiss temperature and T_1 is the temperature below which deviations from Curie-Weiss behavior set in.

In our group, Abel[21] sometime ago used Eq. (3) to successfully fit $\varepsilon'(T)$ data for $KTaO_3$ at 1 bar and elevated pressure. The fits showed that B and T_o decrease with pressure as expected[17] whereas T_1 increases. The increase in T_1 with pressure can be qualitatively understood in that increasing pressure confines the polarizable ions to smaller volumes, thus raising the energy of the lowest quantum level, and hence T_1. A more rigorous test of Eq. (3) was provided by Muller and Burkard[7] for their 1 bar $\varepsilon'(T)$ data on $SrTiO_3$. They found that the fit misses the data significantly in different temperature regimes depending on the choice of fitting parameters.

The second period of interest in quantum effects in ferroelectricity was a period (about 1975-1985) of extensive theoretical[2-4] and experimental[7-11,15,19] research. The theoretical work, due largely to Schneider and Thomas and their collaborators,[2-4] examined rigorously quantum suppression of the FE transition and the response at and near the displacive limit, $T_c^q = 0$ K, on the basis of lattice dynamic models solved within the framework of both classical and quantum statistical mechanics. Among the results of the models are the following:[2-4,11]

- The transition temperature is given by $T_c(S) \propto (S - S_{min})^{1/\phi}$, where S is a general interaction parameter which is proportional to the mean square fluctuations of the ionic displacements, and S_{min} is the value of S for $T_c^q = 0$ K. Near $T_c^q = 0$ K, $\phi = 2$, whereas for large S, i.e., in the classical regime, $\phi = 1$.
- ε' is given by $\varepsilon' - \varepsilon'_\infty = C(T - T_c)^{-\gamma_T}$ where in the classical limit $\gamma_T = 1$, and at the quantum displacive limit $\gamma_T = 2$.

A good deal of the experimental work on quantum ferroelectrics during this time period was aimed at testing theoretical predictions. To do so it is necessary to relate S to some measurable variable such as pressure, p, and chemical composition, x. As a first approximation, it may be assumed that, for small changes in lattice parameter a, $S \propto a$.[8,11] It can also be assumed that for relatively small changes in p and x, a is a linear function of p and x. Under these assumptions S is proportional to p and x, and we can replace S and S_{min} by x and x_c and p and p_c in the above equations. Here p_c is the value of p at $T_c^q = 0$, which corresponds to $S = S_{min}$.

Experimental[8-11,19] results have generally confirmed the above predictions. Work as a function of composition at 1 bar focused on KTN and KNaT,[8-10] whereas work as a function of pressure at fixed composition has examined a wide range of materials including KTN and KNaT,[11] KH_2PO_4 and $NH_4H_2PO_4$,[11,16] K_2SeO_4[19] and SbSI.[22]

The third period of renewed interest in quantum effects in ferroelectrics is the present. It is motivated by continued advances in modeling[6] and experimental results, primarily the work of Kleeman and collaborators[12,13,23,24] on $Sr_{1-x}Ca_xTiO_3$. On the modeling side, Salje and co-workers[6] have used a modified Landau theory to determine the behavior of the phase diagram in the quantum regime. To account for the influence of pressure or composition on the free energy and phase behavior, a term which couples these variables to the order parameter was introduced. The treatment leads to the following equation for the pressure dependence of the transition temperature (for a second-order transition)

$$T_c(p) = \frac{\theta_S}{\coth^{-1}\left[\coth\left(\frac{\theta_S}{T_c}\right) - kp\right]}, \qquad (4)$$

with a similar equation for $T_c(x)$ with x replacing p. Here θ_S characterizes the temperature of the crossover between the classical and quantum regimes and k is a constant. Equation 4 yields a good fit[6] of experimental data on KH_2PO_4, KTN, $SrTiO_3$ and SbSI.

RESULTS ON A MODEL SYSTEM: CHEMICALLY SUBSTITUTED POTASSIUM TANTALATE (KTaO$_3$)

Niobium-substituted KTaO$_3$, or KTN, in the low Nb limit has been an important system for studying quantum effects in perovskites. Hochli[8,10] and Rytz[9] and their collaborators have investigated this system as a function of composition at 1 bar, and we[25] have investigated fixed compositions at high pressure. We have now extended the pressure studies to KTN samples doped with Ca and Ba. We present below some of the results on a KTN crystal with 2.3 at. % Nb and doped with 0.055 at. % Ca and on a nominally pure KTaO$_3$ (KT) crystal doped with 0.056 at. % Ba. The samples were investigated by dielectric spectroscopy with measurements of the real (ε') and imaginary (ε'' or the dielectric loss tanδ) parts of the dielectric constant performed as functions of frequency ($10^2 - 10^6$ Hz), temperature and hydrostatic pressure. Helium was the pressure transmitting medium. Some measurements were performed under a biasing dc electric field to the sample.

Figure 2 shows the cubic structure of KT and the various possible defects produced by Nb^{5+} and Ca^{2+} and Ba^{2+} substitutions. It is now well established that in substituting for Ta^{5+}, Nb^{5+} occupies an off-center position along the <111> direction leading to the many interesting low-temperature properties of KTN.[25,26] The nature of Ca^{2+} and Ba^{2+} dopants is much less clear. In an earlier paper[26] we argued that Ca^{2+} most likely substitutes at the B site; however, a recent first-principles study by Leung[27] suggests that Ca^{2+} may substitute at both the A and B sites. For Ba the large size of Ba^{2+} suggests that the primary substitution is at the A site. Actually, for the purposes of the present study it is simply sufficient to note that both Ca^{2+} and Ba^{2+} substitution produce dipolar entities which greatly enhance the polarizability, and thereby ε', of the host KTN lattice.

Experimentally, Nb^{5+} and Ca^{2+} substitutions produce remarkable changes in the dielectric properties of KT as discussed elsewhere.[26] Modest pressure has a strong influence on the $\varepsilon'(T)$ response as shown in Fig. 3. The first major effect is a FE-to-Relaxor crossover.[26] Below ~ 6 kbar the crystal transforms into a FE phase, but above this pressure the transition is to a relaxor state with its characteristic strong frequency dispersion. The second major effect in Fig. 3 is the vanishing of the transition, the feature of interest for the present purposes. At 9.2 kbar there is no evidence in the $\varepsilon'(T)$ data, or in $\varepsilon''(T)$, of any transition or impending transition. Rather, the data indicate that the crystal is a quantum PE at these conditions with ε' independent of T below ~ 8 K.

Analysis of the $\varepsilon'(T)$ data for this crystal shows agreement with theoretical predictions. In particular, Fig. 4 shows (at 10^5 Hz) plots of log $1/(\varepsilon' - \varepsilon'_\infty)$ vs. log ($T - T_c$), where T_c corresponds to the peak value of ε' in both the FE and relaxor phases, and $T_c = 0$ for the 9.2–kbar data. Because of the increased losses associated with the hopping of the Ca-related defect,[26] the $\varepsilon'(T)$ data above, T_c extend over a limited range only, but nevertheless the results in Fig. 4 show the predicted behavior. At 1 bar and 8.3 kbar T_c is sufficiently far away from the quantum displacive limit

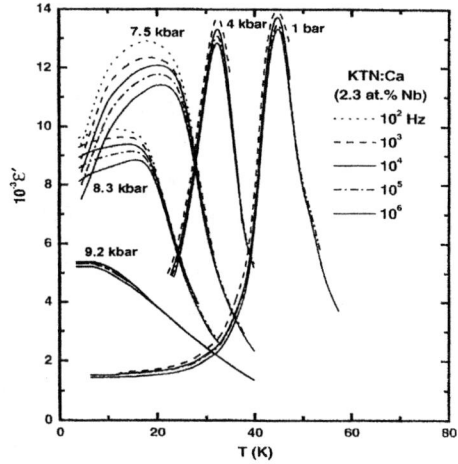
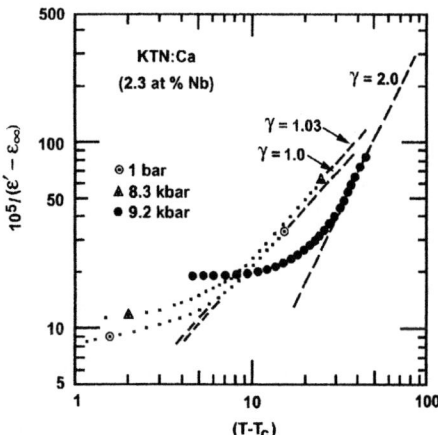

Fig. 3 Temperature dependence of ε' of KTN:Ca at different frequencies and pressures.

Fig. 4 Log-log plot of $(\varepsilon' - \varepsilon'_\infty)^{-1}$ vs. reduced temperature for KTN:Ca at 10^5 Hz and different pressures.

($T_c = 0$ K) so that $\gamma_T \simeq 1.0$; however, 9.2 kbar is very close to this limit, and we see the emergence of a distinct $\gamma_T = 2.0$ regime in the response, as predicted. The flat response in the data emphasizes the quantum paraelectric nature of the crystal at 9.2 kbar.

Because the characteristic energies are small in the quantum regime, dc biasing electric fields have a strong influence on the dielectric response. Such fields stabilize the local potential well of the off-center ion. The results are a shift of the transition temperature, or $\varepsilon'(T)$ peak, to higher temperatures with increasing field, suppression of the magnitude of ε' and a large reduction in the dielectric loss, $\tan\delta$. Some results are shown in Figs. 5 and 6. At 5.6 kbar, the transition is to a normal FE state at low temperatures. The 0.6 kV/cm field further stabilizes the FE phase, and the thermal hysteresis between the field heating (FH) and field cooling (FC) scans emphasizes the first-order nature of the transition. At 8.3 kbar (not shown), however, the transition is to a relaxor state, and the biasing field restabilizes the FE phase while shifting T_c to higher temperatures. In the quantum PE state, the 9.2-kbar data in the inset in Fig. 6 the main effect of the biasing field is a relatively large suppression of the magnitude of ε'. At higher fields than the 0.85 kV/cm in Fig. 6 it is likely that the relaxor phase will emerge.

Figure 7 shows the dielectric response of the Ba-doped KT crystal. Several things are immediately noticeable. First is the quantum PE nature of the response in the low temperature constant ε' region extending over a relatively wide temperature range. Secondly, we note the large enhancement of ε' compared with that of pure KT shown in Fig. 3. This is associated with high polarizability of the Ba dipolar entity. In computer simulations on Ca-doped $SrTiO_3$, Kleemann et al[13] find a large peaking of the susceptibility at the site of the Ca dopant. This peaking averages out over the whole sample leading to a large enhancement in ε'. Undoubtedly, a similar behavior occurs in KT:Ba. Thirdly, in Fig. 7, we note the fairly large suppression of ε' with pressure. This is the expected behavior for a soft-mode system.[17]

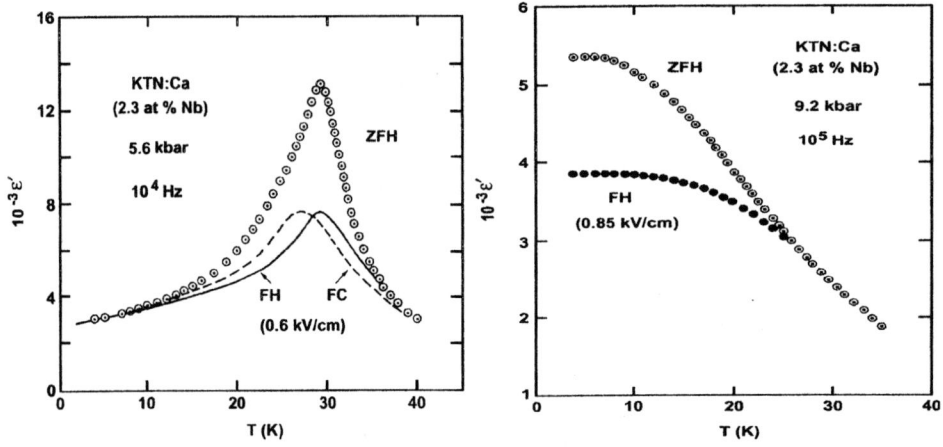

Fig. 5 Influence of a bias field on $\varepsilon'(T)$ of KTN:ca at 5.6 kbar

Fig. 6 Influence of a bias field on $\varepsilon'(T)$ of KTN:Ca at 9.2 kbar.

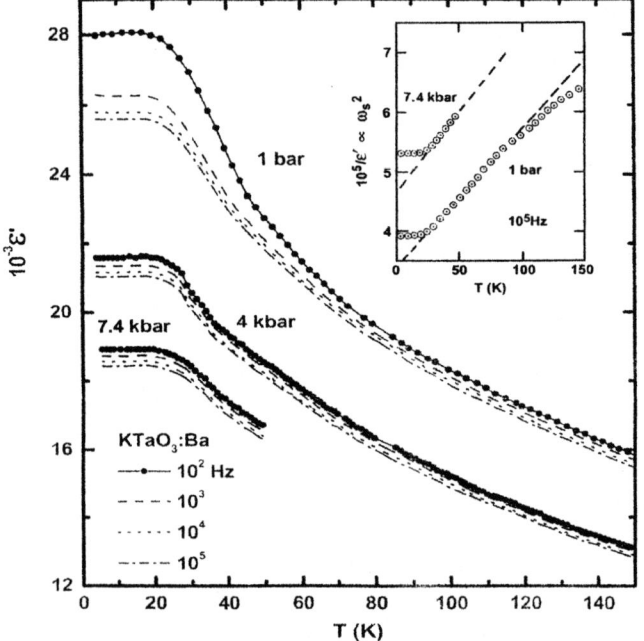

Fig. 7 Temperature dependence of ε' of KT:Ba at different pressures and frequencies. The inset shows ε'^{-1} vs. T plots.

The inset in Fig. 7 shows $1/\varepsilon'$ vs. T plots for KT:Ba. Above the low temperature flat response $\varepsilon'(T)$ follows a Curie-Weiss law over a limited temperature range above which deviations set in. We believe that the Curie-Weiss regime is dominated by the response of the host lattice, whereas the high temperature deviations are related to the Ba doping and the increased conductivity of the sample.

UNIQUE MANIFESTATION OF THE NON-EQUILIBRIUM NATURE OF THE GLASS TRANSITION

One of the most distinctive pressure effects is the finding that, for the pressure-induced suppression of the glassy state, the kinetic glass transition temperature appears to vanish with a finite slope,[25] i.e., dT_m/dP is finite as $T_m \rightarrow 0~K$. This effect is in marked contrast with the behavior of FE and AFE crystals where it is found that the transition temperatures (T_c and T_N) vanish with an infinite slope, i.e., $dT_{c,N}/dP \rightarrow -\infty$ as $T_{c,N} \rightarrow 0~K$. We believe that the behavior of the glasses is a unique manifestation of the non-equilibrium (or glassy) nature of the transition, a point that we made earlier[25] but is worth repeating.

That the above conclusion is so, follows from the third law of thermodynamics according to which the difference in entropy (ΔS) between the two phases involved in an *equilibrium phase transition* goes to zero at $T \rightarrow 0~K$. The consequence of this fact is that $dT_c/dP \rightarrow -\infty$ as $T_c \rightarrow 0~K$. This can be readily seen from both the Clausius-Clapeyron equation and the Ehrenfest equation for a second-order transition.[25] The observation that T_m vanishes with a finite slope for the glasses can then be taken to be a strong indication of the non-equilibrium nature of the kinetic glass transition. The finite slope as $T_m \rightarrow 0~K$ results from the residual configurational entropy in the glassy phase. This is a general result evidence for which comes also from the spin-glass literature.[25]

CONCLUDING REMARKS

Quantum fluctuations, or zero-point motions, strongly influence *FE* properties. Among their manifestations are: new critical exponents, the suppression of T_c below its classical value, the ultimate vanishing of T_c ($T_c = 0~K$ is the Quantum Displacive Limit), and the formation of a quantum *PE* state. We have pointed out and demonstrated that pressure is the "clean" variable for reducing T_c to 0 K and studying these quantum effects. We have also shown that electrical bias has a strong influence on the properties in the quantum regime where characteristic energies are small. Finally, we discussed a unique manifestation of the glass transition in the quantum regime revealed by pressure experiments.

ACKNOWLEDGEMENTS

This work was supported by the Division of Materials Sciences, Office of Basic Energy Sciences, United States Department of Energy and by Sandia's Research

Foundations under Contract No. DE-AC04-AL85000 at Sandia National Laboratories and by the Division of Materials Sciences under Contract No. DE-AC05-96OR-22464 at Oak Ridge National Laboratory

REFERENCES

1. Barrett, J. H., *Phys. Rev.* **86**, 118 (1951).
2. Opperman R., and Thomas, H., Z., *Phys. B* **22**, 387 (1975).
3. Schneider, T., Beck, H., and Stoll, E., *Phys. Rev. B* **13**, 1123 (1975); *B* **12**, 5198 (1975).
4. Morf, R., Schneider, T., and Stoll, E., *Phys. Rev. B* **16**, 462 (1977).
5. Schmeltzer, D., *Phys. Rev. B* **28**, 459, 1983; *B* **29**, 2815 (1984).
6. Hayward, S. A., and Salje, E. K. H., *J. Phys. Cond. Matt.* **10**, 1421 (1998) and references therein. See also Perez-Mato, J. M., and Salje, E. K. H., *J. Phys. Cond. Matt.* **12**, L 29 (2000).
7. Muller, K. A., and Burkard, H., *Phys. Rev. B* **19**, 3539 (1979).
8. Hochli, U. T., and Boatner, L. A., *Phys. Rev. B* **20**, 266 (1979).
9. Rytz, D., Hochli, U. T., and Blitz, H., *Phys. Rev. B* **22**, 359 (1980).
10. Hochli, U. T., *Ferroelectrics* **35**, 17 (1987).
11. Samara, G. A., *Physica B* **150**, 179 (1988). See also Samara, G. A., in *High Pressure in Research and Industry*, Proc. 8th AIRAPT Conf. Uppsala, Sweden, 305 (1981); G. A. Samara, Proc. Int. Symp. On Solid State Phys. Under Pressure, S. Minomura, ed. (Reidel, Boston), 183 (1985).
12. Kleeman, W., Dec, J., and Westwanski, B., *Phys. Rev. B* **58**, 8985 (1998).
13. Kleeman, W., Dec, J., Wang, Y. G., Lehnen, P. and Prosandeev, S. A., *J. Phys. Chem. Solids*, **61**, 167 (2000).
14. Bednorz, J. G., and Muller, K. A., *Phys. Rev. Lett.* **52**, 2289 (1984).
15. Muller, K. A., *Jap. J. Appl. Phys.* **24**, Supplement 24-2, 89 (1985), and references therein.
16. Samara, G. A., *Phys. Rev. Lett.* **27**, 103 (1971); see also *Phys. Rev. B* **8**, 2033 (1973) and *Ferroelectrics* **22**, 925 (1979).
17. Samara, G. A., and Peercy, P. S., in *Solid State Physics*, edited by H. Ehrenreich, F. Seitz and D. Turnbull, Vol. **36**, p. 1, Academic Press, NY (1981).
18. Salce, B., Gravil, J. L., and Boatner, L. A., *J. Phys. Cond. Matt.* **6**, 4077 (1994).
19. Samara, G. A., Massa, N. E., and Ullman, F. G., *Ferroelectrics* **36**, 335 (1981).
20. Slater, J. C., *Phys. Rev.* **78**, 748 (1950).
21. Abel, W. R., *Phys. Rev. B* **4**, 3697 (1971).
22. Samara, G. A., *Ferroelectrics* **9**, 209, (1975).
23. Bianchi, U., Kleeman, W., and Bednorz, J. G., *J. Phys. Cond. Matt.* **6**, 1229 (1994).
24. Wang, Y. G., Kleeman, W., Dec, J. and Zhong, W. L., *Europhysics Letters* **42**, 173 (1998); *Ferroelectrics* **229**, 39 (1999).
25. Samara, G. A., *Ferroelectrics* **117**, 347 (1991).
26. Samara, G. A., and Boatner, L. A., *Phys. Rev. B* **61**, 3889 (2000).
27. Leung, K., *Phys. Rev. B* **63**, (in press).

Bound Charge Diffusion and Polar Nanocluster Dynamics in Proton Glass Crystals

V. Hugo Schmidt

Physics Department, Montana State University, Bozeman, MT 59717

Abstract. Slater proposed that KH_2PO_4 (KDP) has 2 polar and 4 nonpolar (with higher energy E_b) H_2PO_4 H-bond configurations. Takagi postulated H_3PO_4 and HPO_4 groups with still higher energy E_a. Via intrabond H transfer, Takagi pairs form, diffuse independently, and recombine or else annihilate with new partners, allowing H-bond reconfiguration. In both KDP-type ferroelectrics and in proton glasses such as $Rb_{1-x}(NH_4)_xH_2PO_4$ (RADP), at higher temperatures there are many Takagi groups. They have short life and are likely to annihilate with new partners, leading to nearly Debye-like dielectric relaxation. Below the ferroelectric transition in KDP, each Takagi pair member requires energy E_b to diffuse one step farther from its partner. This energy bias leads to their rapid recombination. In proton glasses with intermediate x, there is no ferroelectric or antiferroelectric transition. Well above the freezing range, the Takagi concentration falls rapidly with temperature and the Takagi group lifetime and path to annihilation grow long. The energy landscape of the annihilation path must be level on the average. However, the random energy steps $\pm E_b$ or 0 make the highest barrier encountered be proportional (on average) to the square root of the diffusion path length. As temperature falls, this diffusion inhibition makes recombination more likely than annihilation with new partners. The Takagi pair lifetime becomes shorter, but the fluctuations resulting from their diffusion become biased, so their contribution to dielectric and NMR relaxation weakens. In this same temperature range, the Takagi group population temperature dependence grows weak, because many Takagi groups become stranded, finding no annihilation partner. These stranded groups cause relaxation in their vicinity but no relaxation elsewhere, so the crystal is nonergodic on the time scale concerned. 2D NMR shows the mean (averaged over the crystal) time required for a given nucleus to see a change caused by such diffusion.[1] We propose 4D NMR[2] to test whether this time is different in different crystal nanoregions because of nonergodicity, and to find the time required to restore ergodicity. Another proposed application of 4D NMR is for proton glass crystals with small x, for which ferroelectric nanoclusters coexist with disordered proton glass regions at low temperatures.[3] We expect that 4D NMR can find the size and lifetime of such polar nanoclusters. Such experiments could also study polar nanocluster size and lifetime in relaxor ferroelectric and related high-piezoelectric-strain crystals.

INTRODUCTION

Structural Considerations

Polarization dynamics in ferroelectric and related materials is of considerable interest for theory and applications. In most such materials the microscopic mechanism for polarization change is difficult to characterize, but in H-bonded crystals such as proton glasses there is a simple mechanism that has been widely accepted for nearly half a century.

To explain this mechanism, we first review the structure of KH_2PO_4 (KDP) and its isomorphs. The phosphate ions are arranged topologically in a diamond lattice. The cubic symmetry is broken because adjacent phosphate ions are linked by an O-H···O bond that is nearly perpendicular to the c axis, and because potassium ions alternate with phosphate ions in chains running along c. This gives a tetragonal structure in the paraelectric (PE) phase above the ferroelectric (FE) transition temperature T_c.

Neutron diffraction showed that the hydrogen bonds in the FE phase are asymmetric, with the protons around a given phosphate ion obeying the Pauling ice rules. The first rule is that exactly one proton is in each of the four bonds linking a phosphate to its four neighbors. The second rule is that each of these four protons is in an off-center position in its bond, in such a way that each phosphate ion has two close protons and two far protons in these four bonds. These two rules alone allow a great deal of disorder, or entropy, in the proton configuration, and such disorder occurs in normal ice at all temperatures. In KDP in the FE phase, the protons are ordered. Relative to the c axis, the close protons are at the bottom (top) of the phosphate ion in an up (down) polarized FE domain.

In $NH_4H_2PO_4$ (ADP) the structure is isomorphic to that of KDP except that each ammonium ion forms four N-H···O bonds to four neighboring phosphate ions. The low-temperature phase is antiferroelectric (AFE) and neutron diffraction shows that the O-H···O bonds are ordered with one close proton at the top, and one at the bottom, of each phosphate ion. There are four such ordered arrangements, corresponding to the four types of AFE domains.

In mixed "proton glass" crystals such as $Rb_{1-x}(NH_4)_xH_2PO_4$ (RADP), for $0.22 < x < 0.74$ there is no FE or AFE transition. Instead, the dynamics slow down with decreasing temperature and a spread in time constants develops, with the system finally becoming nonergodic. This behavior is similar to that of spin and structural glasses, hence the "proton glass" name. Outside this x range, for lower x there is coexistence of FE and PE phases, and for higher x the AFE and PE phases coexist. This results from quenched inhomogeneity in the local ammonium/rubidium concentration ratio.

Considerable experimental evidence supports the idea that in the PE phases of both KDP and ADP, and at all temperatures in proton glass, the ice rules are still obeyed in a dynamic sense. This means that at a given instant, almost all phosphate ions will have two close O-H···O-bond protons, but which of the two protons are close will change with time. We now examine the mechanism for this change.

Dynamical Process

Slater[1] based his brilliant theory for the FE phase transition in KDP on Pauling's ice rule hypothesis, which had not yet been verified for ice, let alone KDP. His theory postulated that of the six arrangements of two protons in the four bonds around a phosphate ion, those two giving FE configurations have zero energy, while the other four have a positive energy ε_0. His model predicted an abrupt FE transition, with any polarization possible exactly at T_c, so one could say the transition is frozen at a tricritical point. His model provided no mechanism for proton rearrangement, except highly improbable *en masse* movement.

Takagi[2] proposed that the ice rules could be broken, at a cost of energy $\varepsilon_1 \gg \varepsilon_0$ for each HPO_4 and H_3PO_4 group created (now called Takagi groups in his honor). His motivation was to explain the observed somewhat rounded and gradual increase of polarization below T_c. A more important result of his proposal was its provision of a dynamic mechanism, consisting in creation of Takagi group pairs, their diffusion, and eventual recombination or annihilation. We now examine this mechanism.

The creation process is described by the reaction

$$H_2PO_4 + H_2PO_4 \leftrightarrow HPO_4 + H_3PO_4, \tag{1}$$

where the reverse arrow describes the recombination or annihilation process. A Takagi group can *in effect* diffuse by means of proton intrabond transfer, according to the reactions

$$HPO_4 + H_2PO_4 \leftrightarrow H_2PO_4 + HPO_4 \text{ and } H_3PO_4 + H_2PO_4 \leftrightarrow H_2PO_4 + H_3PO_4. \tag{2}$$

We distinguish recombination of two original pair partners from annihilation with new partners. The creation, diffusion (or drift in an electric field), and eventual annihilation induced us to call this process "bound charge semiconduction [3]," where "*bound*" refers to the fact that the actual charge motion is confined to protons moving *within* their bonds. However, the Takagi group carries a fractional positive or negative charge, and so will drift in an electric field. This process is not to be confused with protonic semiconduction, which occurs in all H-bonded crystals but with much higher activation energy.

MODEL

Basic Considerations

We are concerned with the carrier density and mobility, which can be deduced from dielectric, NMR and other measurements. To determine the density, we must equate the creation and annihilation rates. In equilibrium, the relation

$$4N\nu \exp(-2\varepsilon_1/kT) - n/t_l = 0 \tag{3}$$

must hold, where $2N$ is the density of creation sites (of O-H···O bonds), ν is the attempt frequency, n is the Takagi group density, and t_l is the mean carrier lifetime.

This lifetime is governed at high temperature by diffusion distance to annihilation, and at low temperature by random barriers that must be crossed on the path to annihilation. Expressions for each case will now be derived.

The distance-limited lifetime t_d is based on the approximation that barriers can be neglected at "high" temperature. Each step of the Takagi group to a new site is assumed to require time ν^{-1}. The topology of the H-bond network (over short distances) is such that a Takagi group has three choices for its next site, namely its previous site or two new sites. The double branching of this "Cayley tree" or "Bethe lattice" allows the carrier a high probability of escape from its original partner, once it has diffused a few steps away. Its probability of finding an annihilation partner is roughly n/N for each step, so its mean time to annihilation is $t_d = N/n\nu$. Inserting this t_d for t_l in Eq. (3) yields

$$n(hi\text{-}T) = 2N\exp(-\varepsilon_1/kT). \tag{4}$$

This result is familiar from intrinsic semiconductors; the carrier concentration activation energy is half the band gap (here the carrier pair creation energy).

This equation, if valid at low temperature, would predict negligible carrier concentration and nearly no carrier-mediated dynamics, in contradiction to experiment. A more correct approach at low temperatures is to consider the random barriers encountered on the annihilation path. What is the origin of these barriers? In KDP-type crystals at low temperatures, about half the H_2PO_4 groups are polar and have energy 0, while the other half are nonpolar and have the Slater energy ε_0. Accordingly, one step of a Takagi group may raise or lower the configurational energy by ε_0 or may leave it unchanged. We consider a typical basic diffusion process of four steps with energy changes represented by $\wedge$, that is, with a basic barrier of height ε_0. In following an annihilation path of N/n steps, if the path on average is level, the highest barrier encountered is in the range $\varepsilon_0(N/4n)^{1/2}$. The average path must be level, otherwise the configurational energy would increase or decrease with time as it exchanges energy with the phonon reservoir. We can say that the diffusing Takagi group encounters a fractal energy landscape.

What is the lifetime t_b as thus limited by barriers? As a rough approximation, it can be determined entirely by the highest barrier, so that $t_b = \nu^{-1}\exp[(N/4n)^{1/2}\varepsilon_0/kT]$. We now have two lifetime expressions, t_d and t_b, for the carrier lifetime t_l. The longer of the two at any given temperature should be closer to the correct one, so it seems a good approximation to set

$$t_l = t_d + t_b = N/n\nu + \nu^{-1}\exp[(N/4n)^{1/2}\varepsilon_0/kT]. \tag{5}$$

Inserting this t_l into Eq. (3) provides the following transcendental expression for finding the carrier concentration n at any temperature,

$$4N\nu\exp(-2\varepsilon_1/kT) - n/\{N/\nu n + \nu^{-1}\exp[(N/4n)^{1/2}\varepsilon_0/kT]\} = 0. \tag{6}$$

The high-temperature limit, found by ignoring t_b, appeared already in Eq. (4).

The low-temperature limit, found by ignoring t_d in Eq. (5), gives the following two equivalent transcendental equations for n(lo-T),

$$\exp\{[-2\varepsilon_1 + (N/4n)^{1/2}\varepsilon_0]/kT\} = n/4N, \quad -2\varepsilon_1 + (N/4n)^{1/2}\varepsilon_0 = kT\ln(n/4N). \tag{7}$$

From the second expression it is seen that as T goes to zero, n goes to

$$n(zero\text{-}T) = (\varepsilon_0/4\varepsilon_1)^2 N. \tag{8}$$

What is the crossover temperature T_x at which the n(lo-T) expression of Eq. (7) becomes more accurate than the n(hi-T) expression of Eq. (4)? This T_x can be estimated by setting the n(hi-T) and n(zero-T) expressions from Eqs. (4) and (8) equal at T_x, and solving for T_x,

$$2N\exp(-\varepsilon_1/kT_x) = (\varepsilon_0/4\varepsilon_1)^2 N, \quad T_x = \varepsilon_1/k\ln(32\varepsilon_1^2/\varepsilon_0^2). \tag{9}$$

To find a numerical value for T_x and other parameters, we choose ε_0 and ε_1 values that are typical for deuterated FE and AFE KDP-type crystals and deuteron glass crystals such as $Rb_{1-x}(ND_4)_x D_2 PO_4$ (DRADP), namely $\varepsilon_0/k = 140$ K and $\varepsilon_1/k = 940$ K. With these values, T_x from Eq. (9) becomes 129 K as the temperature at which the Takagi group concentration stops falling exponentially with temperature and start to level out. Insertion of these energy parameters into Eq. (6) provides the fractional Takagi group population n/N, whose inverse N/n is the approximate number of diffusion steps from creation to annihilation. A semilogarithmic plot of N/n *vs.* 1/T appears in Fig. 1.

Distribution In Time Constants

The value of N/n is significant in finding the dielectric relaxation time distribution, which should be about the same as the correlation time distribution for the electric field gradient (efg) fluctuations seen by deuterons (an important parameter for deuteron NMR). The maximum value of this time, τ_{max}, is approximately the mean time required for a Takagi group to traverse its path of N/n steps from creation to annihilation. The justification for this statement is that in this time τ_{max} there are n Takagi groups per unit volume that each on the average have visited N/n sites, so about N sites (all the sites in the unit volume) have a chance to rearrange during this time. For NMR, this means that the deuterons have lost memory of their original off-center positions in their bonds, so this is the correlation time for the efg fluctuations seen by them. For dielectric relaxation, this means that all the deuterons have a chance to redistribute their intrabond positions in response to, say, an electric field step.

From this discussion and the text associated with Eq. (5), we see that $\tau_{max} = t_1$ as given in Eq. (5). A semilogarithmic plot of τ_{max} *vs.* 1/T appears in Fig. 2. At higher temperatures, τ_{max} increases exponentially with 1/T as T decreases because the Takagi concentration n falls exponentially with 1/T, while the mobility is nearly independent of T. At lower temperatures, roughly below T_x, the concentration temperature dependence is becoming weak, but the mobility is developing temperature dependence because kT is becoming small compared to ε_0. Accordingly, τ_{max} keeps increasing even below T_x, in fact with nearly the same slope in Fig. 2 as seen above T_x.

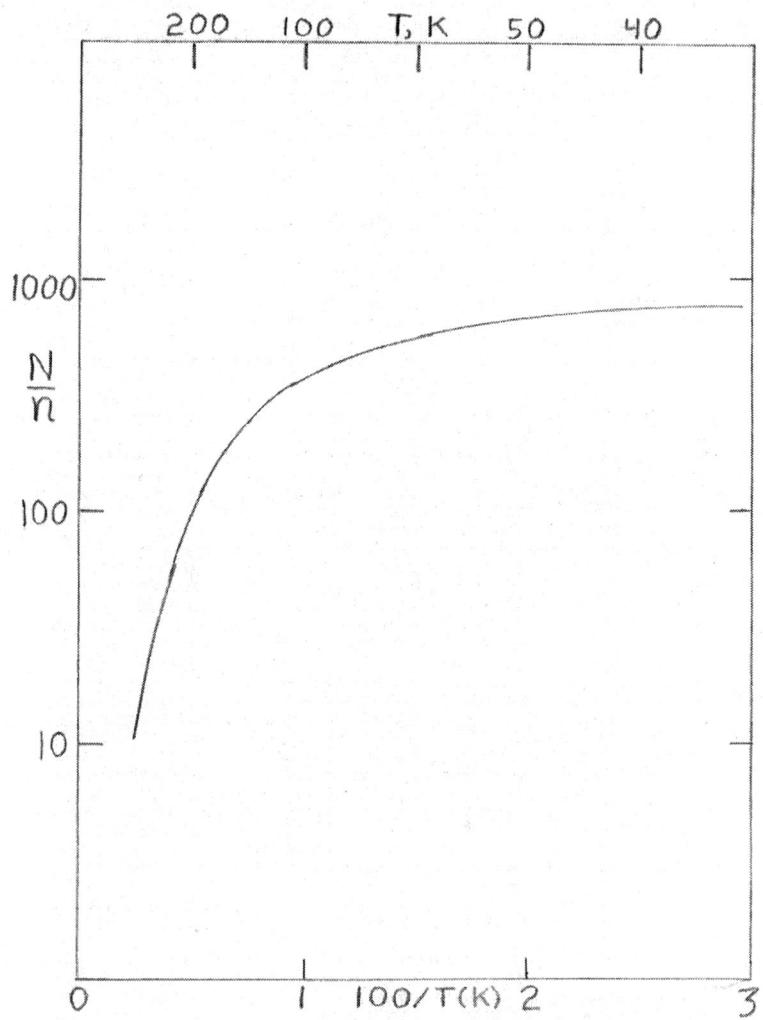

FIGURE 1. Inverse carrier concentration.

We now discuss the origin of this relaxation time spread. In the high-temperature limit, diffusion relaxes deuteron sites at a constant rate after applying a step electric field change, so there should only be one time constant, $\tau = t_l = t_d$. To find this τ, we divide the concentration $2N$ of initially unrelaxed deuterons by the rate $vn = 2\nu N \exp(-\varepsilon_1/kT)$. At lower temperatures, as the random barriers come into play, a time constant spread develops because diffusing Takagi groups usually encounter only lower barriers at early times after a field step, while higher barriers are encountered at later times. Thus the rate of relaxing new sites decreases with time, with a resulting time constant spread, until t_l is reached. At this time, essentially all sites are relaxed, so t_l is the maximum value τ_{max} in the time constant spread, as was discussed above.

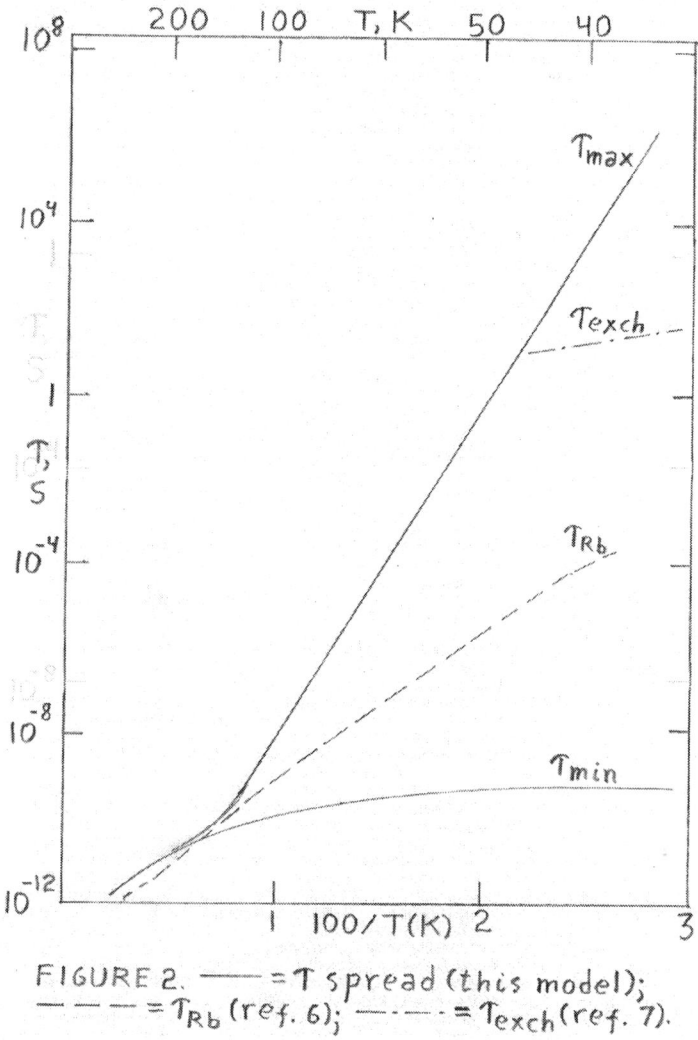

FIGURE 2. ——— = τ spread (this model); – – – = τ_{Rb} (ref. 6); — · — · = τ_{exch} (ref. 7).

To find the minimum time constant τ_{min} of this spread, we note that for an exponential decay, the time constant τ equals the initial value divided by the initial decay rate. For a spread in time constants, this is a good definition for τ_{min}. To find this τ_{min}, we divide the concentration $2N$ of initially unrelaxed deuterons by the rate vn at which they initially relax if we assume that the initial relaxation rate is unaffected by barriers. At temperatures above T_x, n is given approximately by Eq. (4), so

$$\tau_{min}(\text{hi-T})=2N/2\nu N\exp(-\varepsilon_1/kT)=\nu^{-1}\exp(\varepsilon_1/kT). \tag{10}$$

The straight-line portion of τ_{min} at high temperature in Fig. 2 corresponds to Eq. (10).

The minimum time constant τ_{min} in Fig. 2 remains finite at low temperature, but the maximum value τ_{max} is heading for infinity. At what temperature T_e is τ_{max-e} so long that the system ceases being ergodic on the experimental time scale? We can use the fact that n has weak temperature dependence at low temperature, so that in Eq. (5) for $t_l(=\tau_{max})$ we can substitute n(T=0) from Eq. (8) for n(T_e). We then find

$$\tau_{max-e}=\nu^{-1}\exp\{[N/4n(T=0)]^{1/2}\varepsilon_0/kT_e\}=\nu^{-1}\exp(2\varepsilon_1/kT_e), \tag{11}$$

from which we obtain

$$T_e=2\varepsilon_1/k\ln(\nu\tau_{max-e}). \tag{12}$$

If we choose $\varepsilon_1/k=940$ K (as above), $\nu=10^{12}$/s, and $\tau_{max-e}=3600$ s=1 hr as the experimental time, we obtain $T_e=52$ K. This temperature agrees quite well with two values found by different groups. The value 47 K for DRADP with x=0.60 was found from field cooling-zero field heating [FC-ZFH] experiments by Levstik et al. [4]. The value 39 K for the As-for-P substituted crystal DRADA with x=0.28 was found from FC-ZFH and ZFC-FH experiments by Pinto et al. [5]. The effective time for both experiments was about 15 minutes, and T_e is only a weak function of τ_{max-e}.

COMPARISON WITH NMR RESULTS

We can compare deuteron position correlation time results predicted above with those found from two NMR experiments. First, the ^{87}Rb spin-lattice inverse relaxation rate measurements of Korner et al. [6] in DRADP with x=0.50 at f=98.163 MHz show nearly ideal Bloembergen-Purcell-Pound minima for both the $\Delta m=1$ and $\Delta m=2$ transitions. At the minima, the correlation times τ_{Rb} for the efg fluctuations seen by the Rb nuclei due to the deuteron intrabond transfers are $1/2\pi f$ and $1/4\pi f$ respectively for those transitions. The values of τ_{Rb} at other temperatures are found by multiplying (at lower temperature) or dividing (at higher temperature) the above τ_{Rb} values by the ratio of the inverse rate value at the given temperature to ½ the value at the minimum. These values are shown on Fig. 2. They agree quite well with our calculated τ at high temperature, and fall near the middle of the broad τ range at lower temperature. What remains to be calculated is the effect of multiple nearby deuterons (6 at nearly equal distances) on the efg fluctuation spectrum, and how the effective τ_{Rb} is related to the broad τ distribution at lower temperatures. Another consideration, also outside the scope of this paper, is that at intermediate temperatures, between T_e and T_x, many Takagi pairs recombine rather than annihilate with new partners, with long enough lifetimes and diffusion paths to have a significant effect on NMR relaxation.

The other NMR experiment involves two-dimensional (2D) exchange NMR measurements by Dolinšek et al. [7] in DRADA with x=0.32. The τ_{exch} values from their Fig. 8 plotted in Fig. 2 are considerably larger than the τ_{Rb} values shown in Fig. 2 for the same temperature range, but still fall within our calculated τ envelope. The calculation of the expected τ_{exch} based on this τ envelope has not yet been done.

PROPOSED 4D EXCHANGE NMR EXPERIMENTS

2D exchange NMR is able to measure exchange times directly, so they need not be inferred from spin-lattice relaxation time data. It is also possible to find whether or not the exchange process has a spread in relaxation times. If there is such a spread, 2D exchange NMR cannot determine whether the nuclei that exchange slowly during one time period tend to exchange more slowly in a subsequent period. To investigate this question, Schmidt-Rohr and Spiess [8] developed a 4D exchange NMR technique and applied it to study glassy dynamics in poly(vinyl acetate). Their Fig. 1 that contrasts the 2D and 4D pulse sequences is reproduced in Fig. 3. In the 4D sequence, the first 3 pulses produce an echo due only to those nuclei whose clusters have not reoriented significantly. The 4th occurs at the echo peak and selects these "slow" nuclei which are then subjected to a 2D pulse sequence. These selected slow nuclei were found to remain slow for a time about two orders greater than the mean exchange time.

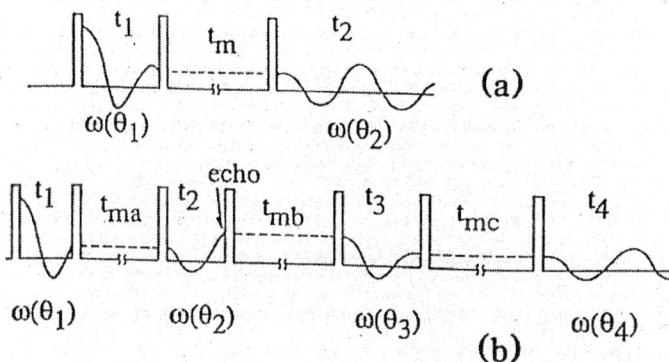

FIG. 3. Schematic pulse sequences for exchange NMR experiments: (a) 2D experiment, (b) 4D experiment. 90° pulses and a typical signal from an isotropic sample are indicated. (From ref. 8.)

This technique could be used for ^{87}Rb NMR in proton or deuteron glasses if the efg changes caused by the hydrogen intrabond motion are large enough to allow selection of slow nuclei. For deuteron NMR, there are only two efg values, and the echo process will be only partially successful in selecting slow nuclei. Still, it is likely that useful results could be obtained.

What kinds of questions can be addressed by 4D exchange NMR in proton or deuteron glasses, or in other ferroelectric materials where local variations such as nanoclusters occur? For proton or deuteron glass compositions far from FE or AFE phases in the phase diagram, the crystal may be homogeneous in the time average. A 2D exchange NMR set of experiments might find a spread in relaxation times at very low temperature because Takagi groups are locally trapped on the experimental time scale. Then only those nuclei in the vicinity of these Takagi groups will see effects of the intrabond exchange. A 4D experiment could determine whether at later times these Takagi groups migrate to other parts of the crystal, finally restoring ergodicity.

If the composition is such that FE or AFE clusters occur, there will be an energy bias against penetration of these clusters by Takagi groups. If the cluster boundaries are not eroded by Takagi groups, then these are quenched clusters and a 4D experiment will conclude that the crystal is inhomogeneous on all time scales at the given temperature. From measurement of the time scale for restoring homogeneity, together with a suitable model, it may be possible to estimate cluster size. Similar considerations apply to possible 4D exchange NMR investigations of nanoclusters in relaxor and relaxor-based ferroelectrics.

CONCLUSIONS

Our model for Takagi group effective diffusion can explain the wide spread in relaxation times and hydrogen intrabond exchange times observed in proton and deuteron glasses. Model predictions are consistent with 1D and 2D NMR results for intrabond transfer correlation times for deuteron glasses reported by two other groups. They also agree with the temperature for onset of nonergodicity found by two groups for two different deuteron glass crystals. 4D exchange NMR experiments are proposed for investigation of spatial inhomogeneities in exchange rates in proton and deuteron glasses and relaxor-type ferroelectrics.

ACKNOWLEDGMENTS

This work was supported by National Science Foundation Grant DMR-9805272.

REFERENCES

1. Slater, J. C., *J. Chem. Phys.* **9**, 16 (1941).
2. Takagi, Y., *J. Phys. Soc. Jpn.* **3**, 273 (1948).
3. Schmidt, V. H., *J. Molecular Structure* **177**, 257-264 (1988).
4. Levstik, A., Filipič, Kutnjak, Z., Levstik, I., Pirc, R., Tadić, B., and Blinc, R., *Phys. Rev. Lett.* **66**, 2368-2371 (1991).
5. Pinto, N. J., Ravindran, K., and Schmidt, V. H., *Phys. Rev. B* **48**, 3090-3094 (1993).
6. Korner, N., Pfammatter, Ch., and Kind, R., *Phys. Rev. Lett.* **70**, 1283-1286 (1993).
7. Dolinšek, J., Zalar, B., and Blinc, R., *Phys. Rev. B* **50**, 805-821 (1994).
8. Schmidt-Rohr, K., and Spiess, H. W., *Phys. Rev. Lett.* **66**, 3020-3023 (1991).

Nonexponential Relaxation in Piezoelectric PVDF

Gary W. Bohannan[1]

Department of Physics
Montana State University
Bozeman, Montana 59717

Abstract. Polymers are known for displaying nonexponential relaxation to both electrical and mechanical stresses. A numerical method for integrating fractional differential equations has made it possible to verify the connection between a power-law ac permittivity function and power-law decay currents resulting from application of dc step voltages. The responses show significant dependence on the duration of the applied pulse. The history dependence predicted by the simulations has been confirmed experimentally. The computational techniques should be applicable to a wide range of systems exhibiting nonexponential relaxation.

INTRODUCTION

Virtually all dielectric materials display power-law relaxation. [1,2] Ferroelectric materials are no exception to this rule. [3,4] The power-law behavior of the "universal dielectric response" is shared with a vast array of other responses in materials. [5]

While there is some disagreement as to whether the power-law descriptions are fundamental to the underlying physics or just coincidental curve fitting functions, the use of power-law descriptions have been accepted within mainstream dielectric spectroscopy for many years. [6]

The fractional calculus is the natural language of power-law dynamics. With recently developed methods for numerical integration of fractional differential equations, it is now possible to simulate power-law responses. The purpose of this article is to demonstrate, computationally and experimentally, that, at least for one material, the power-law description really does model multi-scale polarization dynamics. That is, we will predict long-time relaxation up to an hour after application of an on-off dc field based on ac permittivity measurements covering 20 Hz to 300 kHz.

---

[1] Supported by a Fellowship from the Montana Space Grant Consortium, NSF Grant DMR-9805272 and NASA EPSCoR Grant NCC5-240.

THE EQUIVALENT CIRCUIT MODEL

We are looking for a model to guide the development of a constitutive relation of the form

$$\mathbf{D}(\omega) = \epsilon_0 \, \epsilon(\omega) \mathbf{E}(\omega), \tag{1}$$

relating the electric displacement, $\mathbf{D}$, to the applied external electric field $\mathbf{E}$ via ϵ, the relative permittivity. In SI units, ϵ_0, the permittivity of free space, converts between intensive and extensive units.

The relative permittivity may be found from the impedance, $Z(\omega)$, according to

$$\epsilon(\omega) = \frac{1}{i\omega C_C \, Z(\omega)}, \tag{2}$$

where C_C is the capacitance of the empty test cell.

A candidate form of a fractional permittivity equation may be obtained by starting with an ideal equivalent circuit for a system with *dc* conductivity as shown in Figure 1.

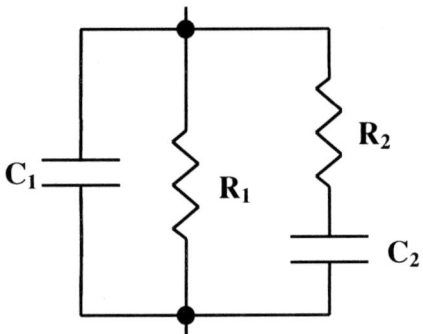

FIGURE 1. An ideal element RC circuit incorporating conductivity

The impedance for the circuit is

$$Z(\omega) = \frac{1}{i\omega C_1 + \dfrac{1}{R_1} + \dfrac{1}{R_2 + \dfrac{1}{i\omega C_2}}}, \tag{3}$$

which can be transformed to a permittivity, $\epsilon(\omega) = \epsilon'(\omega) + i\epsilon''(\omega)$, according to Equation 2 to obtain

$$\epsilon(\omega) = \frac{\dfrac{C_1 + C_2}{C_c} + \dfrac{\tau}{R_1 C_c} + \dfrac{C_1}{C_c}(i\omega\tau) + \dfrac{R_2}{R_1}\dfrac{C_2}{C_c}(i\omega\tau)^{-1}}{1 + i\omega\tau} \tag{4}$$

where $\tau = R_2C_2$.

By replacing the resistive elements with constant phase elements (CPE) according to the prescription of Cole and Cole [6,7]:

$$R_1 \to R_1(i\omega\tau)^{-\gamma}, \quad R_2 \to R_2(i\omega\tau)^{-\alpha},$$

we obtain

$$\epsilon(\omega) = \frac{\epsilon_s + \epsilon_\infty(i\omega\tau)^\delta + \epsilon_c(i\omega\tau)^\beta + \epsilon_c(i\omega\tau)^\nu}{1 + (i\omega\tau)^\delta}, \tag{5}$$

where

$$\tau = R_2C_2, \quad \epsilon_s = \frac{C_1 + C_2}{C_c}, \quad \epsilon_\infty = \frac{C_1}{C_c}, \quad \epsilon_c = \frac{\tau}{R_1C_c},$$

$$\delta = 1 - \alpha, \quad \beta = -1 + \gamma, \quad \nu = \gamma - \alpha.$$

The CPE represents a linear element with a frequency independent phase shift somewhere between that of a resistor and that of a capacitor, i.e. $0 < \alpha < 1$ and $0 < \gamma < 1$. Note that this model reduces to the classical Cole-Cole model when $\epsilon_c = 0$.

Taking the Laplace transform of Equation (5), we obtain the fractional differential form

$$\mathbf{D}(t) + \tau^\delta {}_0D_t^\delta \mathbf{D} = \left(\epsilon_s + \epsilon_\infty \tau^\delta {}_0D_t^\delta + \epsilon_c \tau^\beta {}_0D_t^\beta + \epsilon_c \tau^\nu {}_0D_t^\nu\right) \epsilon_0 \mathbf{E}. \tag{6}$$

The fractional time derivative operator symbol ${}_aD_t^q$ and the related operator ${}_aS_t^q$, to be used in what follows, are defined by [8,9]

$${}_aD_t^q f(t) = \lim_{N \to \infty} \left\{ \left[\frac{t-a}{N}\right]^{-q} \frac{1}{\Gamma(-q)} \sum_{j=0}^{N-1} \frac{\Gamma(j-q)}{\Gamma(j+1)} f\left(t - j\left[\frac{t-a}{N}\right]\right) \right\}, \tag{7}$$

and

$${}_aS_t^q f(t) = \lim_{N \to \infty} \left\{ \left[\frac{t-a}{N}\right]^{-q} \frac{1}{\Gamma(-q)} \sum_{j=1}^{N-1} \frac{\Gamma(j-q)}{\Gamma(j+1)} f\left(t - j\left[\frac{t-a}{N}\right]\right) \right\}. \tag{8}$$

These differ only by the limits of the sum. The symbol ${}_aS_t^q$ is introduced here to allow separation of the value at t from the rest of the sum covering past history. For computational purposes we drop the $N \to \infty$ limit to use a finite number of equally spaced points. We also set $a = 0$ by convention. The fractional derivative is only defined over an interval and does not generally have meaning when evaluated at $t = a$. [8]

¿From here we follow the approach outlined in the discussion of fractional relaxation in [10], and rearrange Equation (6) as

$$\tau^\delta \, _0D_t^\delta \mathbf{D} = -\mathbf{D}(t) + \left(\epsilon_s + \epsilon_\infty \tau^\delta \, _0D_t^\delta + \epsilon_c \tau^\beta \, _0D_t^\beta + \epsilon_c \tau^\beta \, _0D_t^\beta\right) \epsilon_0 \mathbf{E}. \tag{9}$$

Applying the operator $\tau^{-\delta} \, _0D_t^{-\delta}$ to both sides

$$\mathbf{D}(t) = -\tau^{-\delta} \, _0D_t^{-\delta} \mathbf{D}$$
$$+ \left(\epsilon_s \tau^{-\delta} \, _0D_t^{-\delta} + \epsilon_\infty + \epsilon_c \tau^{\beta-\delta} \, _0D_t^{\beta-\delta} + \epsilon_c \tau^{\nu-\delta} \, _0D_t^{\nu-\delta}\right) \epsilon_0 \mathbf{E}. \tag{10}$$

We can isolate $\mathbf{D}[N]$ in terms of the previous $N-1$ values of $\mathbf{D}$,

$$\mathbf{D}[N] = \left(1 + \left(\frac{dt}{\tau}\right)^\delta\right)^{-1} \left\{ -\tau^{-\delta} \, _0S_t^{-\delta} \mathbf{D}\right.$$
$$\left. + \left(\epsilon_s \tau^{-\delta} \, _0D_t^{-\delta} + \epsilon_\infty + \epsilon_c \tau^{\beta-\delta} \, _0D_t^{\beta-\delta} + \epsilon_c \tau^{\nu-\delta} \, _0D_t^{\nu-\delta}\right) \epsilon_0 \mathbf{E}\right\}. \tag{11}$$

This form allows us to predict the values of the electric displacement, often referred to as the polarization, resulting from arbitrary applied fields based on *ac* permittivity measurements. This includes prediction of the transient responses, both at short and long times, even though the permittivity measurements are made in steady state, post-transient, conditions.

APPLICATION TO PIEZOELECTRIC PVDF

Poly(vinylidine fluoride) (PVDF) $(CH_2CF_2)_n$ is a polymer containing microcrystals embedded in an amorphous polymer matrix. The material studied was 28 μm thick, with a silver electrode layer of approximately 7 μm applied to each side by the manufacturer, Measurement Specialties (formerly known as AMP). The material was stretched and poled under heat and high electric field during the manufacturing, so that the material was ferroelectric and piezoelectric. Permittivity measurements of several samples from different manufacturing batches of the PVDF sheets, cut in 4.7 cm × 4.7 cm and 9.6 cm × 9.6 cm squares, were obtained under zero applied mechanical stress.

After initial measurements of the *ac* permittivity and of the relaxation currents following application of on-off *dc* voltage pulses, we realized that the available electrometer did not have the sensitivity to confirm the predicted discharging currents for the smaller samples. For this reason, a Faraday cage was constructed to hold the larger samples of 9.6 cm × 9.6 cm. We found that sandwiching two layers of the material together with a thin layer of adhesive would reduce the low frequency piezoelectric response and increase the current signal to the level necessary. With the high voltage sides placed to the inside of the sandwich, the chance for external atmosphere charge contamination was reduced. The lamination process kept the adhesive out of the electric field but allowed the two sheets to respond in parallel. The lamination did not change the permittivity or resonance characteristics in any measurable way from that of single sheets.

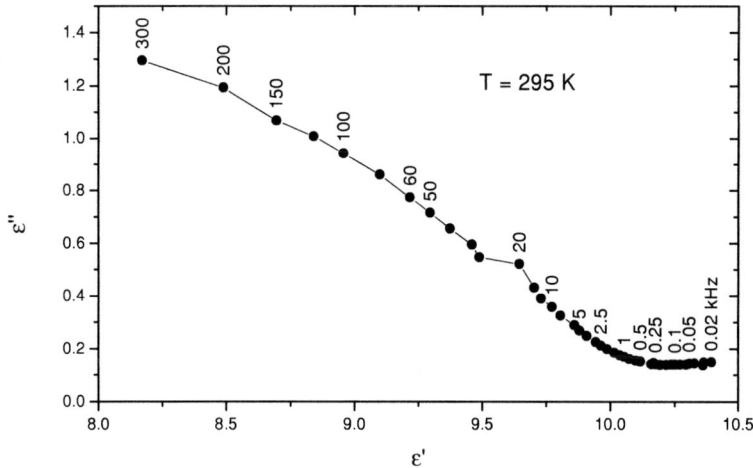

FIGURE 2. Cole-Cole plot for a typical PVDF sample at room temperature. Note the "glitch" due to the resonance around 10 to 20 kHz.

The data of Figure 2 were fit to Equation (5). The data from 12 to 20 kHz were given lower weight due to an electro-mechanical resonance. The time constant is $\tau = 173$ ns, the exponents are $\delta = 0.603$ and $\nu = 0.524$ and the coefficients are $\epsilon_\infty = 3.75, \epsilon_s = 9.32, \epsilon_c = 0.465$. The underlying exponents of the model are $\alpha = 0.397$ and $\gamma = 0.921$. The results of the fit are shown in Figure 3. The model predictions extrapolated to much lower frequencies are shown as well. The time domain experiments will confirm this extrapolation.

The resonance was more pronounced in some of the material tested, but it was above 5 kHz in all cases and will be ignored in what follows. While the values of the coefficients ϵ_i, the time constant τ, and the exponent δ varied slightly, the exponent most responsible for the long-time transient response, β, did not change from batch to batch, nor did it vary by size of the sample tested.

Interpretation of the fitting parameters for the model of Equation (5) leads to a somewhat surprising result. The value of the exponent γ in the CPE replacing R_1 in Figure 1 is 0.921, nearly unity. If it were unity, then R_1 would have been replaced by a capacitor. Exponents less than 0.5 would be more "resistor-like." As it is, this CPE is primarily a charge storage element. The exponent α is in the typical range reported for polymers.

PREDICTION OF LONG-TIME DECAY CURRENTS

The parameter values were used in a numerical simulation of Equation (11). The charging and discharging currents predicted for a 100 second pulse are shown in Figure 4. The model clearly predicts that the discharging current matches the charging current for a short period, then drops off with an exponent one larger

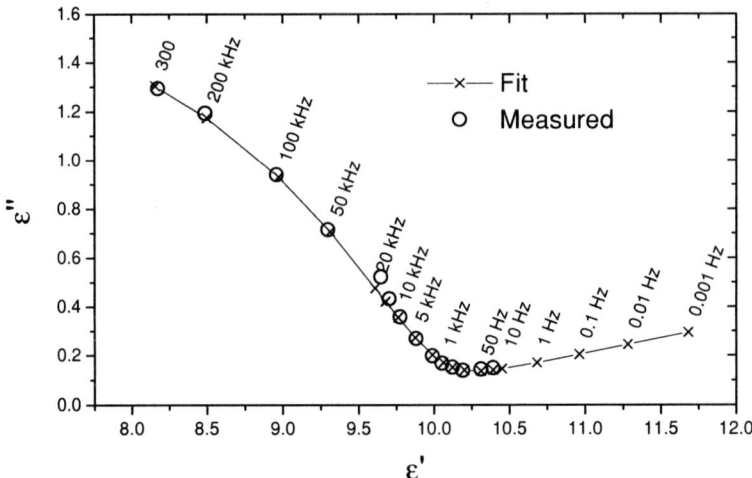

FIGURE 3. Parameter fit for PVDF sample at room temperature. The time constant is $\tau = 173$ ns, the exponents $\delta = 0.603$ and $\nu = 0.524$ and the coefficients $\epsilon_\infty = 3.75, \epsilon_s = 9.32, \epsilon_c = 0.465$. The underlying exponents of the model are $\alpha = 0.397$ and $\gamma = 0.921$. The model has been extrapolated to very low frequency. Solid lines have been added as aids to the eye.

than that of the charging current.

The simulation indicates that virtually all of the charge transferred during the charging cycle is recovered during discharge. A net 0.054% of the "polarization" remained after 4000 seconds. To within the accuracy of the available data, integrating the current from the 4000 second mark to infinity is exactly that remaining polarization. With this, we can say that PVDF is a perfect insulator in that all of the charge applied across it is recovered (after an infinite time). The extrapolated *dc* conductivity is zero. This does not mean that there is no energy loss. In fact, compared with polystyrene, PVDF is quite lossy in this regard. The energy loss is due to the phase shift in the CPE.

The numerical simulation is done in relative units, so the predicted current in amperes is obtained by multiplying the scaled current of Figure 4 by $C_c \times$ Voltage. The predicted current for this sample is shown in Figure 5.

Figure 6 shows the results of three tests. In each case, the electrodes were shorted together for several hours prior to application of the voltage to ensure the zero initial condition required by the model. In the first test, a discharging curve after applying 100 volts for an hour was obtained. The results are plotted as circles on the figure. In the second test, a pulse of 100 volts for 100 seconds was applied and the discharge current was then recorded. The squares on the figure show the results of this measurement. The third test applied a pulse of intermediate length, 10 minutes. The crosses on the figure record this test. Note that the longer the charging time, the longer time the discharging current follows the charging curve

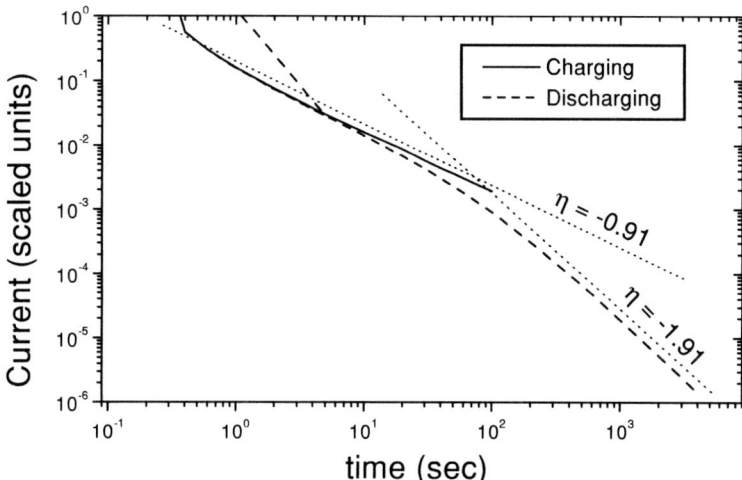

FIGURE 4. Predicted charging and discharging currents for PVDF sample at room temperature using Equation (5) and the parameters from Figure 3. The two asymptotic exponents are also shown. Current in amps is found by multiplying the displayed value by $C_c \times$ Voltage.

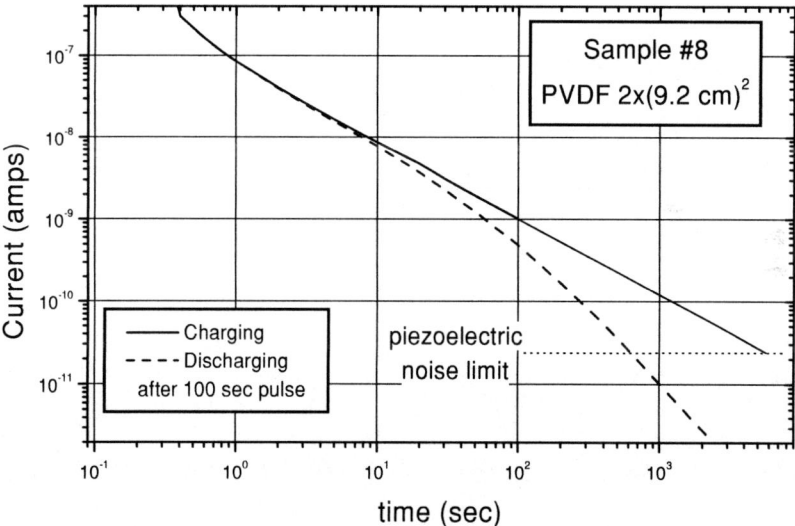

FIGURE 5. Predicted charging and discharging currents for PVDF sample, $2 \times (9.6 \text{ cm} \times 9.6 \text{ cm})$, at room temperature.

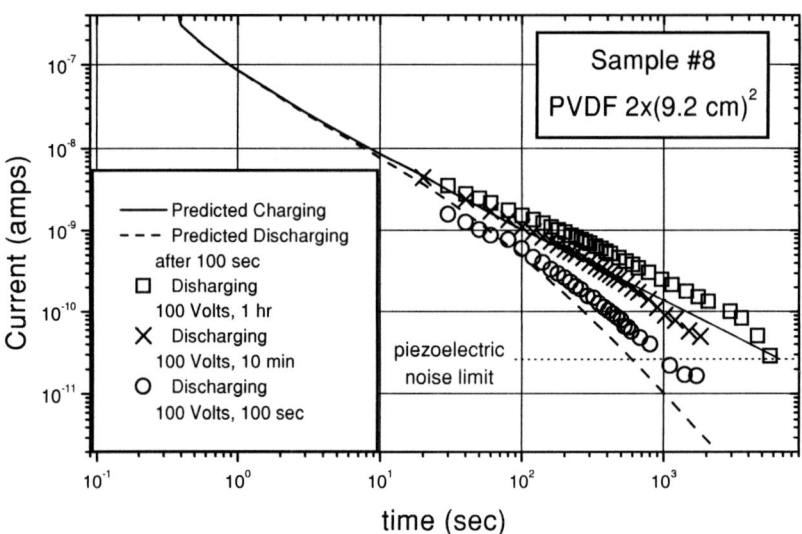

FIGURE 6. Measured discharging currents for PVDF sample, 2 × (9.6 cm × 9.6 cm), at room temperature. Solid and dashed lines are predicted values, symbols are measured data.

slope.

The deviation of the measured results from the predicted values in Figure 6 are greatly distorted due to the log-log plot. The currents predicted by the numerical simulation, and those measured experimentally, exhibit the separation caused by previous histories predicted by Jonscher [1].

DISCUSSION AND SUMMARY

A complete description of the numerical method and its error analysis would be far beyond the scope of these proceedings. It is worth noting, however, that the time domain simulation is most sensitive to the exponents δ and β. These are the most sensitive parameters in the curve fitting as well. Additionally, the method is quite insensitive to variation in the time step, dt, which could be varied by a factor of 10 to 100 with no noticeable variation in the output. The method is surprisingly stable and robust.

To place these results in perspective, the initial ac measurements were made over the range of 20 Hz to 300 kHz. All of these measurements represent dynamics in the fraction-of-a-second time scale. The permittivity curve was fit to this limited data and then extrapolated by five decades in frequency. These parameters were input to a numerical simulation that used a 0.2 second time step. This numerical time step is over one million times larger than the underlying time constant τ. In all this, we ignored the effects of the resonance and made no effort to account for the now sizable piezoelectric response from application of 100 volts in the relaxation current testing versus 0.5 to 1.0 volts used in the *ac* testing. There was no accounting for any nonlinearity and no accounting of the effects of using hand thrown switches. Clearly, automating the data collection process would produce more accurate data by removing the potential biases of human observation.

One of the reasons we were able to do this at all was that the resonance was isolated and well above the low frequency dispersion. Undoubtedly there were other interesting fast dynamics occurring on the application and removal of the field. The simulation was able to skip over these, at some loss of accuracy, to produce a quantitative prediction of the long-time transient response.

Part of the difference between the predicted curves and the measured may well be the loss of accuracy in the time scale of the numerical calculation. With a time step of over a million times the underlying time constant, there is very likely a decay in the accuracy of the time evolution. It is anticipated that further comparison with experiment will provide feedback to the development of improved numerical schemes.

It is by no means assured that all of the differences between the predicted and measured currents are the fault of the computational model. There may well be some additional ultra-slow process that was not discovered in the *ac* testing between 20 Hz and 300 kHz. Such an ultra-slow response is found in KH_2PO_4 (KDP), for example. [11] Nonetheless, any such additional processes must be minimal in order to achieve the degree of accuracy demonstrated here.

The computational methods used here should allow further investigation into nonexponential dynamics of other materials as well. In particular, the new high-strain single crystals [12] will more than likely display both charge and displacement creep in applications where slow non-zero biased operation is necessary. Modeling of the dynamics of a variety of materials may also assist in developing a fundamental understanding of the underlying source of nonexponential dynamics.

REFERENCES

1. A.K. Jonscher. *Dielectric Relaxation in Solids*. Chelsea Dielectric Press, London, 1983.
2. S. Westerlund and L. Ekstam. Capacitor theory. *IEEE Trans. Dielectrics and Electrical Insulation*, 1, 1994.
3. A. Isnin and A. K. Jonscher. Dielectric response of some ferroelectrics at "low" frequencies I. *Ferroelectrics*, 210:47–65, 1998.
4. A. K. Jonscher and A. Isnin. Dielectric response of some ferroelectrics at "low" frequencies II. *Ferroelectrics*, 210:67–81, 1998.
5. K. L. Ngai. Evidences for universal behavior of condensed matter at low frequencies/long times. In T.V. Ramakrishnan and M. Raj Lakshmi, editors, *Non-Debye Relaxation in Condensed Matter: Proceedings of a Discussion Meeting, Bangalore*, pages 23–191. World Scientific Publishing Co. Pte. Ltd., 1987.

6. J. R. Macdonald. *Impedance Spectroscopy*. John Wiley & Sons, New York, 1987.
7. S. Cole and R. Cole. Dispersion and absorption in dielectrics. *J. Chem. Phys.*, 9:341–351, 1941.
8. K. Oldham and J. Spanier. *The Fractional Calculus*. Academic Press, New York, 1974.
9. I. Podlubny. *Fractional Differential Equations: An Introduction to Fractional Derivatives, Fractional Differential Equations, to Methods of their Solution and some of their Applications*, volume 198 of *Mathematics in Science and Engineering*. Academic Press, San Diego, CA, 1999.
10. T. Nonnenmacher and W. Glöckle. A fractional model for mechanical stress relaxation. *Phil. Mag. Lett.*, 64:89–93, 1991.
11. K. Okada, H. Sugié, and K. Kan'no. Extremely slow response of the polarization to the external field in KH_2PO_4. *Phys. Lett.*, 44A:59–60, 1973.
12. K. Uchino. High electromechanical coupling piezoelectrics: relaxor and normal ferroelectric solid solutions. *Solid State Ionics*, 108:43–52, 1998.

Computer Simulations of Domain Pattern Formation in Ferroelectrics

Rajeev Ahluwalia and Wenwu Cao

Materials Research Lab, The Pennsylvania State University, PA 16802, USA

Abstract. We study domain pattern formation in ferroelectrics based on a 2-D time-dependent Ginzburg Landau theory. The model includes electrostrictive and elastic effects in the form of a long-range interaction between the polarization fields that is obtained by eliminating the strain fields subject to the elastic compatibility constraint. We simulate a 2-D square to rectangle transition that has four equivalent polarization states in the ferroelectric phase. Starting from an unstable paraelectric state, we simulate the domain pattern evolution in the absence of external electric field. For the case without defects, the final pattern is a twinned state that has only head to tail (uncharged) domain walls. However, for the case with randomly distributed dipolar defects, head to head and tail to tail charged walls are observed. These results are in accordance with recent experiments where charged domain walls have been observed.

It is now well established that the ferroelectric transformation is accompanied by the formation of domains of the low temperature phase states. Ferroelectric materials that have only two polarization states usually form strain free $180°$ domains. However for multiple degenerate systems, such as $BaTiO_3$ and PZT, the electrostrictively generated strain is responsible for the creation of $90°$ domain walls. It is important to understand the mechanism of formation of these domain structures as physical properties like the dielectric constant and hysteresis are mainly governed by motion of the domain walls.

A useful technique to study the domain pattern formation in phase transitions is the time-dependent Ginzburg Landau (TDGL) theory. In the context of ferroelectric phase transitions, this approach has been sucessfully used to study the formation and growth of domains [1-4]. In this article, we use a TDGL model to study pattern formation in a model 2-D ferroelectric system that has four degenerate polarization states. We also include the electrostrictive coupling of polarization with the strain and the contributions due to randomly distributed dipolar defects.

The 3-D free-energy including strain and electrostrictive coupling has been given before [5]. For simplicity, we consider a 2-D square to rectangle ferroelectric transition in the absence of an external field with free-energy given as

$$F = \int d\vec{r}[f_l + f_g + f_{el} + f_{es} + f_d] \tag{1}$$

where f_l is a Landau free energy density given as

$$f_l = \frac{\alpha_1}{2}(P_x^2 + P_y^2) + \frac{\alpha_{11}}{4}(P_x^4 + P_y^4) + \frac{\alpha_{12}}{2}P_x^2 P_y^2 \tag{2}$$

Here, f_l is identical to the free energy density used in earlier works [4,5]. The gradient energy f_g is given as

$$f_g = \frac{g_1}{2}[(\frac{\partial P_x}{\partial x})^2 + (\frac{\partial P_y}{\partial y})^2] + \frac{g_2}{2}[(\frac{\partial P_x}{\partial y})^2 + (\frac{\partial P_y}{\partial x})^2] + g_3(\frac{\partial P_x}{\partial x})(\frac{\partial P_y}{\partial y}). \tag{3}$$

The free energy contribution due to randomly distributed dipolar defects is given as $f_d = -\vec{E}_d \cdot \vec{P}$, where $\vec{E}_d(\vec{r}) = -\vec{\nabla} V_d(\vec{r})$. The potential V_d represents a configuration of randomly placed defect dipoles given as

$$V_d(\vec{r}_i) = \sum_j^{n_d}[\frac{1}{|\vec{r}_i - (\vec{r}_j + \vec{\delta})|} - \frac{1}{|\vec{r}_i - (\vec{r}_j - \vec{\delta})|}]q_0(\vec{r}_j) \tag{4}$$

Here q_0 represents the coarse-grained charge and δ the displacement associated with the defect dipole centered at $\vec{r}_j$. The elastic energy of the system can be written in terms of the bulk strain $\phi_1 = (\eta_{xx} + \eta_{yy})/\sqrt{2}$, the deviatoric strain $\phi_2 = (\eta_{xx} - \eta_{yy})/\sqrt{2}$ and the shear strain $\phi_3 = \eta_{xy} = \eta_{yx}$. Here η_{ij} is the linear elastic strain tensor. The elastic free energy can then be written as

$$f_{el} = \frac{a_1}{2}\phi_1^2 + \frac{a_2}{2}\phi_2^2 + \frac{a_3}{2}\phi_3^2 \tag{5}$$

where a_1, a_2 and a_3 are linear combinations of second order elastic constants. Similarly, the electrostrictive energy in terms of the electrostrictive constants q_1, q_2 and q_3 is given as

$$f_{es} = -q_1\phi_1(P_x^2 + P_y^2) - q_2\phi_2(P_x^2 - P_y^2) - q_3\phi_3 P_x P_y \tag{6}$$

Following the methodology used in earlier work [4], the elastic and electrostrictive contributions can be expressed as an effective long-range interaction between the polarization fields by eliminating the strain fields subject to the elastic compatibility conditions. The effective long-range interaction in fourier space is then given as

$$F_{eff} = \frac{q_2^2}{2a_2} \int d\vec{k} H(\vec{k}) |\Gamma_2(\vec{k})|^2 \tag{7}$$

$$H(\vec{k}) = h_1^2(\vec{k})/\alpha + h_2^2(\vec{k}) - 2h_2(\vec{k}) + h_3^2(\vec{k})/\beta \tag{8}$$

and $\Gamma_2(\vec{k})$ is the fourier transform of $P_x^2 - P_y^2$. The quantities h_1, h_2 and h_3 are given as $h_1(\vec{k}) = k^2 Q(\vec{k})$, $h_2(\vec{k}) = \{1 - (k_x^2 - k_y^2)Q(\vec{k})\}$ and $h_3(\vec{k}) = -\sqrt{8} k_x k_y Q(\vec{k})$, where

$$Q(\vec{k}) = \frac{(k_x^2 - k_y^2)}{(k^4/\alpha + (k_x^2 - k_y^2)^2 + 8k_x^2 k_y^2/\beta)} \tag{9}$$

where $\alpha = a_1/a_2$ and $\beta = a_3/a_2$. Here, we should remark that the above interaction has been obtained by assuming no coupling of the polarization with bulk and shear strains, i.e., $q_1 \to 0$ and $q_3 \to 0$. In order to catch the essential physics and reduce the computational difficulties, the coupling to bulk and shear strains are not included in this work. Based on experimental results, the bulk strain is small in most of the ferroelectric materials and the shear strain does not exist at all the single domain regions for the case under study. Thus we believe that this assumption will not introduce significant errors in the study of domain pattern formation.

In order to study the dynamics of domain pattern formation, we make use of the time-dependent Ginzburg-Landau equations (TDGL). We introduce rescaled variables as follows: $t = (t^*/|\alpha_1|L)$, where L is the kinetic coefficient, $\vec{r}^* = \vec{r}/\phi$ ($\phi = \sqrt{g_1/a|\alpha_1|}$), where a is a dimensionless constant. The polarization is transformed as $P_x = P_R u$ and $P_y = P_R v$, where $P_R = \sqrt{|\alpha_1|/\alpha_{11}}$ is the remnant polarization of the homogeneous state without elastic effects. With this set of rescaled parameters, the TDGL equations are given as

$$\begin{aligned}u_{,t^*} &= u - u^3 - duv^2 + au_{,x^*x^*} + bu_{,y^*y^*} + cv_{,x^*y^*} + \varepsilon_x \\&\quad - \gamma u \int d\vec{k}^* H(\vec{k}^*) \Gamma(\vec{k}^*) \exp(-i\vec{k}^* \bullet \vec{r}^*) \\v_{,t^*} &= v - v^3 - dvu^2 + av_{,x^*x^*} + bv_{,y^*y^*} + cu_{,x^*y^*} + \varepsilon_y \\&\quad + \gamma v \int d\vec{k}^* H(\vec{k}^*) \Gamma(\vec{k}^*) \exp(-i\vec{k}^* \bullet \vec{r}^*)\end{aligned} \tag{10}$$

Here $\vec{\varepsilon}$ is the rescaled electric field due to the randomly distributed defect dipoles, with $q_0^*(\vec{r}^*) = [q_0(\vec{r})/P_R \phi^2 |\alpha_1|]$. In the rescaled equations, the constant

$d = (\alpha_{12}/\alpha_{11})$, $\gamma = (q_2^2/\alpha_{11}a_2)$, $b = (g_2/|\alpha_1|\phi^2)$ and $c = (g_3/|\alpha_1|\phi^2)$. $\Gamma(\vec{k}^*)$ is the fourier transform of $u^2 - v^2$.

To study the domain patterns, equation (10) is discretized using finite differences on a 128×128 grid with periodic boundary conditions. The space discretization step is set as $\Delta x^* = \Delta y^* = 1$ and the time interval $\Delta t^* = 0.02$. The parameters chosen for the simulation are $d=2$ and $\gamma = 0.05$. The gradient coefficients are chosen as $a=b=c=10$ and the elastic parameters are $\alpha=\beta=1$. We first describe the simulation of domain pattern formation for the case without dipolar defects ($\varepsilon_x = \varepsilon_y = 0$), starting from small amplitude initial conditions corresponding to a paraelectric phase. Figure 1 displays the time evolution of domains for the defect free case. The direction of the arrows in figure 1 is the polarization direction and the magnitude is proportional to the length. In figure 1(a) ($t^* = 0.5$) we can clearly see the appearance of domains with non-zero polarization. However, these domains are still randomly oriented. In figure 1(b) ($t^* = 12.5$), four distinct kind of polarization domains corresponding to the minnima of equation (2) can be observed. Note that at this stage both charged as well as uncharged domain walls are observed. We can also observe that the domain walls are beginning to get oriented along [11] or [1$\bar{1}$]. This is due to the long-range anisotropic interaction which prefers alignment along the 45° directions. As we go further in time, the domain walls move in order to get rid of the head to head and tail to tail configurations. As we can observe in figure 1(c)($t^*=125$), head to tail domain walls aligned along [11] or [1$\bar{1}$] are dominant. Finally, in figure 1(d) ($t^*=500$), we can see that we have a twinned pattern that has only head to tail uncharged domain walls.

Next, we describe our results for pattern evolution for the case with defects. For the defect field $\vec{\varepsilon}$, we take q_0^* to be uniformly distributed in the interval $[0.01,0.03]$. The charge separation $\vec{\delta}^*$ can take the value $(\pm c,0)$ or $(0,\pm c)$, where c is a random number uniformly distributed in the interval $[0.08,0.1]$. The defect field is initialized by selecting random points on the discrete grid. We choose the number of defects to $n_d = 327$ (2% of the total number of grid points). The initial conditions for the polarization fields are the same as in the defect free case. In figure 2(a)($t^* = 0.5$), we can see the early time snapshot where most of the system is still paraelectric. It is clear that randomly distributed dipoles locally try to align the polarization. These defects strongly influence the dynamics and the eventual state. In figure 2(b)($t^* = 25$), we can see domains of all four degenerate states coexisting. Interestingly, the number of charged and 180° walls for this case is much more compared to the defect free case at similar stages of the evolution. We believe that this is due to the dipolar defects which locally influence the domain pattern and lead to a pinning of the domains, which makes the domain dynamics slower. The system tries to get rid of charged walls, as can be seen by comparing figure 2(b) and figure 2(c)($t^* = 250$). However, in this case the system gets arrested in a metastable state

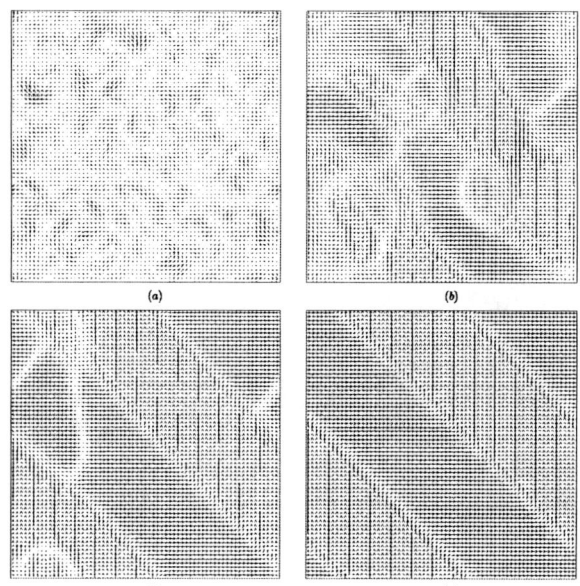

Figure 1. Domain evolution for defect free case

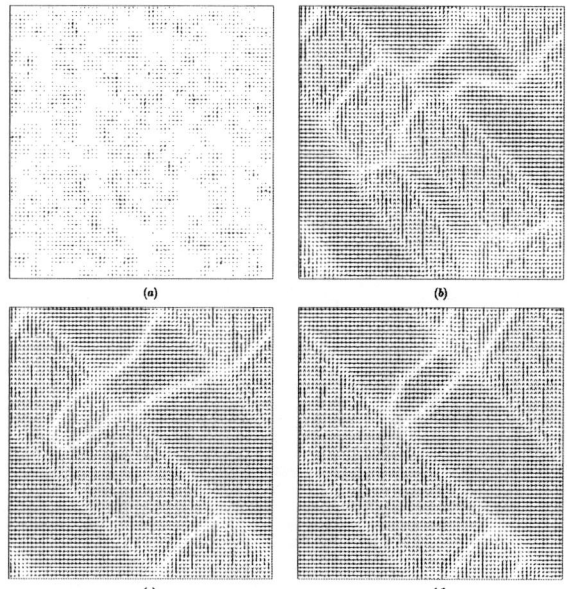

Figure 2. Domain evolution for case with defects

that has charged as well as uncharged walls. This is clear from figure 2(d)($t^* = 1000$), which is the final state as running the simulation longer does not change the domain pattern noticeably.

To conclude, we have made a study of domain pattern formation in ferroelectrics based on the TDGL approach and incorporating elastic coupling. Domain structures evolve with time with domain walls oriented along the [11] or [1$\bar{1}$] directions. This wall direction is preferred to ensure the strain compatibility at the interfaces. For the case with defects, we find that in addition to the uncharged 90° walls, charged walls and 180° walls are stabilized due to the dipolar defects. These results are agreement with recent experiments on PZN-PT single crystals [6]. Although, our model is phenomenological, we have been able to simulate the ferroelectric microstructure qualitatively based on a coarse grained picture. Unfortunately, not all coefficients in the Landau theories are experimentally measured which makes it difficult to give more quantitative predictions. However, our results have demonstrated that the TDGL theory can be used to simulate domain related phenomena in ferroelectric systems if all parameters are measured experimentally.

ACKNOWLEDGEMENT

We acknowledge the support of Office of Naval Research.

REFERENCES

1. S. Nambu and D.A. Sagala, Phys. Rev. B 50, 5838(1994)
2. H.L Hu and L.Q. Chen, Materials. Science. Engineering A 238, 182(1997)
3. W. Cao, S. Tavener and S. Xie, J. Appl. Phys. 86, 5793(1999)
4. R. Ahluwalia and W. Cao, Phys. Rev. B 63, 012103(2001)
5. W. Cao and L. E. Cross, Phys. Rev. B 44, 5(1991)
6. J. Yin and W. Cao, J. Appl. Phys. 87, 133(2000)

Temperature-dependent behavior of $PbSc_{1/2}Nb_{1/2}O_3$ from first Principles

Eric Cockayne and Benjamin P. Burton

Ceramics Division, Materials Science and Engineering Laboratory, National Institute of Standards and Technology, Gaithersburg, MD 20899-8520

Laurent Bellaiche

Physics Department, University of Arkansas, Fayetteville, AR 72701

Abstract.
  We study the ferroelectric phase transition in $PbSc_{1/2}Nb_{1/2}O_3$ (PSN) using a first-principles effective Hamiltonian approach. Results for PSN with NaCl-type ordering of Sc and Nb on the B sites shows that a Pb-centered effective Hamiltonian is appropriate for the ordered cell. We obtain a complete effective Hamiltonian for ordered PSN that has the form of effective Hamiltonians for simple perovskites. We modify the effective Hamiltonian to include the effects of Sc-Nb disorder by adding one additional term giving the effective force on the Pb site due to the nearest neighbor B-site cations. Monte Carlo simulations of our model shows that disordered PSN has a lower Curie temperature than ordered PSN, in agreement with a previous first-principles study based on a virtual crystal approach. No evidence for relaxor behavior is seen in our model.

INTRODUCTION

  $PbSc_{1/2}Nb_{1/2}O_3$ (PSN) crystallizes in a perovskite-type structure. When quenched from high temperatures, PSN has a disordered arrangement of Sc and Nb on the perovskite B-sites. Below the order-disorder transition temperature of 1480 K, however, the Sc and Nb ions order into a NaCl-type arrangement [1]. Disordered $PbSc_{1/2}Nb_{1/2}O_3$ (PSN) exhibits relaxor behavior, at least in a limited temperature range [2], while perfectly ordered PSN is believed to be a normal ferroelectric [3].
  The relatively simple structure of PSN, the existence of both order-disorder and ferroelectric transitions, and the contrasting properties of the ordered and disordered phase make it a good candidate system for a unified first-principles effective Hamiltonian approach. Progress toward such a Hamiltonian was made by Hemphill et al. [4] (HBGV), who used the approach of Bellaiche et al. [5] to create a first-principles effective Hamiltonian for PSN. This Hamiltonian contained two parts:

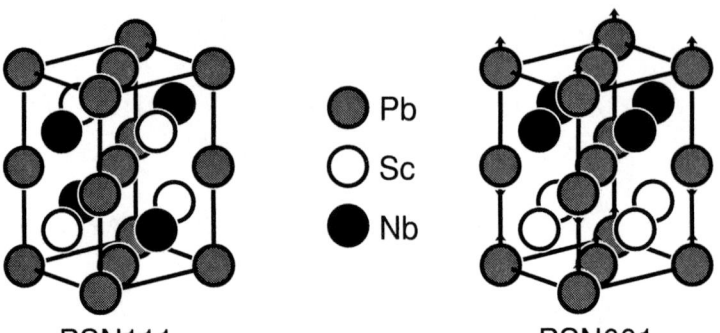

FIGURE 1. $PbSc_{1/2}Nb_{1/2}O_3$ supercells mentioned in this work. For clarity, only the cations are shown. The PSN111 cell is the experimental one. The hypothetical PSN001 supercell is used to compute the local field coupling parameter in the effective Hamiltonian. Relaxation of Pb ions is shown, to scale.

an "average" effective Hamiltonian H_{ave} designed to reproduce the behavior of a $PbSc_{0.5}Nb_{0.5}O_3$ virtual crystal [6] and a local part H_{loc} depending on the arrangement of the Nb and Sc. Simulations of this model yield a lower Curie temperature for disordered PSN than for ordered PSN, in contrast with experiment [3].

The virtual crystal approach, however, does not include all of the physics of the system. For example, one does not easily get the local forces on Pb and O due to changing the B-site cation arrangement. In this study, we take a complementary approach: we start with an effective Hamiltonians for ordered PSN. With the long-term goal of producing an effective Hamiltonian that is valid for any B-site cation configuration, we use first-principles results for a hypothetical PSN supercell with 001-type ordering of Sc and Nb to estimate the effective local field on the Pb due to the nearest neighbor B-site cations. The field coupling parameter can then be applied to any configuration of Sc and Nb. Adding this parameter to our model, we again find, as did HBGV, that disordered PSN has a lower T_c than ordered PSN.

FIRST-PRINCIPLES CALCULATIONS

The NaCl-type ordered PSN structure is shown in Figure 1. It is henceforth labeled PSN111 because this superstructure adds (111)/2-type superlattice peaks to the simple perovskite diffraction pattern.

We perform all calculations on high symmetry cells based on the ideal perovskite structure. The volume per ABO_3 formula was set to the experimental value at room temperature, giving a pseudolattice constant of 4.0828 Å [7] All calculations are performed using VASP (the Vienna ab initio simulation package [8–11]). VASP is a code for plane-wave pseudopotential density functional theory calculations. We

TABLE 1. Born effective charges (in $|e|$) and electronic dielectric constant in PSN111 and the PbSc$_{0.5}$Nb$_{0.5}$O$_3$ virtual crystal.

	$Z^\star_{Pb}$	$Z^\star_{Sc}$	$Z^\star_{Nb}$	$Z^\star_{O_\parallel}$	$Z^\star_{O_\perp}$	ϵ_∞
PSN111	3.85	5.13	7.09	−5.01	−2.48	7.18
PSN-vc	3.90	\multicolumn{2}{c}{$Z^\star_B = 6.44$}	−5.29	−2.54	7.16

used ultrasoft Vanderbilt-type pseudopotentials [12] as supplied by G. Kresse and J. Hafner [13]. The total number of valence electrons used was 14 for Pb, 9 for Sc, 11 for Nb, and 6 for oxygen. All of our calculations were done using a plane-wave energy cutoff of 337.4 eV. We used the local density approximation (LDA) for the exchange-correlation energy. Brillouin zone integration was obtained by calculating Kohn-Sham wavefunctions for a set of **k** points in the Brillouin zone, positioned so as to be equivalent to an $8 \times 8 \times 8$ Monkhorst-Pack grid for a primitive perovskite cell.

Before making any perturbations to compute force constants or electric properties, the atoms were allowed to relax in a symmetry-preserving manner. In PSN111, the O atoms between Sc and Nb moved 0.046 Å away from the Sc toward the Nb.

As described in [14,15], we used the King-Smith and Vanderbilt [16] method to calculate the polarizations of appropriately perturbed PSN111 cells and from this, the Born effective charge tensors of the ions. To calculate the electronic contribution to the dielectric tensors, we used a modification of the method of Bernardini, Fiorentini, and Vanderbilt [17,18]. Full details are given in [14,15].

After obtaining the symmetry-preserving equilibrium structure for PSN111, we displaced each symmetry-independent ion in each Cartesian direction in turn by ±0.01 Å and calculated the induced forces. From finite differences, we determined the force constant matrices. We diagonalized the corresponding dynamical matrices to obtain the normal mode frequencies and eigenvectors. Note that the 20-atom cell shown in Figure 1 has a special property. For a simple perovskite structure, this cell is the smallest supercell which allows the phonons at the Γ, X, M and R points of the cubic Brillouin zone to be determined simultaneously. For complex perovskites ordered in a manner commensurate with this cell, such as PSN111, it allows the equivalent set of phonons to be computed for direct comparisons with simple perovskites.

EFFECTIVE HAMILTONIAN AND TEMPERATURE-DEPENDENT PROPERTIES OF PSN

In Table 1, we give the principal components of the Born effective charges. The Born effective charges show the usual anomalies, especially for the B site cations and for O in the direction of the B-O bonds. The Nb effective charge of +7 in PSN111 is similar to that of Nb in other complex perovskites containing Nb [19], and less than the typical value for Nb of about +9 in simple ANbO$_3$ perovskites [20,21]. The Sc effective charge is as anomalous as the Nb. The PSN111 reference structure has Fm$\bar{3}$m symmetry and thus an isotropic electronic dielectric tensor, whose value is 7.18.

TABLE 2. Symmetry labels and frequencies (in cm^{-1}) for modes at Γ and X for PSN111 in the high-symmetry $Fm\bar{3}m$ double perovskite structure. For comparison with phonon results for simple perovskites, the breakdown of each mode into the symmetry label(s) of a simple perovskite are shown in parentheses. Note that the A site is chosen as the center of symmetry for the simple perovskite labels, while Nb is the center of symmetry for the double perovskite labels. Imaginary frequencies indicate instabilities.

Label	ν	Label	ν	Label	ν
$\Gamma_1(R_{2'})$	830	$X_1(X_{2'} + M_4)$	92	$X_{4'}(X_3)$	167
$\Gamma_{12}(R_{12'})$	588	$X_1(X_{2'} + M_4)$	740	$X_5(X_{5'} + M_5)$	51 i
$\Gamma_{15}(\Gamma_{15} + R_{25'})$	122 i	$X_1(X_{2'} + M_4)$	807	$X_5(X_{5'} + M_5)$	184
$\Gamma_{15}(\Gamma_{15})$	0	$X_2(M_3)$	162 i	$X_5(X_{5'} + M_5)$	268
$\Gamma_{15}(\Gamma_{15} + R_{25'})$	190	$X_{2'}(X_1 + M_{3'})$	137	$X_{5'}(X_5 + M_{5'})$	78 i
$\Gamma_{15}(\Gamma_{15} + R_{25'})$	350	$X_{2'}(X_1 + M_{3'})$	240	$X_{5'}(X_5 + M_{5'})$	78
$\Gamma_{15}(\Gamma_{15} + R_{25'})$	574	$X_{2'}(X_1 + M_{3'})$	424	$X_{5'}(X_5 + M_{5'})$	150
$\Gamma_{15'}(R_{25})$	168 i	$X_{2'}(X_1 + M_{3'})$	571	$X_{5'}(X_5 + M_{5'})$	193
$\Gamma_{25}(\Gamma_{25})$	118	$X_3(M_2)$	601	$X_{5'}(X_5 + M_{5'})$	364
$\Gamma_{25'}(R_{15})$	41 i	$X_{3'}(M_{2'})$	17 i	$X_{5'}(X_5 + M_{5'})$	565
$\Gamma_{25'}(R_{15})$	324	$X_4(M_1)$	377		

The infrared phonon frequencies and symmetry labels for PSN111 are shown in Table 2, along with the label(s) that the corresponding phonon eigenvector would have if the structure were a simple perovskite.

The first step toward creating an effective lattice Hamiltonian for a ferroelectric is to identify the lattice instabilities and to find a local basis for them. As shown in Table 2, PSN111 has three instabilities at Γ and four instabilities at X. Following the choice of HBGV, we based our effective Hamiltonian on the eigenvectors and eigenvalues of the force constant matrix, rather than the dynamical matrix. Note that the numbers and symmetries of lattice instabilities are the same whether one chooses to diagonalize the dynamical matrix or the force constant matrix.

In Table 3, we show the symmetry labels, force constant matrix eigenvalues, and contributions of each kind of atomic motion to the force constant matrix eigenvectors for all of the instabilities in PSN111.

The instabilities fall into 2 types: (1) those that are mainly Pb-displacive and (2) oxygen octahedral tilting. Because Pb-displacive instabilities have greater dispersion and because they include the Γ_{15} instability associated with the ferroelectric transformation, we choose a Pb-centered basis for the effective lattice Hamiltonian. Recognizing that fluctuations of the octahedral tilting may have an effect on the phase transformation thermodynamics, we nonetheless ignore them here for simplicity.

TABLE 3. Force constant matrix symmetry labels, eigenvalues (in eV/Å^2), and contributions of various atomic motions to eigenvectors for PSN111 instabilities, and for the PSN111 X_1 mode used in creating the effective Hamiltonian.

Label	Eigenvalue	Pb %	Sc %	Nb %	O$_\parallel$ %	O$_\perp$ %
Γ_{15}	−2.9063	63.3	1.0	0.5	0.7	34.5
$\Gamma_{15'}$	−1.6517	0.0	0.0	0.0	0.0	100.0
$\Gamma_{25'}$	−1.0008	79.1	0.0	0.0	0.0	20.9
$X_{5'}$	−2.4229	63.3	0.3	3.8	0.2	32.4
X_2	−1.5439	0.0	0.0	0.0	0.0	100.0
X_5	−1.4448	78.4	0.0	0.0	0.0	21.6
$X_{3'}$	−0.2185	100.0	0.0	0.0	0.0	0.0
X_1	6.2604	95.3	0.0	0.0	4.7	0.0

The effective Hamiltonian has the same form that is used for simple perovskites [22]:

$$U = U_0 + U_\text{onsite} + U_\text{local} + U_\text{dipole-dipole} + U_\text{strain} + U_\text{strain coupling}, \quad (1)$$

where the structure is described as a set of local distortion amplitudes $\{\vec{\xi}_i\}$ on the Pb sites. The strain and strain coupling terms include the effects of inhomogeneous strain, as detailed in Ref. [23], and for which an additional local distortion amplitude variable $\{\vec{u}_i\}$, centered on B sites, was added to the model. The only modification that we made in the effective Hamiltonian for PSN111 compared with the form in [22] was to include full anharmonicity in the sixth order local distortion term; *i.e.* replace the $D|\vec{\xi}_i|^6$ term with $D|\vec{\xi}_i|^6 + D'|\vec{\xi}_i|^2(\xi_{ix}^4 + \xi_{iy}^4 + \xi_{iz}^4) + D''(\xi_{ix}^6 + \xi_{iy}^6 + \xi_{iz}^6)$.

The forms of each term and the calculations used to determine the coefficients are given in Reference [22]. On a technical point, we represent the $X_{5'}$, X_5, and X_1 instabilities of Table 3 as unmixed (simple perovskite label) $M_{5'}$, $X_{5'}$ and $X_{2'}$ patterns of $\{\vec{\xi}_i\}$, respectively, in calculating the effective Hamiltonian parameters.

Having a model for ordered PSN, we now modify this model in order to study PSN with B-site disorder. We add a single term to our model for ordered PSN in order to incorporate the effects of the local "field" on the Pb ion. Our local field term has the same form that was used in the random field Potts model of Quian and Bursill [24]:

$$U_\text{local field} = x \sum_i (\sum_j^{nnB} \hat{e}_{ji} \sigma_j) \cdot \vec{\xi}_i, \quad (2)$$

where $\hat{e}_{ji}$ is the unit vector in the direction from site j to site i and σ_j is +1 for Nb and −1 for Sc. The simplest PSN supercell that gives a local field at the B sites is the 001-ordered cell shown in Figure 1. Under symmetry-preserving relaxation, each Pb moves 0.259 Å away from the 001 planes containing Nb toward the 001 planes containing Sc, and each O between Sc and Nb moves 0.072 Å toward the Nb.

TABLE 4. Effective Hamiltonian parameters for PSN. Units are eV per five atom cell, except for Z^*, which is in eÅ, and ϵ_∞, which is dimensionless.

A	-0.8112	a_L	17.24	C_{11}	128.3
B	752.2	a_T	-4.131	C_{12}	38.08
C	542.0	b_L	-0.0340	C_{44}	122.4
D	-1.702×10^4	b_{T1}	-0.0340	g_0	-50.60
D'	6.474×10^4	b_{T2}	1.224	g_1	-136.5
D''	-5.963×10^4	c_L	-0.4300	g_2	-212.6
E	5.708×10^4	c_T	-0.8601	f	1.117
Z^*	25.53	ϵ_∞	7.18	x	-3.041

This relaxation in PSN001 is similar to the X_1 mode in PSN111. We projected the relaxation pattern in PSN001 onto the X_1 mode of PSN111, and then found the value of x which, in the effective Hamiltonian for 001 cation ordering, gave local fields of sufficient strength to reproduce the magnitude of relaxation observed in the fully first-principles computation. The coefficients for the effective Hamiltonian are given in Table 4.

The temperature-dependent behavior of our effective Hamiltonian is determined by the usual Monte Carlo procedure. We simulate the structure and properties on a $10 \times 10 \times 10$ supercell representing 5000 atoms. For disordered PSN (PSN-dis), we randomly occupy the B sites with Sc or Nb with probability 0.5.

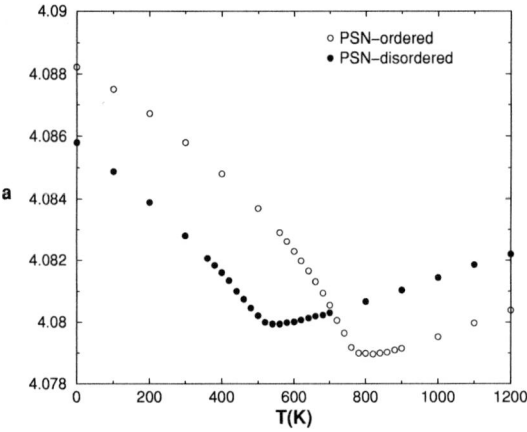

FIGURE 2. Lattice parameter a (in Å) as a function of temperature in PSN model.

At a series of temperatures between 0 K and 1200 K, we calculated the lattice parameters and angles and the polarization. Dielectric properties at T = 0 are calculated by applying an electric field term to the model and measuring the change in polarization with applied field. For $T > 0$, the dielectric properties are calculated

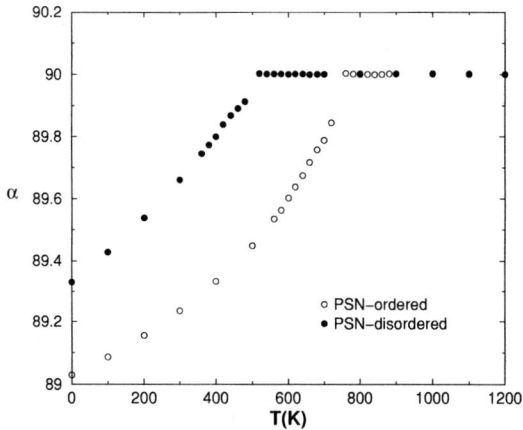

FIGURE 3. Rhombohedral angle α as a function of temperature in PSN model.

through Monte Carlo averaging of polarization fluctuations [25,26]:

$$\epsilon_{\alpha\beta} = (\epsilon_\infty)_{\alpha\beta} + \frac{V}{\epsilon_0 k_B T}(<P_\alpha P_\beta> - <P_\alpha><P_\beta>) \quad (3)$$

Owing to asymmetry of the disordered cell, the physical properties measured along each axis were slightly different. In the thermodynamic (large cell) limit, and also in the ensemble average, all axes become equivalent. We take a subensemble average by averaging the results for each axis.

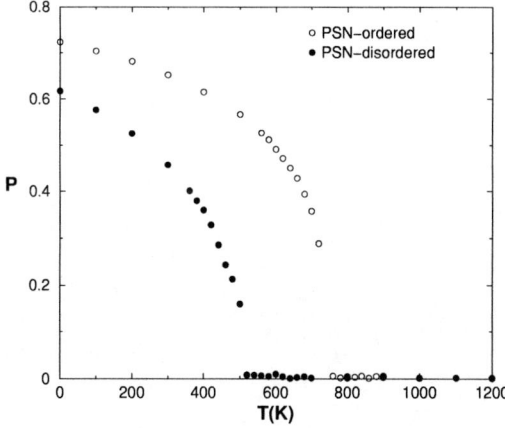

FIGURE 4. Electric polarization P (in C/m^2) as a function of temperature in PSN model.

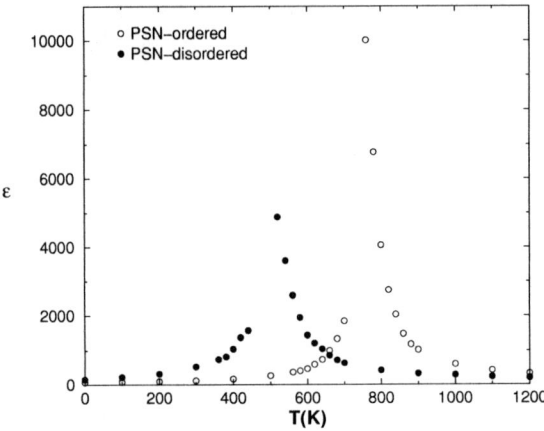

FIGURE 5. Dielectric constant $\epsilon = (Tr[\overleftrightarrow{\epsilon}])/3$ as a function of temperature in PSN model.

The results are shown in Figures 2-5. They are consistent with ferroelectric transitions from a high-temperature cubic phase to a low-temperature rhombohedral phase in both the ordered and disordered cells.

DISCUSSION

Our model gives $T_c \approx 760$ K for ordered PSN111 and $T_c \approx 520$ K for disordered PSN (PSN-dis). These temperatures are lower than the values of 1265 K and 944 K, respectively, computed by HBGV. About 80 K of the decrease in T_c is due to our inclusion of a positive f parameter, which acts as an artificial +2.6 GPa pressure [27]; the rest is due to differences in the effective Hamiltonians. We find that disordering in PSN lowers T_c by about 32 %, in very good agreement with HBGV, who predict a 25 % lowering. Experimentally, on the other hand, Malibert et al. [3] measure $T_c \approx 350$ K for PSN111 and a *higher* $T_c \approx 400$ K for PSN-dis. Further work is needed to understand the source of the discrepancies in the magnitudes and trends in T_c.

The total polarization and lattice distortion at low temperatures are less for PSN-dis that for PSN111. The dielectric constants for both PSN111 and PSN-dis peak at the corresponding transition temperatures. The dielectric constants above T_c fit the Curie-Weiss law $\epsilon = \epsilon_\infty + C/(T - T_c)$ quite accurately, and yield a Curie-Weiss constant C of about 1.7×10^5 for PNS111 and 1.1×10^5 for PSN-dis. Below T_c, the dielectric behavior follows the Curie-Weiss law over a smaller temperature range, and C is higher for the *disordered* system.

There is no clear signature in these results that PSN-dis is a relaxor ferroelectric, such as a spread of high dielectric constant over a range in temperature in which

TABLE 5. Force constant matrix eigenvector for lowest-frequency polar mode in PSN and in the PbSc$_{0.5}$Nb$_{0.5}$O$_3$ virtual crystal.

	u_{Pb}	u_{Sc}	u_{Nb}	$u_{O_\parallel}$	$u_{O_\perp}$
PSN111	0.7959	0.1401	0.0982	−0.4151	−0.0848
PSN-vc	0.7420	$u_B = 0.2439$		−0.4369	−0.0904

the system is nonpolar. One should note, however, that our calculation is for the static dielectric constant and experimentally, the dielectric peak is sharpest at low frequencies. Additionally, for the 10 × 10 × 10 supercell, we estimate that finite size effects are significant within 60 K of T_c. Experimentally, however, the relaxor effects in disordered PSN occur over a narrower range. [2]. HBGV did find a broad peak in the dielectric constant for their model for PSN-dis, possibly because they included a term in their model for the local stress field produced by Sc/Nb size mismatch. Further study will be needed to show which factors are necessary for relaxor behavior.

An important issue in the development of effective lattice Hamiltonians for ferroelectric alloys is whether or not the microscopic properties of a system with fixed composition vary significantly with the configuration. To investigate this issue, we compared microscopic properties computed by HBGV for virtual PSN with the PSN111 properties that we calculated. The results are shown in Tables 1 and 5. The similarities are encouraging and suggest that it is valid to start with the effective Hamiltonian for either the virtual crystal or the PSN111 ordered cell and then add a local field term or terms to encompass the effects of cation configuration.

CONCLUSIONS

We have used first-principles computations to compute the electronic dielectric constant, Born effective charges, and lattice dynamics of ordered PbSc$_{1/2}$Nb$_{1/2}$O$_3$. ¿From these results, we extract an effective Hamiltonian for ordered PSN. In order to model PSN with Sc-Nb disorder, we modify the effective Hamiltonian to include the local field on the Pb ions due to the local arrangement of Sc and Nb ions. Simulations of this model show that the ferroelectric phase transition in disordered PSN takes place at lower temperature than that in the ordered phase, in contrast with experiment. We find significant differences in the Curie-Weiss constants of ordered and disordered PSN in our model, but no evidence for relaxor behavior.

ACKNOWLEDGMENTS

L.B. thanks the financial assistance provided by the Arkansas Science and Technology Authority (Grant N99-B-21), the Office of Naval Research (Grant N00014-00-1-0542) and the National Science Foundation Grant (DMR-9983678).

REFERENCES

1. Stenger, C.G.F., and Burggraaf, A.J., *Phys. Stat. Sol. A* **61**, 275-285 (1980).
2. Chu, F., Reaney, I.M., and Setter, N., *J. Appl. Phys.* **77**, 1671-1676 (1995).
3. Malibert, C., Dkhil, B., Kiat, J.M., Durand, D., Bérar, J.F., and Spasojevic-de Biré, A., *J. Phys.: Condens. Matter* **9**, 7485-7500 (1997).
4. Hemphill, R., Bellaiche, L., Garcia, A., and Vanderbilt, D., *Appl. Phys. Lett.* **77**, 3642-3644 (2000).
5. Bellaiche, L., Garcia, A., and Vanderbilt, D., *Phys. Rev. Lett.* **84**, 5427-5430 (2000).
6. In the *virtual crystal approximation*, the Sc and Nb pseudopotentials V_{Sc} and V_{Nb} are replaced by a single B-site pseudopotential V_B which is effectively $0.5\ V_{Sc} + 0.5\ V_{Nb}$.
7. Perrin, C., Menguy, N., Suard, E., Muller, Ch., Caranoni, C., and Stepanov, A., *J. Phys. Cond. Matt.* **12**, 7523-7539 (2000).
8. Kresse, G., and Hafner J., *Phys. Rev. B* **47**, 558-561 (1993).
9. Kresse, G., Thesis, Technische Universität Wien, 1993.
10. Kresse, G. and Furthmüller, J., *Comput. Mat. Sci.* **6**, 15-50 (1996).
11. Kresse, G. and Furthmüller, J., *Phys. Rev. B* **54**, 11169-11186 (1996).
12. Vanderbilt, D., *Phys. Rev. B* **41**, 7892-7895 (1990).
13. Kresse, G., and Hafner, J., *J. Phys.: Condens. Matter* **6**, 8245-8257 (1994).
14. Cockayne, E., and Burton, B.P., *Phys. Rev. B* **62**, 3735-3743 (2000).
15. Cockayne, E., submitted to *J. Appl. Phys.*
16. King-Smith, R.D. and Vanderbilt, D., *Phys. Rev. B* **47**, 1651-1654 (1992).
17. Bernardini, F. Fiorentini, V., and Vanderbilt, D., *Phys. Rev. Lett.* **79**, 3958-3961 (1997).
18. Bernardini, F., and Fiorentini, V., *Phys. Rev. B* **58**, 15292-15295 (1998).
19. Bellaiche, L., Padilla, J., and Vanderbilt, D., *Phys. Rev. B* **59**, 1834-1839 (1999).
20. Resta, R., Posternak, M., and Baldereschi, A., *Phys. Rev. Lett.* **70**, 1010-1013 (1993).
21. Wang, C.-Z., Yu, R., and Krakauer H., *Phys. Rev. B* **54**, 11161-11168 (1996).
22. Rabe, K.M., and Waghmare, U.V., *Ferroelectrics* **194**, 119-134 (1996).
23. Zhong, W., Vanderbilt, D., and Rabe, K.M., *Phys. Rev. B* **52**, 6301-6312 (1995).
24. Qian, H., and Bursill, L.A., *Int. J. Mod. Phys. B* **10**, 2027-2047 (1996).
25. Garcia, A., and Vanderbilt, D., "Temperature-Dependent Dielectric Response of $BaTiO_3$ from First Principles", in *First Principle Calculations for Ferroelectrics*, edited by R.E. Cohen, AIP Conference Proceedings 436, Woodbury, NY,1998, pp. 53-60.
26. Rabe, K.M., and Cockayne, E., "Temperature-Dependent Dielectric and Piezoelectric Response of Ferroelectrics from First Principles", in *First Principle Calculations for Ferroelectrics*, edited by R.E. Cohen, AIP Conference Proceedings 436, Woodbury, NY,1998, pp. 61-70.
27. This parameter was adjusted so that the lattice parameter of the model for disordered PSN agrees with experiment at room temperature.

First Principles and Semi-empirical Calculations of Atomic and Electronic Structure for the (100) and (110) Perovskite Surfaces

E. Heifets[a], R.I. Eglitis[b], E.A. Kotomin[b,c], and G.Borstel[b]

[a]*Carnegie Institution of Washington, 5251 Broad Branch Rd., N.W. Washington D.C. 20015 and California Institute of Technology, MS 252-21, Pasadena CA 91125*
[b]*Department of Physics, University of Osnabrueck, D-49069 Osnabrueck, Germany*
[c]*Institute for Solid State Physics, University of Latvia, 8 Kengaraga str., Riga LV-1063, Latvia*

Abstract. We present and discuss results of the calculations for $BaTiO_3$ and $SrTiO_3$ surface relaxation with different terminations using a semi-empirical *shell model* (SM) as well as *ab initio* methods based on Hartree-Fock (HF) and Density Functional Theory (DFT) formalisms. Using the SM, the positions of atoms in 16 near-surface layers placed atop a slab of rigid ions are optimized. This permits us determination of surface rumpling and surface-induced dipole moments (polarization) for different terminations of the (100) and (110) surfaces. We also compare results of the *ab initio* calculations based on both HF with the DFT-type electron correlation corrections, several DFT with different exchange-correlation functionals, and hybrid exchange techniques. Our SM results for the (100) surfaces are in a good agreement with both our *ab initio* calculations and LEED experiments. For the (110) surfaces O-termination is predicted to be the lowest in energy.

INTRODUCTION

Thin films of ABO_3 perovskite ferroelectrics are important for many high tech applications including high capacity memory cells, catalysis, optical waveguides, integrated optics applications, substrates for the high T_c cuprate superconductor growth, etc. [1-4] where surface structure and quality are of primary importance. In this paper, we calculate the atomic structure of the $SrTiO_3$ and $BaTiO_3$ (100) and (110) surfaces for the ideal cubic phases. It should be noted that at all temperatures bulk $SrTiO_3$ exhibits paraelectric properties, despite the antiferrodistortive (AFD) transition at 105 K to a tetragonal phase in which the oxygen octahedra have rotated in opposite directions in neighboring unit cells [5]. In contrast, iso-structural $BaTiO_3$ undergoes several phase transitions from paraelectric to ferroelectric phases as the temperature decreases.

The $SrTiO_3$ (100) surface relaxation has been characterized by means of low energy electron diffraction (LEED), reflection high-energy electron diffraction (RHEED), and medium energy ion scattering (MEIS) measurements [6-10]. Recently, several *ab initio* [11-17] and shell model (SM) [18-20] studies were published for the (100) surface of $BaTiO_3$ and $SrTiO_3$ crystals. Here we perform much more detailed SM studies for both crystals with different terminations, supported by *ab initio* calculations.

The (110) surface only recently became a subject of intensive experimental investigations, focusing mainly on $SrTiO_3$, and using STM, UPS, XPS, Auger spectroscopies, and LEED [21]. Here we report the first simulations of (011) surfaces of $BaTiO_3$ and $SrTiO_3$ crystals. We performed these (110) surface simulations using SM.

METHOD

In the present study we studied a two-dimensional slab of cubic $SrTiO_3$ and $BaTiO_3$ crystals by means of the SM [22] as realized in the MARVIN computer code [23]. To study the surface relaxation, we optimized the atomic positions in several (varied from one to 16) near-surface planes, placed into the electrostatic field of the slab (simulated by 20 additional planes whose atoms were fixed in their perfect lattice sites). The number of these additional planes was chosen to reach a convergence of the crystalline field in the surface planes.

In our slab calculations we simulated Ti- and Sr(Ba)-terminated (100) surfaces as well as Ba-, Ti- and O-terminated (110) surfaces. For each termination surface, modification was characterized by the surface rumpling (s), interplane distances between top metal and the second crystal layers (d_{12}), and between the second and the third crystal layer (d_{23}). Our calculations of the interplane distances are based on the metal ion (Ti or Sr) displacements from unrelaxed planes, which are known to be much stronger electron scatterers than O ions [6]. In SM calculations, atoms from one to 16 near-surface planes were allowed to relax in order to achieve the minimum total energy. As a result, we obtain the optimized slab geometry and dipole moments caused by core and electron shell displacements from regular lattice sites. The surface energy was calculated as $E_s = E_{tot} - E_1 - n\, E_{bulk}$, where E_{tot} is the total energy for a slab of n relaxed planes placed on the rigid substrate, E_{bulk} the total energy per bulk unit cell, and E_1 the interaction energy between relaxed slab and rigid substrate.

To check our SM results, we also performed *ab initio* calculations based on HF with different DFT *posteriori* electron correlation corrections to the total energy (including generalized gradient approximation, GGA, Perdew-91, and Lee, Yang, Parr (LYP)) as well as the Kohn-Sham equation with a number of exchange-correlation functionals (LDA, PBE, PWGGA), including *hybrid* HF-DFT exchange functionals (B3LYP, B3PW) [24]. For this purpose we used the CRYSTAL-98 computer code [24] using the Gaussian-type basis set, and we optimized atomic positions in several top layers of $SrTiO_3$ slab consisting of 7 planes terminated by Ti and O atoms on both sides of the slab.

In both SM and ab initio simulations we used a single slab and had no periodicity along z axis. All interactions across the slab were summed directly. Ewald procedure was applied only for summation of Coulomb interactions in two dimensions along the slab. Therefore, our calculations took into account the depolarization field.

MAIN RESULTS

A. The (100) Surfaces

Table 1 gives the displacement magnitudes for atomic cores and shells as found in SM calculations for three top layers nearby the surface. Our calculations show that Ti^{4+}, Sr^{2+}, and O^{2-} ions in the planes close to the surface reveal different displacements from their perfect crystalline sites. One can also see the difference between the SrO- and TiO_2-terminated surfaces. In most cases the surface ions are displaced inwards, whereas the ionic displacements in the second layer point outwards from the crystal. In particular, on the SrO-terminated surface of $SrTiO_3$, the surface ions shift inwards by 7% of the bulk lattice constant (a_0=3.89 Å) whereas in the third layer the displacements of Sr ions are reduced to 1.4 %. Similarly, on the BaO-terminated surface of $BaTiO_3$, the surface ions shift inward by 3.7% of bulk lattice constant (a_0=3.96 Å) and by 0.5% in the third layer. Ti ions in the second layer of the $SrTiO_3$ crystal are displaced outwards by ≈1.6%. In $BaTiO_3$ this value is ≈1.3%. Inward displacement of Ti in the fourth layer in both crystals is ≈ 0.2%. The cores of O ions in the first plane in both crystals are displaced outward, but their shells are displaced inward, which means strong O atom polarization. *Both* O cores and shells in the second (TiO_2) layer relax outwards the crystals.

Very similar trends in ionic displacements are observed for the Ti-terminated surface (Table 1). Inward displacements of the surface Ti ions are ≈ 3% in $SrTiO_3$ (≈ 2.7 % in $BaTiO_3$) and outward displacements of Sr ions in the second layer are nearly the same in magnitude. The O ions are displaced inwards in the top layer; again we can see the opposite displacement directions for the O cores and shells in the second plane.

Table 1. SM relaxation of the uppermost three layers in the Sr- and Ti-terminated (100) $SrTiO_3$ surfaces. Totally 16 near-surface planes were allowed to relax. Ionic displacements are in percents of a_0=3.89Å (the bulk crystal lattice parameter). Positive (negative) displacements mean the direction outwards (inwards) the surface. Numbers in brackets are results of previous *ab initio* plane wave with pseudopotentials calculations[12].

Sr-terminated				Ti-terminated			
Layer	Ion	Type	Δz(%)	Layer	Ion	Type	Δz(%)
1	Sr^{2+}	Core	-7.10(-5.7)	1	Ti^{4+}	core	-2.96(-3.4)
		Shell	-5.03			shell	-2.88
	O^{2-}	Core	1.15 (0.1)		O^{2-}	core	-1.73 (-1.6)
		Shell	-3.15			shell	-2.40
2	Ti^{4+}	Core	1.57 (1.2)	2	Sr^{2+}	core	3.46 (2.5)
		Shell	1.53			shell	2.63
	O^{2-}	Core	0.87 (0.0)		O^{2-}	core	-0.21 (-0.5)
		Shell	1.21			shell	1.34
3	Sr^{2+}	Core	-1.42 (-1.2)	3	Ti^{4+}	core	-0.60 (-0.7)
		Shell	-1.10			shell	-0.59
	O^{2-}	Core	0.70 (-0.1)		O^{2-}	core	-0.29 (-0.5)
		Shell	-0.58			shell	-0.43

Table 2. Relaxation of uppermost three layers (per cent of lattice constant) for Ti-terminated SrTiO$_3$ (100) found in the *ab initio* HF with different electron correlation corrections and DFT calculations (see in the text).

				A. Ti-terminated SrTiO$_3$ surface									
N	Ion	[11]	[17]	DFT (Kohn Sham)						Hartree-Fock with *posteriori* corrections			
				LDA	B3LYP	B3PW	BLYP	PBE	PWGGA	HF	HFGGA	HF P91	HFLYP
1	Ti^{4+}	-3.4	-1.79	-2.12	-2.03	-2.19	-2.28	-1.88	-2.31	-2.74	-3.20	-3.19	-3.05
	O^{2-}	-1.6	-0.26	-1.11	-0.72	-0.93	-0.90	-0.57	-1.19	-1.38	-2.20	-2.20	-1.87
2	Sr^{2+}	2.5	4.61	2.21	2.38	2.18	2.64	2.75	2.04	1.91	1.81	1.83	1.87
	O^{2-}	-0.5	0.77	0.07	0.21	0.01	0.12	0.45	0.0	-0.13	-0.15	-0.17	-0.17
3	Ti^{4+}	-0.7	-0.26							-0.26	-0.28	-0.28	-0.28
	O^{2-}	-0.5	0.26							-0.05	-0.13	-0.14	-0.14

				B. Sr-terminated SrTiO$_3$ surface									
N	Ion	[11]	[17]	DFT (Kohn Sham)						Hartree-Fock with *posteriori* corrections			
				LDA	B3LYP	B3PW	BLYP	PBE	PWGGA	HF	HFGGA	HF P91	HFLYP
1	Sr^{2+}	-5.7	-6.66	-8.48	-4.28	-4.29	-2.78	-4.25	-4.28	-2.61	-4.16	-4.13	-3.74
	O^{2-}	0.1	1.02	-2.37	0.64	0.61	2.28	1.02	0.90	1.56	0.41	0.35	0.10
2	Ti^{4+}	1.2	1.79	-0.42	1.16	1.25	1.85	1.27	1.29	0.79	0.48	0.48	0.56
	O^{2-}	0.0	0.26	-0.87	0.85	0.82	1.35	0.73	0.61	0.51	0.23	0.17	0.27
3	Sr^{2+}	-1.2	-1.54							-0.49	-0.69	-0.69	-0.70
	O^{2-}	-0.1	0.26							0.01	-0.25	-0.02	-0.14

Along with SM calculations, we performed *ab initio* calculations for the SrTiO$_3$ surfaces, after first testing them for the bulk properties. LDA calculations underestimate the lattice constant a_0 (by 0.8%) and overestimate the bulk modulus B by 4.5 %. The hybrid B3PW method gives a better result for B (Δ=4%) and only by 0.5% *overestimates* a_0. The HF method without any correlation corrections overestimates a_0 only by 1% but considerably overestimates B (by 16%). Lastly, HF with GGA corrections makes a_0 too small (Δ = -1.5%) but B even larger (Δ= 37%). In other words, it is quite difficult to choose the optimal method reproducing all properties equally well, but the hybrid B3PW method looks like the best one.

Atomic displacements in several top SrTiO$_3$ planes calculated using the *ab initio* methods (see Table 2) are in a good agreement with SM results and discussed *ab initio* plane wave calculations [11, 17]. Both DFT and HF with correlation effects predict Sr displacement on the Sr-terminated surface to be larger than that for Ti atom on the Ti-terminated surface, in agreement with previous plane wave calculations [11,17].

Table 3 gives atomic displacements, the effective static charges (obtained with the Mulliken population analysis), and bond populations between nearest metal and oxygen atoms. Let us use B3PW results for the analysis. The main effect observed well here is strengthening of the Ti-O chemical bond near the surface. Its population on the Ti- terminated surface is 124 me, which is about one and a half times larger than the relevant values in the bulk (82 me). The Ti-O populations between the first and second, the second and third plane, and lastly, the third and fourth planes (124 me, 92 me, and 86 me) also exceed the bulk value. In contrast, the Sr-O populations are very small and even negative which indicates the repulsion. This effect is also well seen from the effective charges: that for Sr is close to the formal ionic charge of 2e, whereas charges for Ti and O atoms are much smaller than the ionic charges of 4e, and -2e, respectively, due to the covalent bonding between them.

Next, we performed SM calculations of the surface energy, E_s, for the relaxed surfaces. Its magnitude saturates when more than 8 near-surface planes are allowed to relax. In the case of Ba-terminated BaTiO$_3$ surface, E_s=1.45 eV/cell, which is only slightly larger then that for the Ti-terminated case (1.40 eV). This tiny difference appears entirely due to the difference in the relaxation energies of the surfaces in both cases. Since the difference is very small, both types of surfaces should co-exist, which is confirmed by both *ab initio* calculations [5] and experiments [6]. For SrTiO$_3$, the Ti-terminated surface energy of 1.37 eV is slightly larger than that for Sr- termination (1.33 eV/cell). The *ab initio* calculations [11, 12] gave quite similar average surface energies (1.26 eV/cell and 1.24 eV/cell for SrTiO$_3$ and BaTiO$_3$, respectively.)

The surface dipole moments for different numbers of relaxed layers oscillate as the number of relaxed near-surface layers increases from one to six. For a larger number

Table 3. Calculated absolute atomic displacements d (in Å), the effective atomic charges Q (in e) and bond populations P between nearest Me-O atoms (in me) for the Ti and Sr- terminations

A. TiO$_2$-terminated SrTiO$_3$ surface						B. SrO-terminated SrTiO$_3$ surface					
			DFT		Hartree - Fock				DFT		Hartree - Fock
N	Ion		B3PW	B3LYP	HF	HFGGA	N	Ion		B3PW	B3LYP	HF	HFGGA
1	Ti^{4+}	d	-0.086	-0.079	-0.107	-0.1245	1	Sr^{2+}	d	-0.168	-0.168	-0.102	-0.162
		Q	2.165	2.197	2.507	2.502			Q	1.829	1.833	1.897	1.894
		P	128	124	146	142			P	-4	-2	-20	-20
	O^{2-}	d	-0.037	-0.028	-0.054	-0.0856		O^{2-}	d	0.024	0.025	0.061	0.016
		Q	-1.241	-1.257	-1.395	-1.400			Q	-1.44	-1.459	-1.581	-1.555
		P	-10	-10	-28	-30			P	166	160	198	208
2	Sr^{2+}	d	0.085	0.094	0.074	0.0705	2	Ti^{4+}	d	0.049	0.046	0.031	0.019
		Q	1.833	1.834	1.890	1.888			Q	2.24	2.282	2.536	2.524
		P	-10	-8	-20	-20			P	56	58	90	84
	O^{2-}	d	0.0004	0.008	-0.005	-0.006		O^{2-}	d	0.32	0.033	0.020	0.009
		Q	-1.297	1.307	-1.418	-1.401			Q	-1.423	-1.434	-1.517	-1.523
		P	92	92	104	104			P	-10	-8	-20	-20
3	Ti^{4+}	d			-0.010	-0.011	3	Sr^{2+}	d			-0.019	-0.027
		Q	2.269	2.313	2.553	2.552			Q	1.857	1.855	1.907	1.906
		P	86	86	114	114			P	-10	-8	-22	-22
	O^{2-}	d			-0.002	-0.005		O^{2-}	d			0.0	-0.01
		Q	-1.363	-1.376	-1.476	-1.476			Q	-1.398	-1.409	-1.508	-1.503
		P	-10	-8	-22	-22			P	80	80	108	108
Bulk	Ti^{4+}	Q	2.272	2.325	2.584	2.50	Bulk	Ti^{4+}	Q	2.272	2.325	2.584	2.50
		P	82	74	112	100			P	82	74	112	100
	O^{2-}	Q	-1.375	-1.392	-1.497	-1.466		O^{2-}	Q	-1.375	-1.392	-1.497	-1.466
		P	-10	-8	-20	-28			P	-10	-8	-10	-28
	Sr^{2+}	Q	1.852	1.852	1.909	1.898		Sr^{2+}	Q	1.852	1.852	1.909	1.898

of relaxed layers these oscillations practically vanish and the dipole moments saturate at the level of 0.1-0.2 eÅ. Note that the same number of layers (six) was found by us as necessary to reach convergence of the crystalline field in the surface region. For both $SrTiO_3$ surfaces and Ti-terminated $BaTiO_3$ surfaces, the dipole moments are negative, whereas for Ba-termination it turns out to be positive (but close to zero). In all cases, a large polarization of ions in the near-surface layers takes place. It manifests itself through the large difference in displacements of cores and shells of the same ions.

A comparison of our SM calculations for the surface structure with the *ab initio* plane-wave studies [11,12], other SM simulations [19] and experimental results [6,7,9] is presented in Table 4. The agreement of our results with the plane-wave ab initio calculations and the experimental data is very good. We observe also only qualitative agreement with other SM calculations for $SrTiO_3$ [18-20] (Table 4). Anyway, all theoretical methods give the same signs for both the rumpling and change of the interplanar distances.

The results of the LEED and RHEED experiments [6-8] presented in the same Table 4 suggest relaxations quite different from calculated quantities. Note also that the results of these two experiments contradict each other (e.g., in the sign of the d_{12} for Sr- terminated surface). Both our simulations and calculations [11] gave values, which are closer to the experimental data [6]. Lastly, it was found in recent MEIS experiments [9] for the Ti-terminated $SrTiO_3$ surface that $s \approx 2$ %, which is close to our result in Table 4.

However, despite a good agreement for the O and Ti atom displacements from the top plane relative to each other, one serious question remains open: both experiments argue that the topmost O atoms move outwards from the surface, whereas all calculations give that in most cases O goes inwards the surface.

B. The (110) Surfaces

The problem with calculating the (110) surfaces of SrTiO3 and $BaTiO_3$ is that they consist of charged planes. If the (110) surface was to be modeled exactly as one would expect after crystal cleavage, it would have an infinite dipole moment perpendicular to the surface which makes the surface unstable. To avoid this problem, in our calculations we removed half of O atoms from the O-terminated surface, the Sr (Ba) atoms from the Ti-terminated surface, and Ti and O atoms from the Sr (Ba)-terminated surface [22] (Fig.1). As a result, we obtain the so-called *type-II* stable surface with charged planes but a zero dipole moment [25]. The initial atomic configuration for the O-terminated surface, where every second surface O atom is removed and others occupy the same sites as in the bulk structure (Fig.1b), we call hereafter *asymmetric* denoted in oncoming Tables as *A*. Since such a removal of half of O atoms disturbs the balance of interatomic forces along the surface, we studied also another, *symmetric* initial surface configuration (denoted as *B*) in which the O atom is placed in the *middle* of the distance between two equivalent O atoms in the bulk (Fig.1a,c).

Table 4 Surface rumpling s, and relative displacements of the three near-surface planes for the Sr(Ba)- and Ti-terminated SrTiO$_3$ and BaTiO$_3$ (100) surfaces (in percents of the bulk lattice parameters).

	SrTiO$_3$					
	Sr-terminated			Ti-terminated		
Method	s	d_{12}	d_{23}	s	d_{12}	D_{23}
Present study (SM)	8.2	-8.6	3.0	1.2	-6.4	4.0
Shell model [19]	4.5	-4.75	1.45	1.1	-3.95	1.2
Ab initio [11]	5.8	-6.9	2.4	1.8	-7.0	3.2
LEED experiment [6]	4.1±2	-5±1	2±1	2.1±2	1±1	-1±1
RHEED experiment [7,8]	4.1	2.6	1.3	2.6	1.8	1.3

	BaTiO$_3$					
	Ba-terminated			Ti-terminated		
Method	s	d_{12}	d_{23}	s	d_{12}	D_{23}
Present study (SM)	2.7	-5.0	1.8	1.8	-4.9	2.5
Ab initio [12]	1.4	-3.7	1.5	2.3	-5.2	2.0

The calculated surface energies for Ba-, Sr-, Ti-terminations of BaTiO$_3$ and SrTiO$_3$ (Table 5) are much larger than those for the (001) surfaces. The O-terminated asymmetric surfaces have considerably lower energies for both BaTiO$_3$ and SrTiO$_3$. These energies turn out to be close to those for the (100) surface. Thus, O-termination should be predominant when crystal is cleaved or grown along the (011) plane. On the other hand, symmetrical O-termination is energetically costly and thus unfavorable.

Table 6 gives the predicted values for surface rumpling and the relative displacements of the two top layers. For the Ti-terminated surface, the rumpling for both SrTiO$_3$ and BaTiO$_3$ (110) surfaces is similar and very large, ≈ 13-14 %. This relaxation is much larger than what we found for the (100) surface. The reduction of relative distances between the first three layers is quite similar for both crystals, being about 4-5 %. Note also that surface ions are strongly polarized.

For the O-terminated surface, when O ions are placed initially into symmetrical (B) surface positions, all near-surface atoms are displaced along the z axis perpendicular to the surface. On the other hand, if O ions in SrTiO$_3$ and BaTiO$_3$ are placed initially in asymmetrical positions (A), atoms reveal also in-plane displacements, i.e., in the direction *parallel* to the surface. As a result, we found another optimized surface structure, with considerably smaller surface energy. In this structure, Ba (Sr) atoms in the second plane are only moderately (≈4 %) shifted outwards. But now the surface O ions tend to move inwards, so that they are displaced along the z axis by as much as 11-14 %. The surface O ions are strongly polarized; the relative core-shell separation is ≈ 4 %. As a result, the first and second planes turn out to be compressed. Their separation is reduced by ≈12 %. The distance between the second and third planes increases by ≈9 % because of the strong O atom inward displacements in the third plane. Note that these O atoms are also strongly polarized. Similar to the (100) surface, the (110) surface-induced dipole moments oscillate as the number of relaxed

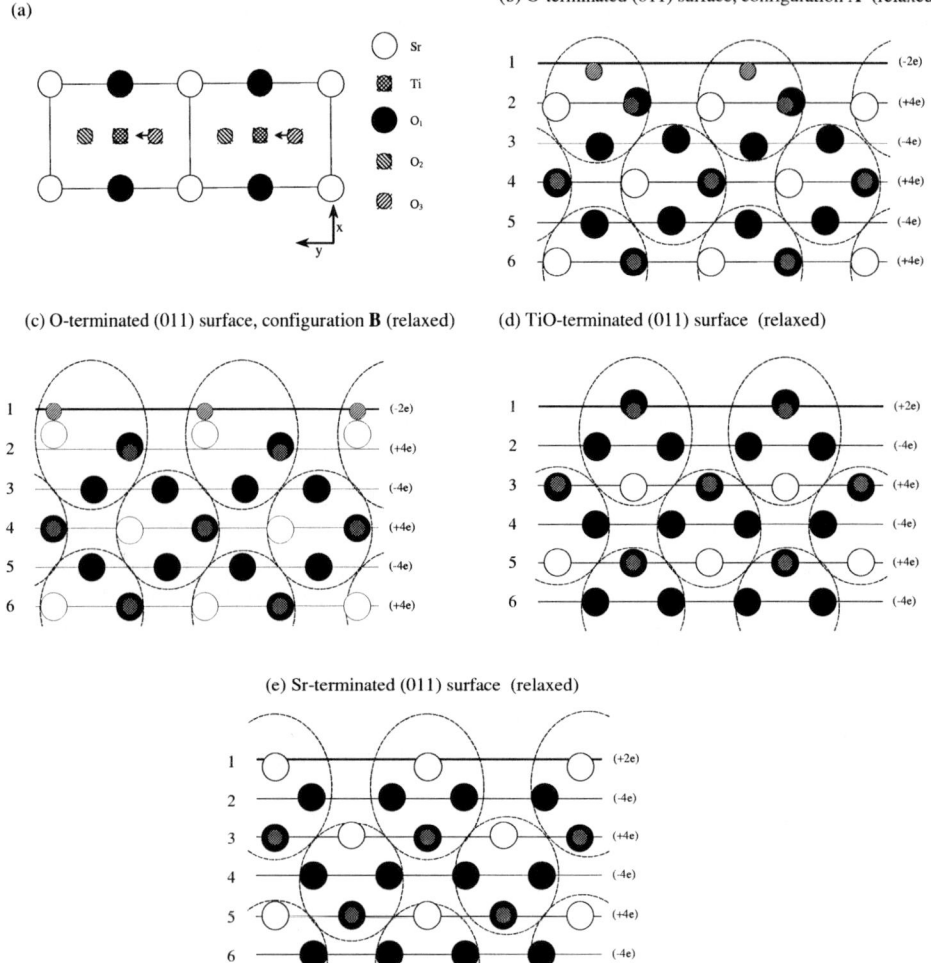

Fig.1. (a) Top view of the (110) O-terminated surface, directions of the O atom displacements are shown by arrows. In our model, we remove atoms O_2 from the O-terminated surface and search for the atomic relaxations when O_3 are placed initially into asymmetric A or symmetric B positions (see text for explanations). Atoms of Ti, Sr, and O_1 lie in the second plane below the O-surface plane. (b,c) Side view of the two possible configurations A and B after relaxation. (d,e) Relaxed TiO- and Sr-terminated surfaces. Dashed ellipses containing 5 atoms in the 3 nearest planes show neutral fragments from which the surface unit cell is built. Numbers in brackets on the right-hand side indicate the corresponding effective charges of related planes in the cell.

Table 5. Surface energies (in eV/unit cell) for (011) BaTiO$_3$ and SrTiO$_3$ surfaces with different terminations calculated by means of SM

Termination	Surface energies	
	BaTiO$_3$	SrTiO$_3$
Ba(Sr)-terminated	4.2	3.4
TiO-terminated	2.3	2.4
O-terminated (A), asymmetric	1.8	1.6
O-terminated (B), symmetric	4.8	3.4

Table 6. Surface rumpling s, and relative displacements of the three near-surface plane for the O- and Ti-terminated SrTiO$_3$ and BaTiO$_3$ (110) surfaces (in percents of the bulk lattice parameter) obtained for the symmetrical initial O position **B**. Numbers in brackets correspond to the asymmetrical O position **A**.

	O-terminated		Ti-terminated		
	Δd_{12}	Δd_{23}	s	Δd_{12}	Δd_{23}
SrTiO$_3$	2.36(-11.83)	-2.73(8.69)	14.47	-4.27	-3.86
BaTiO$_3$	3.34	-3.89	13.38	-5.27	-3.25

near-surface layers increases from one to ten. Then these oscillations vanish and the dipole moment saturates. The values of the surface dipole moments perpendicular to the BaTiO$_3$(110) saturate at the -0.93 e Å for Ba-terminated and -1.61 e Å for Ti-terminated surfaces.

This is in sharp contrast with the results for the (100) surface for which we got dipole moments of 0.07 e Å in the case of the Ba-termination, and -0.19 e Å for the Ti-terminated surfaces. That is, the (110) surface dipole moments and respectively, surface polarizations, are much larger. In contrast, for the O-terminated BaTiO$_3$ in both initial configurations- asymmetric and symmetric- dipole moments are positive and of the same order of magnitude as for the Ba-, Ti-terminations. It is, however, important to stress, that in the asymmetric O configuration A we observe also the strong on-plane polarization p_y caused by the dipole moments much larger than those perpendicularly to the surface. This is a manifestation of the ferroelectric reconstruction predicted for the first time for the (100) surface [18].

The perpendicular surface polarization of Sr- and Ti-terminated SrTiO$_3$ is quite similar to that for the BaTiO$_3$. However, dipole moments for the O-terminated asymmetric SrTiO$_3$ surface reveal strong oscillations even for 16 relaxed layers (similar to those observed above for the surface energy). This is caused by this surface instability with respect to the AFD-type relaxation, which is observed for SrTiO$_3$ at low temperatures. In contrast, p_y dipole moment for the symmetric O configuration saturates rapidly to zero.

DISCUSSION AND CONCLUSION

A comparison of our semi-empirical SM results with *our ab initio* HF and DFT as well as with previous plane wave pseudopotential calculations [11-13] and experimental low-energy electron diffraction [6] studies of SrTiO$_3$ (001) surface clearly demonstrates their good agreement for the *rumpling* and the relative displacements of the second and third planes. We found that the metal-terminated surface relaxations for $SrTiO_3$ and $BaTiO_3$ (110) surfaces are much larger than those for the (100) surfaces. This is in a line with observations for other oxides. Another important prediction of our calculations is that the asymmetric O-termination is the energetically most favorable amongst all (110) surface terminations. The formation of large dipole moments parallel and perpendicular to the relaxed surface even in a cubic phase of perovskite could considerably affect the ferroelectric properties of thin ferroelectric films.

ACKNOWLEDGMENTS

This study was partly supported by DAAD (grant to EK through Osnabrück University) and by ONR grant #N00014-97-1-0052 to R. E. Cohen. Authors are indebted to R. E. Cohen, R. Dovesi, C.R.A. Catlow, F. Cora, S. Dorfman, D. Fuks, and D. Vanderbilt for fruitful discussions.

REFERENCES

1. Noguera C, *Physics and Chemistry at Oxide Surfaces*, Cambridge Univ. Press, N.Y., 1996.
2. Lines M.E., and Glass A.M., *Principles and Applications of Ferroelectrics and Related Materials*, Clarendon, Oxford, 1977.
3. Auciello O., Scott J.F., and Ramesh R., *Physics Today*, July 1998, pp.22-30.
4. Proceedings of the Williamsburg workshop on Ferroelectrics-99, *J. Phys. Chem. Sol.*, **61**, No 2 (2000).
5. Zhong W, and Vanderbilt D., *Phys. Rev.* B **53**, 5047-5055 (1996).
6. Bickel N., Schmidt G., Heinz K., and Müller K., *Phys. Rev. Lett.* **62**, 2009-2013 (1989).
7. Hikita T., Hanada T., Kudo M., Kawai M., *Surf. Sci.* **287/288**, 377-380 (1993).
8. Kudo M., Hikita T., Hanada T., Sekine R., and Kawai M., *Surf. and Interf. Analysis*, **22**, 412-416 (1994).
9. Ikeda A., Nishimura T., Morishita T., and Kido Y., *Surf. Sci.* **433-435**, 520-525 (1999).
10. Nishimura T., Ikeda A., Namba H., Morishita T., Kido Y., *Surf. Sci.* **421**, 273-278 (1999).
11. Padilla J., and Vanderbilt D, *Surf. Sci.* **418**, 64-70 (1998).
12. Padilla J., and Vanderbilt D, *Phys. Rev.* B **56**, 1625-1630 (1997).
13. Meyer B., Padilla J., and Vanderbilt D, *Faraday Discussions*, **114**, 395-405 (1999).
14. Cora F., and Catlow C.R.A., *Faraday Discussions*, **114**, 421-430 (1999).
15. Cohen R.E., *Ferroelectrics* **194**, 323-342 (1997).
16. Fu L., Yashenko E., Resca L., and Resta R., *Phys. Rev.* B **60**, 2697-2703 (1999).
17. Cheng C., Kunc K., and Lee M.H., *Phys. Rev.* B, **62**, 10409-10417 (2000).
18. Ravikumar V., Wolf D., and Dravid V.P., *Phys. Rev. Lett.*, **74**, 960-964 (1995).
19. Prade J., Schröder U., Kress W., and de Kulkarni F.W., *J. Phys: Condens. Matter*, **5**, 1-15 (1993).
20. Tinte S., and Stachiotti M.G., *AIP Conf. Proc* **535** ed. R Cohen, 273-282 (2000).
21. H.Bando H., Aiura Y., Haruyama Y., Shimizu T.,and.Nishihara Y., *J. Vac. Sci. Technol.*, B **13**, 1150-1158 (1995); Szot K., and Speier W., *Phys. Rev.* B **60**, 5909-5920 (1999); .Jiang O.D., and Zegenhagen J., *Surf. Sci.* **425**, 343-350 (1999); .Souda R., *Phys. Rev.* B **60**, 6068-6074 (1999).
22. Heifets E., Kotomin E.A., and Maier J., *Surf. Sci.*, **462**, 19-35 (2000).
23. Gay D.H., and Rohl A.L., *J. Chem. Soc. Faraday Trans.* **91**, 925-935 (1995).
24. Saunders V.R., Dovesi R., Roetti C., Causa M., Harrison N.M., Orlando R. and Zicovich-Wilson C.M., *Crystal-98 User Manual* (University of Torino, 1999).
25. Tasker P.W., *J. Phys. C : Solid State Phys.*, **12**, 4977-4986 (1979).

Accurate construction of transition metal pseudopotentials for oxides

Ilya Grinberg*, Nicholas J. Ramer§, and Andrew M. Rappe*

*Department of Chemistry and Laboratory for Research on the Structure of Matter
University of Pennsylvania, Philadelphia, PA 19104-6323

§ Department of Chemistry, Long Island University - C. W. Post Campus, Brookville, NY 11548

Abstract. We generate a series of Zr pseudopotentials and use them to calculate the properties of $PbZrO_3$, in order to examine the relationship between pseudo-atomic properties and solid-state oxide results. We find that lattice constants and bond lengths within the oxide unit cell are quite sensitive to pseudopotential construction errors, and clear correlations emerge. These trends motivate our identification of two criteria for accurate transition metal pseudopotentials for use in oxide calculations, which are similar to the criteria for metal use. We find that both the preservation of all-electron tail norm and the preservation of all-electron ionization energy are necessary to give good lattice constants for oxides.

INTRODUCTION

Perovskites (oxides with chemical formula ABO_3) are an important class of materials; many common minerals are perovskites, and ferroelectric perovskites such as $PbZr_xTi_{1-x}O_3$ (PZT) have many industrial applications including sonar. The phase diagram of PZT has been extensively studied experimentally and theoretically. Recently a virtual crystal method was used to predict theoretically a compositional phase transition in this material [1]. The changes in structure between the tetragonal and rhombohedral phases of PZT are subtle and are sensitive to temperature, composition and pressure. This makes it a good industrial material, but also presents challenging problems in modeling this material theoretically.

Density functional theory [2,3] (DFT) calculations have been widely used to study PZT and other perovskites with both the local density approximation (LDA) and generalized gradient approximation (GGA) for the exchange-correlation functional. Within DFT, one has a choice of using either all-electron or pseudopotential (PSP) methods. While the former are more accurate, they are much more computationally expensive; therefore many DFT studies of perovskites in recent years have been done using PSPs.

The absence of core electrons in the calculation and the reduced-cutoff plane-wave expansion of the PSP greatly reduce the computational cost of the solid-state calculation as compared those employing all-electron potentials. Modern PSP construction methods insure that the PSP agrees with the all-electron potential outside a specified core radius (r_c) for a given reference configuration. A perfect PSP would be completely transferable, i.e. it would mimic the behavior of the all-electron nucleus and core potential in various local chemical environments, producing solid-state results identical to those of an all-electron calculation. In practice, the transformation of a real, physical all-electron system into an artificial one consisting of PSPs and valence electrons will often introduce errors in the calculation. This can be due to either inaccuracies in the wave function and PSP in the valence region in atomic configurations other than reference or to the omission of the wave function oscillations in the core region. Methods capable of generating transferable PSPs with small plane-wave cutoffs have been developed over the course of the past twenty years [4–10].

PSP DFT calculations involving transition metal elements have become widespread. Nevertheless, some fundamental questions regarding PSP transferability remain unresolved. While it is axiomatic that a PSP must preserve certain all-electron properties to be considered transferable, it is unclear which all-electron properties are crucial. Various criteria for comparing to all-electron results have been proposed, such as agreement between all-electron and PSP eigenvalues along with total energy differences, norm-conservation at the reference configuration and preservation of logarithmic derivatives at r_c [4], and the correct chemical hardness matrix [11]. The agreement between all-electron and PSP eigenvalues and total energy differences are the criteria which are most often used and are generally considered when determining if a given PSP is transferable. However, no clear correlation has been firmly established between these criteria and solid-state results.

Recently, we have discovered that solid-state transition metal lattice constants and bulk moduli are very sensitive to PSP error, with variations in the solid-state DFT results directly correlated to errors in the atomic properties of the PSP [12]. Extending the norm conservation concept of Hamann, Schlüter and Chiang [4], we have shown that errors in the norm of the wave function beyond r_c in configurations other than the reference as well as total energy difference and eigenvalue errors affect solid-state results. Therefore, both the energy differences and wave function norms must be preserved for configurations other than reference in order to obtain correct solid-state properties.

In this work we will examine $PbZrO_3$ (PZ), as a test case to determine the relationship between solid-state results and PSP error for perovskite oxides.

METHODOLOGY

A PSP-DFT calculation of PZ structure involves three PSPs—lead, zirconium and oxygen. The PSPs for lead and oxygen are very accurate and are able to respond to perturbations away from the reference configuration, perhaps due to the presence of semicore-like $5d$ electrons in lead and $2s$ electrons in oxygen. The last $5s$ radial node in Zr is so far from the nucleus that one shell of semi-core states ($4s,4p$) must be included as valence. In the standard nonlocal PSP formalism [5], the use of three non-local channels (s,p,d) for five valence states ($4s,4p,5s,4d$ and $5p$) can introduce a large error in the $5s$ and $5p$ states, even in the reference configuration. While $5p$ error is unimportant ($5p$ states are not populated even in neutral Zr), the $5s$ state is populated in the neutral Zr and loses electrons to oxygen in the perovskite solid. As a result, the zirconium PSP will generally contain much more transferability error than the lead or oxygen PSPs. Since the BO_6 octahedral complex plays a crucial role in determining the characteristics of the perovskite unit cell, and since we expect the Zr PSP to provide most of the PSP error, we focus our investigation on the Zr PSP.

The designed nonlocal (DNL) PSP construction approach of Ramer and Rappe [7] allows us to adjust the amount of norm and total energy difference error in various PSP configurations, while leaving the reference configuration unchanged. We can, therefore, systematically introduce variations in Zr $5s$ norm and total energy differences to gauge the consequences of variations of atomic PSP properties for solid-state calculations.

To understand how norm-conservation and total energy difference agreement affect solid-state results, we examine five different Zr PSPs. All calculations are done using the Perdew-Zunger parameterization of the Ceperley-Alder LDA [13] exchange-correlation functional. We pick the $4s^2 4p^6 5s^0 5p^0 4d^0$ state as the reference atomic configuration. Zirconium has an oxidation number of +4 in $PbZrO_3$, and this atomic configuration will therefore be important in the solid state. Since the Zr–O bond is ionic, the ionization energy of Zr going from neutral to +4 configurations will be crucial.

All PSPs were created from a +4 ionized $4s^2 4p^6 5s^0 5p^0 4d^0$ reference configuration with varying depths and placements of the DNL augmentation operator, $\hat{A}$. Since the +4 state is the reference configuration, norm is conserved for the $4s$, $4p$ and $4d$ states, but it is not conserved for the $5s$ state. The $5s$ wave function tail norm can be adjusted, whereas the other norms are exact in the reference configuration and are not affected by the DNL operator. We therefore focus on the $5s$ norm. For all five PSPs we compute the PSP +4 ionization energy and the $5s$ norm in the valence region ($r \geq 1.7$ Bohr) in the +4 ionized state. For each Zr PSP we carry out DFT calculations to determine the equilibrium lattice constants a and c and the equilibrium volume for the tetragonal phase of PZ. While this phase is not found experimentally, the trends observed in our calculations will be relevant for other phases of PZ and for other perovskites. Therefore, tetragonal PZ is a valid test case for studying PSP effects and it is more tractable than the 40-atom ground

TABLE 1. Pseudopotential (PSP) results for Zr atom. Ionization energy to +4 state (I) and norm of the tail region of the +4 $5s$ state (N_s) are given for an all-electron atom (AE) and for the PSPs described in the text. All energies are in Ry and all lengths in Å. Lattice constants a and c and volume V are given for solid-state calculations of tetragonal PbZrO$_3$ using the specified Zr PSP.

	I	N_s	a	c	V
AE	5.7330	0.8976			
PSP A	5.7612	0.8762	4.0738	4.2222	70.0711
PSP B	5.7443	0.8909	4.0858	4.2378	70.7451
PSP C	5.7344	0.8992	4.0924	4.2458	71.1076
PSP D	5.7274	0.8920	4.1091	4.2719	72.1301
PSP E	5.7202	0.8973	4.1113	4.2745	72.2489

state structure. The results for the five PSPs are found in Table 1.

RESULTS

From the results in Table 1, it is clear that the Zr ionization energy is correlated with the volume of the PZ unit cell. Ionization energy decreases from PSP A to PSP E by about 0.04 Ry and the lattice constants a and c increase by about 1%, increasing the unit cell volume by about 3%. However, the changes in the lattice constants and volume do not depend on ionization energy alone. This can be seen by comparing the results for PSP C, PSP D and PSP E. While the ionization energy deceases by approximately 0.007 Ry from PSP C to PSP D and from PSP D to PSP E, the volume change from PSP C to PSP D is 8.6 times larger than the volume change from PSP D to PSP E. Upon examining the tail norms of the three PSPs, the reason for this dramatic difference becomes clear. The changes in norm of the $5s$ state are approximately equal but opposite in sign going from PSP C to PSP D, and from PSP D to PSP E. Thus in going from PSP C to PSP D, the norm and the ionization energy effects cooperate, significantly increasing the volume of the unit cell. From PSP D to PSP E, the two effects largely cancel, leading to a small increase in volume.

The correlations between atomic properties and solid-state results can be quantified by fitting the DFT calculated PZ volumes to a function of the form

TABLE 2. Fit of PbZrO$_3$ volume dependence on atomic properties. All volumes are in Å^3/ primitive cell. δV_{DFT} is the difference between the equilibrium volume obtained by solid-state calculations using a given pseudopotential (PSP) and the equilibrium volume using PSP A. δV_{Fit} is the difference between equilibrium volume obtained through Eq. 2 using I and N_s values for a given PSP and PSP A. The final column is the percent difference between fitted and DFT δV values.

	δV_{DFT}	δV_{Fit}	% error
PSP A	0.000	0.000	0.0
PSP B	0.674	0.630	-6.9
PSP C	1.037	0.999	-3.8
PSP D	2.059	1.937	-6.3
PSP E	2.178	2.240	2.8

$$V_{\text{PZ}} = aI + bN_s + c \qquad (1)$$

where I is the +4 ionization energy, N_s is the norm of the $5s$ state in the +4 configuration and a, b and c are constants. Fitting to calculated DFT PZ volumes we get:

$$V_{\text{PZ}} = -80.637I - 50.493N_s + 528.9125. \qquad (2)$$

We check the quality of the fit by comparing the volume differences obtained with Eq. 2 with computed DFT volume differences. As shown in Table 2, volume differences are predicted within 7%. We have examined fitting the volume to other atomic variables such as +3 ionization energy and the norm of $4d$ states and $5s$ states in other ionized configurations. All combinations showed a simple linear correlation between ionization energy, tail norm and volume, but the choice of +4 ionization energy and +4 ionized $5s$ norm gave a slightly more accurate fit.

The dependence of PZ volume on atomic properties can be understood by viewing the zirconium-oxygen bond character as a mixture of covalent and ionic bonding. Covalent bonding relies on direct overlap of atomic wave functions, favoring shorter bond lengths and smaller volumes. Ionic bond lengths result from a balance of Coulomb attraction and Pauli repulsion; this balance generally leads to longer bond lengths and larger volume. The proportion of covalent and ionic character in the zirconium-oxygen bond is affected by the size of the tail norms and ionization energies. Decreasing the +4 ionization energy of zirconium decreases the energetic cost of charge transfer from Zr to O, making the bond more ionic. This increases the volume. More diffuse Zr $5s$ wave functions will result in a larger overlap with oxygen atomic orbitals, increasing the hopping element. This favors covalent bonding and leads to shorter bond lengths and smaller volumes.

CONCLUSION

In this study, we have examined the relationship between pseudopotential properties on the atomic level and the results of solid-state DFT calculations in perovskites, taking tetragonal $PbZrO_3$ as a test case. We constructed a family of Zr pseudopotentials with varied total energy difference and norm conservation properties. We then calculated the equilibrium lattice constants and volumes of $PbZrO_3$ to gauge how the variations in atomic-level properties correlate with the results of the solid-state calculations. We find that the volume of the unit cell is sensitive to small changes in ionization energy and norm of the valence states, with variations of 3% in unit cell volume depending on the pseudopotential. Therefore, both of these properties must be enforced in pseudopotential construction in order to reduce pseudopotential error at the solid-state level. We also find that the changes in volume are linear in changes of the two atomic properties, demonstrating that errors in solid-state volume are a predictable consequence of pseudopotential errors.

This work was supported by the Office of Naval Research grant No. N-00014-00-1-0372 and the Air Force Office of Scientific Research, Air Force Materiel Command, USAF, under grant number F49620-00-1-0170. AMR acknowledges the support of the Camille and Henry Dreyfus Foundation. Computational support was provided by the San Diego Supercomputer Center, National Center for Supercomputing Applications and NAVOCEANO MSRC.

REFERENCES

1. N. J. Ramer and A. M. Rappe, *Phys. Rev. B* **62**, R743 (2000).
2. P. Hohenberg and W. Kohn, *Phys. Rev.* **136**, B864 (1964).
3. W. Kohn and L. J. Sham, *Phys. Rev.* **140**, A1133 (1965).
4. D.R. Hamann, M. Schlüter, and C. Chiang, *Phys. Rev. Lett.* **43**, 1494 (1979).
5. L. Kleinman and D. M. Bylander, *Phys. Rev. Lett.* **48**, 1425 (1982).
6. A. M. Rappe, K. M. Rabe, E. Kaxiras and J. D. Joannopoulos, *Phys. Rev. B* **41**, 1227 (1990).
7. N. J. Ramer and A. M. Rappe, *Phys. Rev. B* **59**, 12471 (1999).
8. D. Vanderbilt, *Phys. Rev. B* **41**, 7892 (1990).
9. S. Goedecker, M. Teter, and J. Hutter, *Phys. Rev. B* **54**, 1703 (1996).
10. J. Vackar, A. Simunek, and R. Podloucky, *Phys. Rev. B* **53**, 7727 (1996).

11. M. Teter, *Phys. Rev. B* **48**, 5031 (1993)
12. I. Grinberg, N. J. Ramer and A. M. Rappe, *Phys. Rev. B* in press (2001).
13. J. P. Perdew and A. Zunger, *Phys. Rev. B* **23**, 5048 (1981).

Ferroelectric and Piezoelectric Properties in the Presence of Compositionally Broken Inversion Symmetry

Na Sai, B. Meyer and David Vanderbilt

Department of Physics and Astronomy, Rutgers University, Piscataway, New Jersey 08855-0849

Abstract. We extend our first-principles study of novel ferroelectric perovskite systems in which compositional inversion symmetry is broken [N. Sai, B. Meyer and D. Vanderbilt, Phys. Rev. Lett. 84, 5636 (2000)] by (i) focusing on the piezoelectric response of the triple-cell heterovalent system Ba(Ti-δ,Ti,Ti+δ)O_3, and (ii) studying the strength of the symmetry breaking in the double-cell system (Ba,Sr)(Ti-δ,Ti+δ)O_3 with simultaneous A-site and B-site substitutions. We observe the enhanced piezoelectric response coefficient e_{33} when increasing the compositional parameter δ in the triple-cell system. This enhancement is quite drastic for the metastable minimum, but only modest for the stable minimum, corresponding respectively to the minority and majority wells of the ferroelectric double-well structure. In the double-cell system, we find that the increase in strength of symmetry breaking with the chemical or concentration perturbation is dominated by the term linear in δ, in contrast with the δ^3 dependence found in the triple-cell system. A symmetry-based justification of the dominance of the linear term is provided.

INTRODUCTION

Technologically important materials that display ultra-high piezoelectric response have recently been found in many alloy ferroelectrics or more complex solid-solution systems [1]. Numerous theoretical calculations have been carried out on perovskite oxides such as Pb(Zr,Ti)O_3 or ferroelectric relaxors represented by Pb(Zr,Nb,Ti)O_3 and Pb(Mg,Nb,Ti)O_3 to investigate the origin of the large response [2–5]. It was shown that the ferroelectric and piezoelectric properties exhibit great sensitivity to composition [1,4]. This feature, together with the variety of crystal configurations, has provided this class of materials with an exceptional tuning ability for improving the efficiency of applications.

The potential, however, for designing new materials with tuning freedom has by no means been exhausted. Very recently, we have for the first time reported a theoretical study of ferroelectric systems with compositionally broken inversion symmetry [6]. To summarize briefly, we studied ferroelectric systems of the form

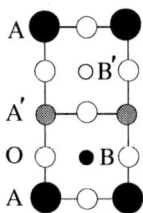

FIGURE 1. Side view of the $(A_{1/2}A'_{1/2})(B_{1/2}B'_{1/2})O_3$ structure. In our study, A and A' are Ba and Sr, and B and B' are either Sc and Nb or virtual Ti-δ and Ti+δ atoms.

$(A_{1/3}A'_{1/3}A''_{1/3})BO_3$ and $A(B_{1/3}B'_{1/3}B''_{1/3})O_3$ with three different cations that alternate layer by layer along the stacking direction. Focusing on the breaking of the inversion symmetry along the ferroelectric direction, we found a great disparity between the effects of heterovalent substitutions, such as in $Ba(Sc_{1/3}Ti_{1/3}Nb_{1/3})O_3$, and isovalent substitutions, such as in $(Ba_{1/3}Sr_{1/3}Ca_{1/3})TiO_3$. The heterovalent symmetry-breaking perturbation is so strong as to completely suppress one of the minima in the ferroelectric double well, while isovalent substitution only weakly affects the ferroelectric behavior. In particular, the study has emphasized a triple-cell model system with cyclically alternating cations Ti, Ti-δ and Ti+δ. Here δ is a continuous parameter, tunable between 0 and 1, that accounts for the nuclear charge difference between a true Ti atom and virtual atoms having slightly less or more nuclear charge. Surprisingly at first sight, the strength of the symmetry breaking in this system was found to scale as δ^3.

It is generally expected that the piezoelectric response will increase for a system of lowered symmetry [7]. Therefore one might expect that it should be possible to enhance certain ferroelectric properties, such as the piezoelectric response, using the enormous freedom allowed for by the type and pattern of imposed compositional order. In the first part of the present paper, we describe previously unreported calculations of the piezoelectric response for the $Ba(Ti-\delta,Ti,Ti+\delta)O_3$ system.

In the second part of our paper, we study a double-unit-cell system denoted by $(A_{1/2}A'_{1/2})(B_{1/2}B'_{1/2})O_3$ as shown in Fig. 1. In contrast with the triple-cell system, inversion symmetry in the double-cell system is broken by simultaneous substitutions on the A- and B-sites. As before, the spontaneous polarization direction is assumed to lie along the tetragonal stacking direction. We will show that the strength of the symmetry breaking in the double-cell system has a quite different behavior compared to the triple-cell symmetry-breaking system.

METHOD

Total-energy calculations are based on the Vanderbilt ultrasoft pseudopotential [8] scheme within the local-density approximation. Details of the pseudopotentials can be found in Ref. [9]. A $4 \times 4 \times 2$ Monkhorst-Pack mesh [10] is used for sampling the Brillouin zone in the self-consistent calculations and a 25-Ry plane-wave cutoff is used throughout. Our procedure for locating both local energy minima in the case of a distorted double-well potential has been reported previously in Ref. [6]. To study the piezoelectric response, the polarization P is calculated for each structure using the Berry-phase approach [11,12], with a $4 \times 4 \times 10$ k-point grid used for the k-space integration in the polarization calculations.

RESULTS

A Piezoelectric response of the triple-cell system

The lattice parameters for the Ba(Ti-δ,Ti,Ti+δ)O$_3$ system are chosen so that the volume is 1248.9au^3 and $c/a = 3.036$. We have assumed the ferroelectric order parameter to lie only along the z-axis. In order to study quantitatively the symmetry breaking as a function of δ, we have tracked the ferroelectric ground state configurations at different δ by gradually increasing δ by small amounts. Each calculated relaxed structure at one value of δ is taken as the initial guess for the next structure, and the atomic coordinates are relaxed again. However, there exists a critical value of δ beyond which the secondary minimum vanishes and only one minimum remains locally stable.

The piezoelectric coefficients e_{ij} are defined as derivatives of the polarization with respect to strain at zero macroscopic electric field. Thus, in leading order, $P_i^\epsilon = P_i^s + e_{ij}\epsilon_j$, where P_i^s is the spontaneous polarization of the unstrained system along the i-th Cartesian direction, ϵ_j is the strain tensor, and e_{ij} is the piezoelectric tensor [2] in Voigt notation. It has been shown that the proper piezoelectric response is just related to the variation of the Berry phase with strain deformation [13]; we can therefore compute the proper response directly using

$$e_{ij} = \frac{e}{2\pi\Omega} \sum_\alpha \frac{d\phi_\alpha}{d\epsilon_j} R_{\alpha i} \tag{1}$$

where ϕ_α is the Berry phase in direction α, $R_{\alpha i}$ is the real-space lattice vector, and Ω is the volume. In our calculations, we focus on the principal element e_{33} and introduce several additional strains in the range $|\epsilon_3| \leq 0.01$ into the fully relaxed ground-state structure (reference state), and re-optimize the coordinates. We find that in order to obtain a good fit for every structure within the strain of this range, the variation of the polarization has to be fitted with at least a third-order polynomial in strain, where the linear coefficient gives the piezoelectric constant.

We also separately computed the clamped-ion piezoelectric response that involves only external strains at fixed coordinates. Its magnitude amounts to only 2-3% of the total piezoelectric response and it carries the opposite sign. Therefore, the total piezoelectric response is mainly due to the atomic sublattice displacements, or "internal strain," similar to what has been found in the PT and PZT materials

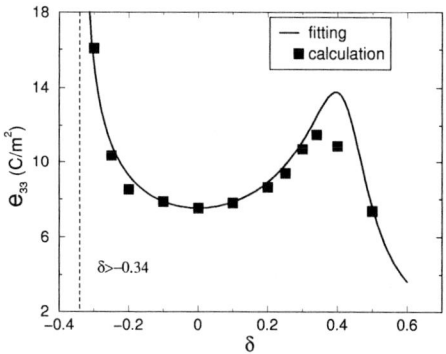

FIGURE 2. Calculated and fitted piezoelectric response e_{33} as a function of δ in the Ba(Ti-δ,Ti,Ti+δ)O$_3$ system.

[2,3]. In the following, we report only the total piezoelectric response including both clamped-ion and internal-strain contributions.

The e_{33} piezoelectric response for both the principal ("majority") and the secondary ("minority") minima are plotted together as function of δ in Fig. 2 (solid squares). For $\delta=0$ the system is just BaTiO$_3$ in the tetragonal phase. We find that the polarization $P_s = 0.24$ C/m^2 is very close to the experimental value of 0.27 C/m^2. e_{33} is found to be 7.5 C/m^2. For positive δ, we plot the piezoelectric response for the energetically-favored principal minimum. The e_{33} coefficient increases by a factor about 1.4 as it reaches a maximum at $\delta \sim 0.4$. Then the value starts to drop and becomes lower than that of BaTiO$_3$ at $\delta = 0$. The response is found to be significantly different for negative δ, corresponding to the secondary minimum; here e_{33} increases sharply for $\delta \leq -0.25$ and appears to diverge as δ approaches some critical value from above. The divergence is well fitted by a power law of the form $e_{33} \simeq A_0(\delta - \delta_c)^{-\gamma}$, where $\gamma = 0.31$, $A_0 = 5.58$, and the critical value δ_c equals -0.34.

An intuitive explanation for the divergence of the piezoelectric response at δ_c is readily available. To produce a large piezoelectric response, the structure needs to have a nearly flat internal energy surface [14]. Thus, in ferroelectrics, the piezoelectric response tends to diverge at the paraelectric-to-ferroelectric phase transition [7]. When δ is tuned in the present system so that the secondary minimum is about to disappear, the saddle point approaches the secondary minimum, the barrier between them becomes very low, and the upward curvature at the secondary minimum is going to zero. Intuitively, this represents the kind of flat energy surface that can provide a large piezoelectric response.

Next, we show that a simple model that takes into account the macroscopic strain can illustrate quantitatively the behavior of the piezoelectric response within a range of δ corresponding to the physical structures. The total energy can be expanded in the ferroelectric soft-mode variable up to fourth order as

$$E = E_0 + E_1 u + E_2 u^2 + E_4 u^4 \qquad (2)$$

where the coefficients E_n are functions of δ, and the zero of u has been chosen to make the $d^3 E/du^3$ term vanish. The leading δ-dependence of the E_1 coefficient is as $a_3 \delta^3$, while E_2 and E_4 have a δ-dependence like $a_0 + a_2 \delta^2$. This behavior follows from the symmetry, which allows simultaneous exchange of $u \leftrightarrow (-u)$ and $\delta \leftrightarrow (-\delta)$ without changing the total energy.

We write the free energy in terms of a single mode of the polarization P that couples to the macroscopic strain ϵ,

$$F(P, \delta, \epsilon) = e_0(\delta) + e_1(\delta) P + e_2(\delta) P^2 + e_4(\delta) P^4 + c\epsilon P^2. \qquad (3)$$

The equilibrium polarization $P_0(\delta)$ is given by $\partial F/\partial P = 0$, so that for $\epsilon \to 0$, the piezoelectric constant is

$$e_{33} = \frac{\partial P_0(\delta)}{\partial \epsilon} = \frac{2c P_0(\delta)}{2e_2(\delta) + 12 e_4(\delta) P_0^2(\delta)}. \qquad (4)$$

The interesting range of δ is when $P_0(\delta)$ is real. Within this range, we substitute the numerically computed coefficients $e_1(\delta)$, $e_2(\delta)$, and $e_4(\delta)$ into Eq. (4) and plot in Fig. 2 (smooth curve) the analytically fitted e_{33} versus δ for both the stable and the metastable minimum. We scale the constant c in Eq. (4) so that the fitted and the calculated e_{33} are equal at $\delta = 0$. Evidently, the two curves exhibit the same behavior. The data and the fit agree well within the δ range of ± 0.25. Discrepancies in e_{33}, however, become more pronounced in the vicinity of the peak that appears at $\delta \approx 0.4$ in the calculations and $\delta = 0.4$ in the fit. The fitted peak is 20% higher than the calculated one.

We also use the above model in order to investigate what has caused the piezoelectric response to peak at $\delta = 0.4$. We calculate the curvature of the free-energy surface $\partial^2 F/\partial^2 P$ at the three stationary points (two minima and saddle point) and plot them in Fig. 3. At $\delta = 0$, the two minima have identical positive curvature. For the case of the metastable minimum, the curvature drops continuously as δ increases and vanishes at $\delta = -0.34$, which is consistent with the divergent piezoelectric response we find at this point. On the positive side of the plot, the curvature drops, forming a valley at $\delta = 0.4$, and then increases again. This result suggests that the behavior of the piezoelectric response can be explained by the nature of the δ-dependence of the curvature. Intuitively, one can say that increasing δ has two effects: a δ^2 tendency that makes e_2 less negative, and a δ^3 symmetry breaking. The first tendency makes both wells more shallow and soft, thus causing e_{33} to increase, as δ first deviates from zero with either sign. But then the δ^3 tendency establishes one of the wells as the principle well, and the e_{33} of this well is eventually suppressed as the growing positive e_2 term causes the remaining single well to deepen and harden with increasing δ.

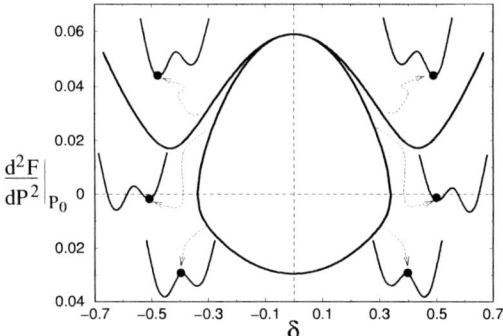

FIGURE 3. Second derivatives of the free-energy surface evaluated at the stationary points (minima or saddle points, as indicated by the sketches) according to the model of Eq. (3). (For $|\delta| > 0.34$ the secondary minimum and saddle point have disappeared.)

B Broken symmetry in the $(A_{1/2}A'_{1/2})(B_{1/2}B'_{1/2})O_3$ system

There are many possible layer sequences that can break the inversion symmetry along the growth direction of a mixed cubic perovskite. Here, we consider a layer sequence that accomplishes this with the minimum repeat period, only 10 atoms per primitive cell. Specifically, we study the $(A_{1/2}A'_{1/2})(B_{1/2}B'_{1/2})O_3$ system shown in Fig. 1, in which there are simultaneous alternating substitutions on both A and B sites.

It has been demonstrated in the case of the triple-cell system that heterovalent substitutions along the order-parameter direction have a much stronger effect on breaking the symmetry than isovalent substitutions. Hence, we first consider an "A-iso-B-hetero" system $(Ba_{1/2}Sr_{1/2})(Sc_{1/2}Nb_{1/2})O_3$ in which the heterovalent perturbation occurs on the B site only. However, for this case we find that the symmetry breaking is already strong enough that there is only a single ferroelectric minimum, similar to what was found for the $Ba(Sc_{1/3}Ti_{1/3}Nb_{1/3})O_3$ triple-cell system [6]. Therefore, in order to understand how the system evolves from the unperturbed to the strongly-perturbed regime, we turn to the system $(Ba,Sr)(Ti+\delta, Ti-\delta)O_3$ where virtual atoms of type $Ti\pm\delta$ have been substituted on the B-sites, and increase δ gradually from 0 to 1.

The equilibrium volume is frozen to equal the sum of the volumes of $BaTiO_3$ and $SrTiO_3$ in their cubic structures [9]. We use $c/a = 1.012$, the same as for the $BaTiO_3$ 5-atom tetragonal cell. The previously-developed search strategy is applied to find the principal and secondary ferroelectric ground states for a series of broken-symmetry systems with δ increasing from system to system. The total energy as a function of the soft mode amplitude u is computed along the line direction that connects the two energy minima as shown in Fig. 4. When δ exceeds 0.3 the second minimum has already vanished, indicating a stronger symmetry breaking compared

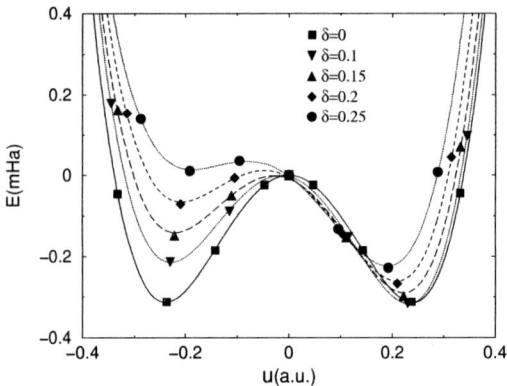

FIGURE 4. Total energy vs. soft-mode amplitude for different δ in the (Ba,Sr)(Ti+δ, Ti-δ)O$_3$ system.

to the triple cell system. Hence, only the energy profiles concluding with $\delta = 0.25$ have been plotted.

Once again we analyze the results for the total energy as a function of the soft mode variable u via the polynomial expansion of Eq. (2). The δ-dependent coefficients E_1, E_2 and E_4 are plotted as functions of δ in Fig. 5. For comparison, we have also plotted the corresponding coefficients for the Ba(Ti-δ,Ti,Ti+δ)O$_3$ system in the same figure. The linear coefficient, which is a primary measure of the symmetry breaking (we name it the "symmetry breaking force"), clearly shows the leading dependence on δ to be linear in the double-cell system while it is as δ^3 in the triple-cell system. Indeed, a much stronger symmetry-breaking effect has been observed at very small δ in the present system. However, it is also indicated that further enhancement of the symmetry breaking with increasing δ will be much milder in this system than in the triple-cell system. The E_2 and E_4 coefficients show similar dependence on δ in the form of $a_1 + a_2\delta^2$ as in the triple-cell system. However, the intercept $E_2(\delta) = 0$ gives $\delta = 0.37$ and the critical value for the existence of a second energy minimum is only $\delta < 0.28$.

To compare the differences between the symmetry-breaking behavior in the two systems, it is instructive to consider the forces on the atoms while they sit at their ideal positions. We call this pattern of forces the "cubic force vector" $\vec{f}_c$. In general, $\vec{f}_c$ can be expanded in powers of δ,

$$\vec{f}_c = \vec{f}_c^{(0)} + \vec{f}_c^{(1)}\delta + \vec{f}_c^{(2)}\delta^2 + \vec{f}_c^{(3)}\delta^3 + \cdots, \quad (5)$$

where $\vec{f}_c^{(0)}$ is the force vector at $\delta = 0$. Note that $\vec{f}_c^{(0)}$ is nonzero in this system, so that the unit vector $\hat{f}_c = \vec{f}_c/|\vec{f}_c|$ can be expanded as

$$\hat{f}_c = \hat{f}_c^{(1)} + a_2\hat{f}_c^{(2)}\delta + a_3\hat{f}_c^{(3)}\delta^2 + \cdots. \quad (6)$$

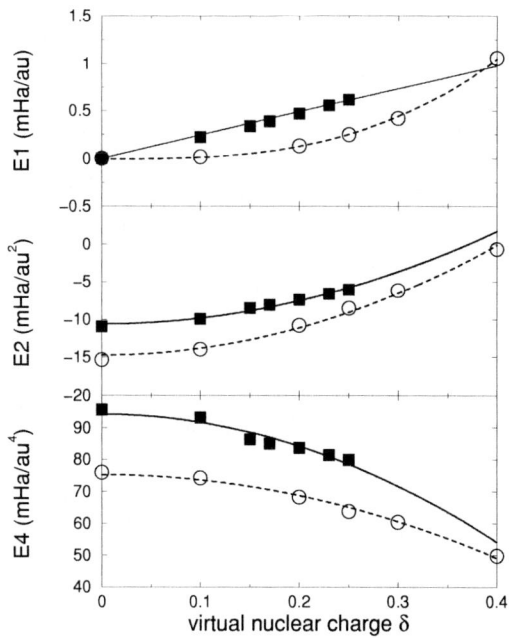

FIGURE 5. Total energy expansion coefficients in Eq. (2) as functions of δ for both the (Ba,Sr)(Ti+δ, Ti-δ)O$_3$ system (solid squares) and Ba(Ti-δ,Ti,Ti+δ)O$_3$ system (empty circles).

In Ref. [6], we identified the C_{3v} symmetry group with the Ba(Ti-δ,Ti,Ti+δ)O$_3$ system. Note that C_{3v} is *not* the point-group symmetry; instead, it refers to the 6-element symmetry group of the $\delta=0$ system, generated by 2-fold mirror reflections along z and 3-fold translations along z. In this C_{3v} group, the ferroelectric mode unit vector $\hat{\xi}$ transforms according to irrep A_2, while the perturbation pattern associated with δ belongs to irrep E. Thus, the coupling of the ferroelectric mode and the cubic force vector $\hat{\xi} \cdot \hat{f}_c$ can only have $\hat{\xi} \cdot \hat{f}_c^{(3)} \delta^2$ as the leading term because $\hat{f}_c^{(3)}$ is associated with the product of $E \times E \times E$ that contains the A_2 representation.

In contrast, the double cell system at $\delta = 0$ does not have symmetry with respect to a primitive translation, and so obeys the S_2 symmetry group generated by just the 2-fold mirror reflection along z. Both the ferroelectric mode vector and the perturbation of δ are odd in terms of the mirror operation, so that they both belong to the same irrep A_u. Therefore, a coupling of the form $\hat{\xi} \cdot \hat{f}_c^{(2)} \delta$ is still prohibited, but $\hat{\xi} \cdot \hat{f}_c^{(1)}$ does not have to vanish. Hence, the coupling between the ferroelectric mode and the symmetry-breaking cubic-force vector in the double-cell system is governed by both the terms in $\hat{\xi} \cdot (\hat{f}_c^{(1)} + a_3 \hat{f}_c^{(3)} \delta^2)$.

To illustrate the above analysis, we show in Fig. 6 the coupling between $\hat{\xi}$ and $\hat{f}_c$. The angles are directly calculated from $\cos^{-1}(\hat{\xi} \cdot \hat{f}_c)$. In the bottom panel, the angle is shown for the Ba(Ti-δ,Ti,Ti+δ)O$_3$ system. We observe that the angle decreases

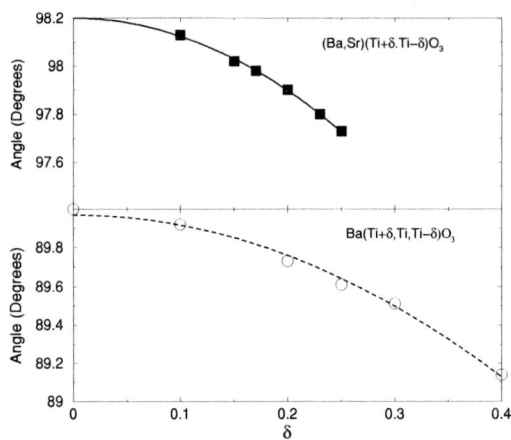

FIGURE 6. Angle between the ferroelectric mode vector $\widehat{\xi}$ and the cubic force vector $\widehat{f}_c$ as a function of δ for the (Ba,Sr)(Ti+δ, Ti-δ)O$_3$ (top panel) and Ba(Ti-δ,Ti,Ti+δ)O$_3$ (bottom panel) systems.

quadratically with increasing δ and intersects $\delta = 0$ at exactly 90°, in agreement with the predicted $\widehat{\xi} \cdot \widehat{f}_c^{(3)} \delta^2$ dependence. In the double-cell system shown in the upper panel, the angle has a similar quadratic dependence on δ, but the intercept is not at 90° but at 98.2°, illustrating that the $\widehat{\xi} \cdot \widehat{f}_c^{(1)}$ term does not vanish in the present system.

CONCLUSIONS

In summary, we studied the evolution of the piezoelectric response in the model ferroelectric system Ba(Ti-δ,Ti,Ti+δ)O$_3$ with compositionally broken inversion symmetry. A divergence in the response is observed for the metastable energy minimum at the critical δ value where this secondary minimum is about to disappear. For the energetically favored minimum, the piezoelectric response is moderately enhanced and peaks in the vicinity of $\delta \sim 0.4$. A simple model based on a free-energy expansion in the zero macroscopic strain limit is used to describe the behavior of these responses. We also extended our studies to the double-cell system (Ba,Sr)(Ti+δ,Ti-δ)O$_3$. We find that the strength of the symmetry breaking depends linearly on δ, providing a different picture from that of the triple-cell system studied previously.

ACKNOWLEDGMENTS

N.S. thanks K.M. Rabe for useful discussions. Support for this work was provided by ONR N00014-97-1-0048 and NSF DMR-9981193.

REFERENCES

1. S.-E. Park and T. R. Shrout, *J. Appl. Phys* **82**,1804 (1997).
2. Gotthard Saghi-Szabo, R.E. Cohen and H. Krakauer *Phys. Rev. Lett.* **80**, 4321 (1998).;ibid. *Phys. Rev. B* **59**, 12771 (1999).
3. L. Bellaiche and D. Vanderbilt, *Phys. Rev. Lett.* **83**, 1347 (1999).
4. L. Bellaiche, A. García, and D. Vanderbilt, Phys. Rev. Lett. **84**, 5427 (2000).
5. R. Hemphill, L. Bellaiche, A. García, and D. Vanderbilt, Appl. Phys. Lett. **77**, 3642 (2000).
6. N. Sai, B. Meyer and D. Vanderbilt, *Phys. Rev. Lett.* **84**, 5636 (2000).
7. E. Cockayne and K.M. Rabe *Phys. Rev. B* **57**, R13973 (1998).
8. D. Vanderbilt, *Phys. Rev. B* **41**, 7892 (1990).
9. R.D. King-Smith and D. Vanderbilt, *Phys. Rev. B* **49**, 5828 (1994).
10. H.J. Monkhorst and J.D. Pack, *Phys. Rev. B* **13**, 5188 (1976).
11. R.D. King-Smith and D. Vanderbilt, *Phys. Rev. B* **47**, 1651 (1993).
12. R. Resta, *Rev. Mod. Phys.* **66**, 899 (1994).
13. D. Vanderbilt *J. Phys. Chem. Solids* **61**, 147 (2000).
14. H.X. Fu and R. Cohen *Nature* **403**, 281 (2000).

New Polaronic-Type Excitons in Ferroelectric Oxides: Nature and Experimental Manifestation

Valentin S. Vikhnin[a], Roberts I. Eglitis[b], Eugene A. Kotomin[b,c], Siegmar E. Kapphan[b] and Gunnar Borstel[b]

[a]A.F. Ioffe Physical Technical Institute, 194021, Saint-Petersburg, Russia
[b]Department of Physics, University of Osnabrück, D-49069 Osnabrück, Germany
[c]Institute for Solid State Physics, University of Latvia, 8 Kengaraga str.,Riga LV-1063, Latvia

Abstract. The current experimental and theoretical knowledge of new polaronic-type excitons in ferroelectric oxides – charge transfer vibronic excitons (CTVE) – is discussed. It is shown that Hartree-Fock-type INDO calculations as well as photoluminescence studies in ferroelectric oxygen-octahedral perovskites confirm the CTVE-concept. Single CTVE as well as a new phase of strongly correlated CTVEs are analysed. It is shown also that polaron- and CTVE-trapping effects including the oxygen vacancy clusters play an important role in polar cluster formation.

INTRODUCTION

Charge Transfer Vibronic Exitons (CTVE) in ferroelectric oxides with mixed ionic-covalent type of chemical bonding consist of the correlated pairs or/and triads of electronic and hole polarons [1-7]. The CTVE case is especially interesting due to the possibility of a strong interaction between its reorienting dipole moment with a soft lattice [2-6]. This is directly connected to a principal problem of real ferroelectrics, namely, with the nature of dipole reorienting centers and their polar clusters which are involved into cooperative phenomena, for instance, phase transitions. It is important to underline that this type of excitons can be considered as proper deep centers, and can be metastable excitations due to a strong equilibrium lattice distortion. That is, they can exist also under certain conditions of a crystal treatment, without any optical pumping. The origin, properties and experimental manifestations of such the localized charge transfer exciton in matrices like ferroelectric oxides is one of the main aims of the present paper. Oxygen-octahedral perovskite-like ferroelectrics as well as related ferroelectric oxides are good examples of the actual solids with the ionic-covalent chemical bonding where a charge transfer is rather strong, on one hand, and vibronic interaction also is not small, on the other hand. Ferroelectric oxides $KTaO_3$ (KTO), $KNbO_3$ (KNO), and solid solution SBN will be considered below as examples of new type excitons of polaronic type (CTVE). Here a review of recent theoretical and experimental investigations of CTVE is presented.

We present here quantitative arguments for the CTVE state appearance which are based on INDO calculations. Namely, we present new data for effective charges and lattice distortion accompanied by CTVE. The experiments which support the CTVE concept are connected with optical studies of ferroelectric oxides (photoluminescence and light absorption data).

It should be underlined that effective ionic charges used here allow us to explain many ferroelectric oxide characteristics. For instance, the calculated absorption band position for F-center in KNO [8] is in a good agreement with experiment.

The same method which was justified before [8-14] is used here for evaluation of the charge transfer values corresponding to excited, CTVE states. Here theoretically calculated charge transfer in CTVE-states becomes rather high due to an important lattice distortion effect which appears as a result of strong vibronic interaction.

Note also that vibronic instability appears already at rather small magnitudes of the charge transfer, but the CTVE equilibrium charge transfer is much higher than this threshold value mentioned above. That is, CTVE state lies deeply within region of distorted lattice states.

Computation Method and its Grounding

Ab initio methods are still cumbersome and time-consuming in the treatment of the electronic and spatial structure of complex systems, especially those with partially covalent chemical bonding, like perovskites. In order to be able to study the relatively complicated case of perovskite solid solutions, there is a need to close the gap between accurate but time consuming *ab initio* methods [15-21] and widely used, simple but not so reliable ad hoc parameter-dependant phenomenological approaches. One possible compromise is to use a semiempirical quantum chemical method that is parameter-dependant, but has parameters which are transferable, for the specific chemical constituents given, and is not subject to adjustment for each new compound in question. An example of such a method is the updated Intermediate Neglect of the Differential Overlap (INDO) method [12,13], which is a semiempirical version of the Hartree-Fock method.

The Fock matrix elements in the modified INDO approximation [12,13] contain a number of semiempirical parameters. The orbital exponent ζ enters the radial part of Slater-type atomic orbitals:

$$R_{nl}(r) = (2\zeta)^{n+1/2}\left[(2n)!\right]^{-1/2} r^{n-1} \exp(-\zeta r) \qquad (1)$$

where n is the main quantum number of the valence shell. We used a valence basis set including 4s, 4p atomic orbitals (AO) for K; 2s, 2p for O; 5s, 5p, 4d for Nb and 6s, 6p, 5d for the Ta atom. The diagonal matrix elements of the interaction of an electron occupying the μ-th valence AO on atom A with its own core are taken as:

$$U_{\mu\mu}^A = -E_{neg}^A(\mu) - \sum_{\nu \in A}(P_{\nu\nu}^{(0)A}\gamma_{\mu\nu} - \frac{1}{2}P_{\nu\nu}^{(0)A}K_{\mu\nu}) \qquad (2)$$

where $P_{\mu\mu}^{(0)A}$ are the diagonal elements (initial guess) of the density matrix, $\gamma_{\mu\nu}$ and $K_{\mu\nu}$ are one-centre Coulomb and exchange integrals, respectively. $E_{neg}^{A}(\mu)$ is the initial guess of the μ-th AO energy. The matrix elements of interaction of an electron on the μ-th AO belonging to the A with the core of another atom B are calculated as follows:

$$V_{\mu}^{B} = Z_{B}\{1/R_{AB} + [\langle\mu\mu/\nu\nu\rangle - 1/R_{AB}]\exp(-\alpha_{AB}R_{AB})\} \quad (3)$$

where R_{AB} is the distance between atoms A and B, Z_B is the core charge of atom B. α_{AB} is an adjustable parameter characterizing the finite size character of the atomic core B and additionally the diffusenes of the μ-th AO. $\langle\mu\mu/\nu\nu\rangle$ is the two-center Coulomb integral. The resonance integral parameter $\beta_{\mu\nu}$ enters the off-diagonal Fock matrix elements for the spin component u:

$$F_{\mu\nu}^{u} = \beta_{\mu\nu}S_{\mu\nu} - P_{\mu\nu}^{u}\langle\mu\mu/\nu\nu\rangle \quad (4)$$

where the μ-th and ν-th AO are centered at different atoms. u is an electron subsystem with α or β spin. $S_{\mu\nu}$ is the overlap matrix between electrons on the μ-th Slater-type AO on the A-atom; the electronegativity of the A-atom $E_{neg}^{A}(\mu)$, defining the μ-th AO energy; the μ-th AO population $P_{\mu\mu}^{(0)A}$, i.e. the 'initial guess' for the diagonal element of the density matrix; the resonance integral $\beta_{\mu\nu}$ entering the off-diagonal Fock matrix element where the μ-th and ν-th AOs are centered at the different A- and B- atoms; and the adjustable exponent α_{AB} characterizing the extended nature of the B-atom core interaction with the electron on the μ-th AO of the A-atom.

In the last decade the INDO method has been succesfully used in the study of bulk solids and defects in many oxides [12,13] and semiconductors [14]. This method has been earlier applied to the study of phase transitions and frozen phonons in pure KNO [9], pure and Li-doped KTO [10], point defects in KNO [8], solid perovskite solutions KNb$_x$Ta$_{1-x}$O$_3$ [22], as well as F centers and hole polarons in KNO [8,11].

The detailed analysis of the development of the INDO parametrization for pure KNO and KTO is given in [9,10]. The INDO method reproduced very well both available experimental data and the results of ab initio LDA-type calculations. In particular, this method reproduces the effect of a ferroelectric instability of KNO due to off-centre displacement of Nb atoms from the regular lattice sites, as well as the relative magnitudes of the relevant energy gains for the [100], [110] and [111] Nb displacements. These are consistent with the order of the stability of the tetragonal, orthorombic and rhombohedral ferroelectric phases, respectively, as the crystal's temperature decreases. This is a very non-trivial achievement since the typical energy gain due to the Nb off-centre displacement is of the order of several mRy per unit cell.

The calculated frequencies of the transverse-optic (TO) phonons at the Γ point in the BZ of cubic and rhombohedral KNO, and the atomic coordinates in the minimum energy configuration for the orthorombic and rhombohedral phases of KNO, are also in good agreement with experiment, thus indicating that a highly successful INDO parametrization has been achieved. The frozen-phonon calculations for T_{1u} and T_{2u} modes of cubic KTO are also in good agreement with experiment. Appreciable covalency of the chemical bonding is seen from the calculated (static) effective charges on atoms (calculated using Löwdin population analysis): 0.62 e for K, 2.23 e for Ta and -0.95 e for O in KTO, which are far from those expected in the purely ionic model (+1 e, +5 e and -2 e, respectively) often used. These charges show slightly higher ionicity in $KTaO_3$ as compared with the relevant effective charges calculated for KNO: 0.54 e for K, 2.02 e for Nb and -0.85 e for O.

The Atomic Structure of Charge Transfer Vibronic Exciton

The important feature of CTVE is the charge transfer effect which is self-consistent with the accompanied lattice distortion. The computations of the electronic and distorted lattice structure of CTVE in KTO-crystal was performed on the basis of semi-empirical Hartree-Fock-type calculations using the Intermediate Neglect of the Differential Overlap (INDO) method. We have also used the periodic (large unit cell) model and have extended the primitive KTO unit cell by a factor of 3x3x3=27. As a result, the following structure of CTVE was obtained (Figure 1): the CTVE has a triad structure [5,6] with rather strong vibronic energy lowering (~ 2.71 eV). This energy lowering is connected with the appearance of a hole polaron on one oxygen ion which has a displacement 5.2 % of the lattice constant towards the first active Ta-ion on which the electronic polaron is partly located.

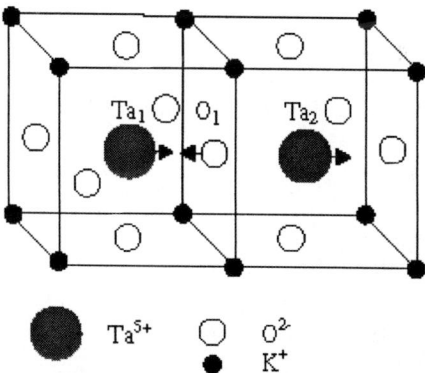

FIGURE 1. The calculated triad structure of the Charge Transfer Vibronic Exciton in $KTaO_3$ crystal. The displacements of three active Ta-O-Ta ions of the triad are shown.

This first Ta-ion has a displacement of 3.1 % towards the active oxygen ion. However, the second active Ta-ion, where the electronic polaron is also partly located, has a displacement of 4.5 % due to a repulsion from the active oxygen ion - in a contrast to the first Ta ion. The final charge distribution between these three Ta-O-Ta ions is the following: The first Ta-ion has the effective charge of 1.74 e, while the second Ta-ion has 2.12 e, and an active oxygen ion – 0.24 e. The energy level of the electronic polaron on the two Ta-ions is by ~ 0.8 eV below the bottom of the conduction band, whereas the energy level of the hole polaron localized on the oxygen ion is ~ 1 eV above the top of valence band of the KTO crystal. A similar triad structure of CTVE was confirmed by consideration of the KNO [7].

We can conclude, that the CTVE has a well-defined triad structure (in the model case under discussion), with rather deep in-gap energy position of electronic states, and with a strong self-consistent lattice distortion.

A New Type Charge Transfer States in Ferroelectric Oxides: INDO-Calculations of Charge Transfer Vibronic Exciton Phase

Special attention should be paid to the *new phase* of the CTVE predicted in ref. [3]. The CTVE phase consists of the strongly correlated CTVEs which are located in *each* unit cell of the crystal. That is, the CTVE phase corresponds to a new state of the crystal which is characterized by a new equilibrium charge transfer as well as a new set of equilibrium lattice displacements.

The first step in the theoretical study of the CTVE phase was done [2,3] using semi-phenomenological models. Namely, first a co-operative Jahn-Teller effect type model of CTVE phase was developed. This model deals with point dipoles corresponding to equilibrium displacements in the CTVE cell as well as with electronic degrees of freedom of CTVE active ions assuming the electronic state degeneracy, or pseudo-degeneracy. These two type degrees of freedom directly interact with lattice polarization and deformation in the linear and quadratic approximations. It was shown that an important role in CTVE formation belongs to a specific vibronic interaction characterized by a direct coupling of charge transfer with lattice distortions. According to these estimates, the CTVE phase energy position [2] was in-gap, rather close to the valence band top.

However, a direct, non-model computation of CTVE phase properties was needed. In the present study the computational analysis was performed by means of the semi-empirical Hartree-Fock-type INDO method. This method was previously used in a treatment of both a single CTVE in incipient ferroelectric KTO [5,6] and in ferroelectric KNO [7] where the existence of CTVE was confirmed and its three-ion structure was obtained.

It is shown in the present work on the basis of INDO-calculations that the novel CTVE phase can exist in the well known incipient ferroelectric KTO. We had used the 3x3x3 supercells in these calculations. The triplet CTVE phase was obtained as a ferroelectric phase with parallel-oriented small polaron electron-hole pairs on O-Ta ions and in-gap states. The total energy lowering per such a O-Ta pair was 2.32 eV. This corresponds to a strong vibronic interaction case. The equilibrium displacements of O-Ta ions in each electron-hole pair are directed towards each other and rather large; 4.33 % for Ta-ion, and 5.62 % for O-ion. Self-consistent charge transfer is also large, -0.77 e, accompanied by the equilibrium effective charge of 1.46 e for Ta-ion, whereas for O ions they are 0.77 e and -0.18 e, respectively. Note that ferroelectrically ordered oxygen ions in the intermediate chains relatively to O-Ta CTVE-chains are much less shifted (0.37 %) in the same direction as oxygen ions in these CTVE- chains (Figure 2).

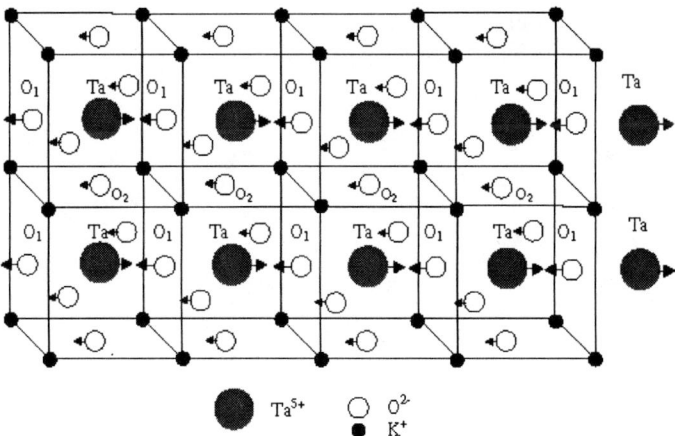

FIGURE 2. The displacement structure in the ferroelectric state of charge transfer vibronic exciton phase.

Another model example, characterized by higher bonding covalency, is KNO. The parameters of ferroelectrically ordered CTVE phase were calculated here on the basis of the same approach as for KTO. Here the total energy lowering per O-Nb pair was 1.99 eV. This also corresponds to a strong vibronic interaction case, which is decreased due small increase in the bond covalency. The equilibrium displacements of O-Nb ions in each polaronic electron-hole pair are also directed towards each other and remain rather large: 4.15 % for the Nb-ion, and 5.28 % for the O-ion. Self-consistent charge transfer is also considerable -0.70 e (the equilibrium effective charge +1.32 e) for the Nb-ion, and +0.70 e (with an equilibrium effective charge -0.15 e) for the O-ion in the CTVE-chains. Oxygen ions in the intermediate chains have 0.35 % shifts which are parallel to the shifts of oxygen ions in the CTVE-chains. This calculation, together with the previous one, confirm the validity of the conclusions about a high stability of the CTVE-phase in ferroelectric oxides.

Novel Type Exciton Manifestation: Photoluminescence Experiments

It should be useful to consider typical luminescence energies induced by CTVE radiative recombination. According to the experiment, we can consider here two characteristic CTVE recombination luminescence energies: the band with a higher quantum energy [23,24] is connected to the "green" luminescence (GL) in KTO, and a more weak with lower quantum energy [25-27, 6] is connected with the "red" luminescence (RL) in KTO.

The GL phenomenon is explained in the present work on the basis of a CTVE trapped on oxygen vacancies [24]. Such a pair of defect centers forms two different states with different local geometries (a linear geometry and right angle geometry). The phonon-induced tunnelling between these states leads to a formation of a new type pseudo-spin centre.

This model allows us to explain why the green luminescence signal is decreasing under oxidation and increasing under reduction of the sample (Figure 3).

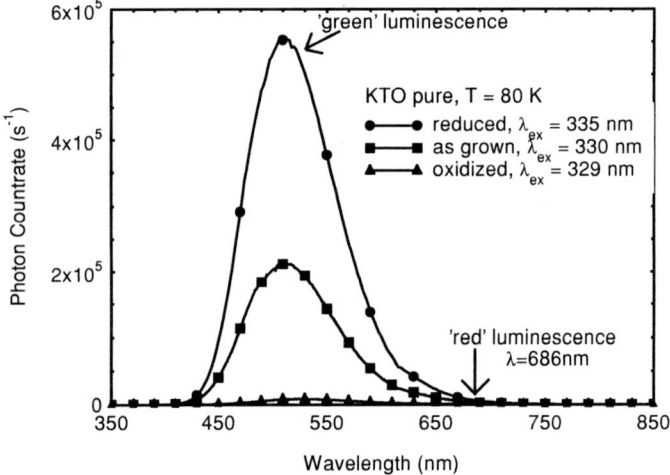

FIGURE 3. The spectrum of green luminescence emission at T=80 K of nominally pure $KTaO_3$, "as grown", after oxidizing, and after reduction in H_2 atmosphere. Spectral position of relatively weak red luminescence emission is also indicated.

The intrinsic nature of the defect as well as good recombination properties for both electrons and holes detected for the green luminescence are explained simultaneously. Asymmetric line shape of green luminescence signal and its temperature dependence can be clearly described in the framework of the contributions of two different sub-bands with rather different (approximately opposite) temperature dependence (Figure 4a). The CTVE-oxygen vacancy model proposed here explains such a behaviour as a result of local configuration instability. Namely, the temperature dependence of the green luminescence line shape is explained on the basis of the first-order local configuration instability. It corresponds to a local transition between two orientational states of the CTVE-oxygen vacancy center; between states with linear geometry, and with right angle geometry. Anharmonic interaction of soft and hard quasi-local modes induces here a local instability effect (under the conditions of the hard mode's quasi-equilibrium temperature averaging).

As it was shown in ref. [27], the RL signal is caused by intrinsic defects. This disappears under any heat treatment, (Figure 4b). Both these properties support the assumption [6] that RL in KTO is the CTVE recombination luminescence.

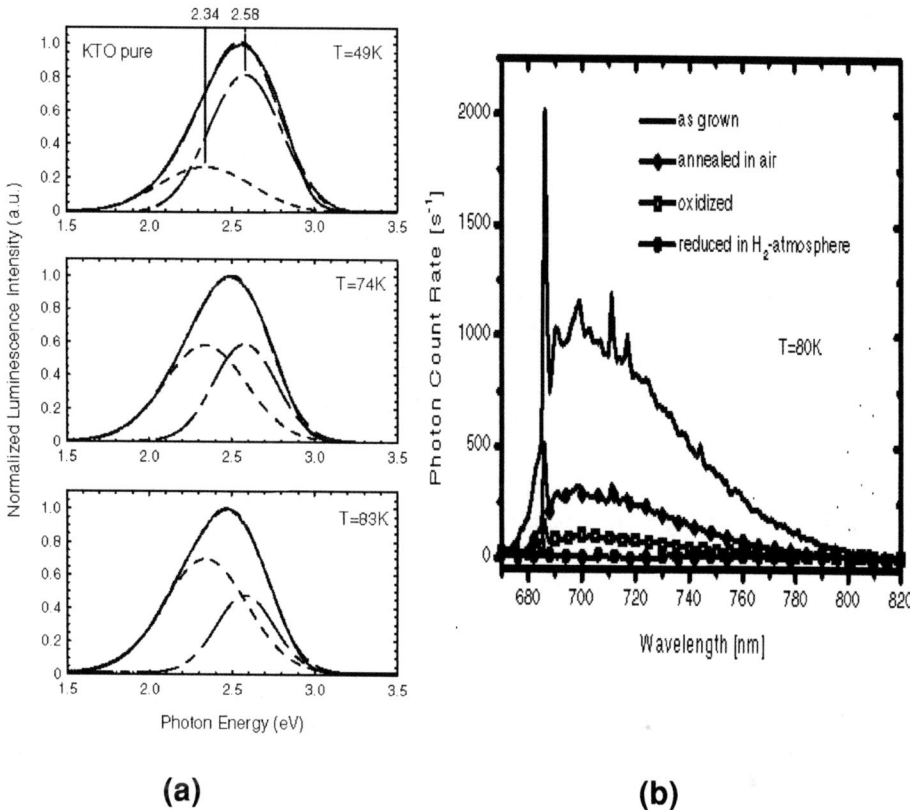

FIGURE 4. a) Spectral distribution of the green luminescence at different temperatures for nominally pure KTaO$_3$. A fit with two gaussian subbands is indicated (dashed lines). Local configuration instability of the first order takes place at T ≈ 74 K. b) The spectrum of red luminescence emission of nominally pure KTaO$_3$ after different type thermal treatment.

In addition, the presence of $(5d^2)Ta^{3+}$ electronic state contribution in RL active state (which was detected in [26] by magneto-optical investigations) can be explained by equilibrium charge transfer in the CTVE discussed above.

Last but not the least, the strong sample dependence of RL could be understood due to the high cross-section of CTVE for trapping by electric (elastic) dipole defects. We assume that a free CTVE exists only in rather pure samples. (Figure 4b).

It should be underlined that in agreement with performed INDO-calculations [7], the energy of a possible electron-hole polaron recombination with CTVE collapse corresponds to a luminescence quantum energy of ~ 2.12 eV in the case of KTO. This value is in a good agreement with experimental value observed for KTO RL [25-27].

An Origin of The Visible Range Absorption in Ferroelectric Oxides: Clusters of Charge Transfer Vibronic Exciton Phase

The open question in the optics of ferroelectric oxides is the question about an origin of visible-range (VIS) light induced absorption bands. They were detected for SBN:Ce, SBN:Cr, as well as in nominally pure SBN [28], and recently in $Ba_{0.77}Ca_{0.23}TiO_3$ [29] (BCT) crystals in approximately the same spectral region (500 – 600 nm), and was accompanied by a near infrared absorption peak induced by small electronic polarons. But the VIS center manifests a rather long life-time, which is much more longer than that for a small electronic polaron. Another important peculiarity of the VIS center is a semi-quantitative conservation of the properties for all experimental situations mentioned above. That is the origin of a VIS center does not direct depend from an actual impurity, but has a more general basis.

The model is essentially based on the CTVE. Namely, we assume, that VIS absorption is related to the states of metastable CTVE clusters of CTVE phase, trapped by impurities (Ce^{3+}, Cr^{3+} in SBN, Ca^{2+} in BCT), or by intrinsic defects (oxygen vacancies, or second component Sr^{2+} ion in SBN). Here specific absorption by CTVE phase clusters appear. It is connected with a light-induced CTVE cluster decay. The presence of characteristic VIS absorption also for nominally pure SBN, and the estimate of the energy quanta for CTVE state decay induced by resonance absorption in the CTVE vibronic state system support the model.

Co-operative Polar Cluster Formation Due To Polaron Or Charge Transfer Vibronic Exciton Trapping on Clusters of Oxygen Vacancies or Strong Off-center Impurities

The oxygen vacancy related model where trapped polarons or/and trapped CTVE undergo hopping motion between different trapping centers – oxygen vacancies – can be rather actual for oxygen-octahedrical perovskites. Note that such type of hopping motion can play an important role for polarization percolation phenomena, additionally to such dipole center reorientations. But the principal result here is an existence of the mechanism of formation of polaronic or CTVE clusters with polar symmetry in the oxygen vacancy regions. If the polar CTVE cluster in the oxygen vacancy region gives rise due to strong ferroelectric CTVE-CTVE correlation [3] of the dipole type CTVE trapped on the vacancies, the situation for polarons trapped on the vacancies is more complicated. We shall use the model where the oxygen vacancy with trapped electronic polaron obeys a non-dipole tetragonal symmetry ground state, while the first low-lying excited state is under the conditions of the Dipole Single Charged Vacancy (DSCV) model [30] and has a reorienting dipole moment corresponding to a polaron localization on one from two B-ions in oxygen-octahedrical perovskite. In spite of even symmetry of the single oxygen vacancy – polaron centre, the DSCV-type state can be seen as a ground state for the oxygen vacancy cluster situation. Namely, this is the case of ferroelectrically correlated DSCV-type states. The main result here is the transition of this ferroelectrically ordered cluster with DSCV-type states from the excited to the ground state due to its energy lowering with center concentration.

Such type energy lowering is due to direct interaction of the cluster mean field with electric dipole moment of the single charged vacancy in the framework of intra-cluster co-operative "Negative-U" effect.

Indirect polaron-polaron interaction is mainly of the dipole-dipole interaction between point electric dipoles located on the polaron sites. This interaction is realized via soft TO-phonons. But non-point "polaron-vacancy" electric dipole moments interact also via low-frequency polar quasilocal mode of a cluster discussed. If the former dipole-dipole interaction has a radius equals to a polarization correlation radius, the latter one has the radius equals the cluster size. Both interactions are ferroelectric-type within correlation radius and create the cluster mean field.

It is important to underline that this cluster mean field of the polarization is increased with the number of single charged vacancies within a cluster on the one hand, and with temperature decreasing, on the other. As a result, the above mentioned intra-cluster transition corresponds to LCI of the first order with co-operative creation of polar cluster consisting in ferroelectrically ordered polarons trapped by oxygen vacancies. The latter behaviour is very similar to polar CTVE-cluster formation [2,3] with co-operative origin of polar cluster state.

The strong off-center impurity case for the diluted solid solution KLT can be considered in the framework of hopping motions of electronic polarons, or CTVE [31] between trapping Li^+ impurity centres (or oxygen vacancies) as the cores. Dipole type CTVE appearance near a strong defect can arise as special kind of dipole-type LCI (with self-consistent charge transfer).

The relaxor properties as well as the dielectric response with two (or more) low frequency dispersion peaks for KLT (due to the action of reoriented dipole-type CTVE, and due to its interaction with reorientations of off-center ion as a defect core) can be grounded here in the natural way. This model of two interacted relaxators was used by us earlier in phenomenological form [32] for the explanation of dielectric response of the weakly doped KTO crystal. It can be actual for the explanation of (i) two relaxation peaks appearance [33] in the KLT low frequency dielectric response as well as (ii) peculiarities of weakly doped KTO dielectric response detected in [32] not only phenomenologically, but also in the framework of CTVE-related microscopical model mentioned above. The reorientations of CTVE can play a role of a relaxator with small activation energy in the second case. Probably, the hopping motion of CTVE between different trapping centres (oxygen vacancies or/and strong off-center Li^+ ions in KLT) can serve a main mechanism of the two peak structure of dielectric response in the first case. Such a hopping motion is controlled by quasi-resonance cross-relaxation of CTVE [31] and exists also at rather low temperatures. The dipole reorientation gives rise here due to transitions to the different orientation states relatively to the initial sites of CTVE-oxygen vacancy, or CTVE-Li^+ centres.

The observation of the green luminescence in KTO and KLT is in agreement with this idea: dielectric response and green luminescence signal both can find the explanation on the basis of "CTVE-oxygen vacancy", or/and "CTVE- Li^+ centres" effect. Moreover, these centres can be responsible for the formation of polar clusters taking part in the polarization percolation phenomena considered in [34]. But here we have to deal not only with the polarization percolation. The percolation due to a transport of real CTVE with their real dipole moments as a result of CTVE cross-relaxation takes place simultaneously with the pure electric dipole reorientations with site conservation. The power-law decay of the probability of CTVE cross-relaxation ($\sim r^{-6}$) changes the situation in the total percolation effect due to its strengthening.

Note in the conclusion that polaron and CTVE hopping motions extend the cluster size with corresponding softening of the polarization percolation conditions.

REFERENCES

1. Vikhnin V.S., *Proc. Est. Acad. Sci., Phys. Math.* **44**, 164-170 (1995).
2. Vikhnin V.S., Ferroelectrics 199, 25-40 (1997); *Z. Phys. Chem.* **201**, 201-213 (1997).
3. Vikhnin V., *Ferroelectrics Lett.* **25**, 27-35 (1999).
4. Vikhnin V.S., Liu H., Jia W. and Kapphan S., *J. Luminescence*, **83-84**, 91-96 (1999).
5. Vikhnin V.S, Liu H., Jia W., Kapphan S., Eglitis R.I., and Usvyat D., *J. Luminescence*, **83-84**, 109-113 (1999).
6. Vikhnin V.S., Eglitis R.I., Kapphan S., ISRF-III, Dubna, Russia, 13-16 June, 2000, Abstract, p. 47.
7. Kotomin E.A., Eglitis R.I., Borstel G., *J.Phys.: Condens. Matter* **12**, L557-L562 (2000).
8. Eglitis R.I., Christensen N.E., Kotomin E.A., Postnikov A.V. and Borstel G., *Phys. Rev. B* **56**, 8599-8604 (1997).
9. Eglitis R.I., Postnikov A.V. and Borstel G., *Phys. Rev. B* **54**, 2421-2427 (1996).
10. Eglitis R.I., Postnikov A.V. and Borstel G., *Phys. Rev. B* **55**, 12976-12981 (1997).
11. Kotomin E.A., Eglitis R.I., Postnikov A.V., Borstel G. and Christensen N.E., *Phys. Rev. B* **60**, 1-5 (1999).
12. Stefanovich E., Shidlovskaya E., Shluger A. and Zakharov M., *Phys. Status Solidi B* **160**, 529-540 (1990).
13. Shluger A. and Stefanovich E., *Phys. Rev. B* **42**, 9664-9673 (1990).
14. Stashans A. and Kitamura M., *Solid State Commun.* **99**, 583-588 (1996).
15. Rabe K.M. and Waghmare U.V., *Phys. Rev. B* **52**, 13236-13246 (1995).
16. Rabe K.M. and Waghmare U.V., *J. Phys. Chem. Solids* **57**, 1397-1403 (1996).
17. Cohen R.E., *Nature* **358**, 136-138 (1990).
18. Cohen R.E. and Krakauer H., *Phys. Rev. B* **42**, 6416-6423 (1990).
19. Zhong W., Vanderbildt D. and Rabe K.M., *Physical Review Letters* **73**, 1861-1864 (1994).
20. Yu. R. and Krakauer H., *Physical Review Letters* **74**, 4067-4070 (1995).
21. Zhong W., King-Smith R.D. and Vanderbildt D., *Physical Review Letters* **72**, 3618-3621 (1994).
22. Vikhnin V.S., Eglitis R.I., Markovin P.A. and Borstel G., *Phys. Stat. Sol. B* **212**, 53-63 (1999).
23. Yamaichi E., Watanabe K, Imamiya K., Ohi K., *J. Phys. Soc. Jpn,* **56**, 1890-1897 (1987); Yamaichi E, Ohno S., Ohi K., *Jap. J. Appl. Phys.* **27**, 583-586 (1988).
24. Vikhnin V.S. and Kapphan S. *Phys. Solid State*, **40**; 834-836 (1998).
25. Grenier P., G. Bernier G., S. Jandl S., B. Salce B., Boatner L.A. *J. Phys.: Condens. Metter* **1**, 2515-2520 (1989).
26. Grenier P., Jandl S., Blouin M., Boatner L.A., *Ferroelectrics,* **137**, 105-111 (1992).
27. Vikhnin V.S., Eden S., Aulich M., Kapphan S., *Sol. St. Commun.*, **113**, 455-460 (2000).
28. Gao M., Kapphan S., Pankrath R., Feng X., Tang Y., Vikhnin V.S. *J. Phys. Chem. Sol.* **61**, 1775-1787, 2000.
29. Wierschem M., Lindemann T., Pankrath R., Kapphan S., Vikhnin V.S. *Appl.Phys.B - Laser and Optics*, 2001, to be published.
30. Prosandeev, S.A., *Sov. Phys. JETP* **83**, 747-754 (1996).
31. Vikhnin V., Liu H., Jia W., *Physics Letters A* **245**, 307-316 (1998).
32. Trepakov, V., Smutny, F., Vikhnin, V., Bursian, V., Sochava, L., Jastrabik, L., Syrnikov, P. *J. Phys. Condens. Matter* **7**, 3765-3777 (1995).
33. Prosandeev S., Trepakov V., Savinov M., Kapphan S, Jastrabik L, private communication.
34. Prosandeev S.A., Vikhnin V.S., Kapphan S. *Eur. Phys. J. B* **15**, 469-474 (2000).

Author Index

A

Ahluwalia, R., 185

B

Bellaiche, L., 62, 191
Bhattacharya, K., 72
Blinc, R., 97, 144
Boatner, L. A., 155
Bobnar, V., 97
Bohannan, G. W., 175
Borstel, G., 201, 228
Burton, B. P., 82, 191
Bussman-Holder, A., 137

C

Cao, W., 185
Caracas, R., 128
Chen, H., 45
Cockayne, E., 82, 191
Cohen, R. E., 11
Colla, E., 45

D

Dalal, N., 137
Dal Corso, A., 107
Davies, P. K., 33
Dawber, M., 1
Dmowski, W., 33

E

Egami, T., 33
Eglitis, R. I., 201, 228

F

Farber, L., 33
Fornari, M., 23

Fu, H., 11

G

George, A. M., 62
Gonze, X., 128
Gregorovič, A., 97
Grinberg, I., 211

H

Heifets, E., 11, 201

J

Jiang, F., 55

K

Kapphan, S., 144
Kapphan, S. E., 228
Kojima, S., 55
Kotomin, E. A., 201, 228
Krakauer, H., 118

L

Larsen, P., 118
Li, J. Y., 72

M

Meyer, B., 218

P

Pirc, R., 97, 144

R

Ramer, N. J., 211
Rappe, A. M., 91, 211
Resta, R., 107
Ríos, S., 1
Rogers, C. L., 91

S

Sai, N., 218
Samara, G. A., 155
Schmidt, V. H., 165
Scott, J. F., 1
Singh, D. J., 23
Suewattana, M., 118

T

Tahan, C., 118
Tkachuk, A., 45

U

Umari, P., 107

V

Vanderbilt, D., 218
Vikhnin, V. S., 144, 228

W

Whatmore, R. W., 1

Z

Zhang, Q., 1
Zhang, S., 118
Zschack, P., 45